POCKET MECHANIC
VEHICLE MANUAL

VOLKSWAGEN LT
LT28 - LT35 - LT46

Diesel Models from 1997-2002

4 Cylinder Diesel, AGK and ATA Engines
5 Cylinder Diesel and Turbodiesel
AGX, AHD, ANJ, APA, BBE, BBF

ISBN: 9781898780786

PM7

03W6/3077

PREFACE

Small though this Workshop Manual is in size, it lacks no detail in covering most of the servicing and repair of the Volkswagen LT vehicles since 1997 with 2.5- and 2.8-litre diesel engines. Depending on the models, the vehicles are marketed as LT28, LT35 and LT46. The manual covers models introduced in 1997 with a five-cylinder 2.5-litre engine, with and without turbocharger and a four-cylinder 2.8-litre engine. Engines have been changed during production and not all engines have the same performance.

Brief, easy-to-follow instructions are given, free from all necessary complications and repetitions, yet containing all the required technical detail and information, and many diagrams and illustrations.

Compiled and illustrated by experts, this manual provides a concise source of helpful information, all of which has been cross-checked for accuracy to the manufacturer's official service and repair procedures, but many instructions have derived from actual practice to facilitate your work. Where special tools are required, these are identified in the text if absolutely necessary and we do not hesitate to advise you if we feel that the operation cannot be properly undertaken without the use of such tools.

The readers own judgement must ultimately decide what work he or she will feel able to undertake, but there is no doubt, that with this manual to assist him, there will be many more occasions where the delay, inconvenience and the cost of having the van off the road can be avoided or minimised.

The manual is called "Pocket Mechanic" and is produced in a handy glove pocket size with the aim that it should be kept in the vehicle whilst you are travelling. Many garage mechanics themselves use these publications in their workshops and if you have the manual with you in the car you will have an invaluable source of reference.

A general faultfinding (troubleshooting) section is included at the end of the manual and all items listed are taken from actual experience, together with the necessary remedies to correct faults and malfunctioning of certain parts.

Although many of the information also apply to the similar "Minivans" (MPVs) of the same manufacturers, we would like to point out that the wiring diagrams at the end of the manual are specifically aimed at the small vans

No liability can be accepted for any inaccuracies or omissions in this workshop manual, although every possible care has been taken to make it as complete and accurate as possible Every care has also been taken to prevent personal injury or damage to equipment when working on the vehicle We have tried to cover all models produced to the day of publication, but are unable to refer to all modifications and changes for certain markets or up-dating of models.

Always insist on fitting genuine Volkswagen Replacement Parts, but make sure to quote the Chassis and/or Engine No. when ordering parts.

The Pocket Mechanic, PO Box 904,
Amersham, HP6 9JA, England

CONTENTS

0. GENERAL INFORMATION

0.0. Introduction

This workshop manual covers the Volkswagen LT models with the following engines:

2.5-litre SDI engine, 75 BHP (55 kW)
2.5-litre TDI engine, 102 BHP (75 kW)
2.5-litre TDI engine, 90 BHP (66 kW) and 109 BHP (80 kW)
2.5-litre TDI engine, 83 BHP (61 kW) and 95 BHP (70 kW)
2.8-litre TDI engine, 125 BHP (92 kW) and 131 BHP (96 kW)
Not all engines are fitted to the various models in any particular country.

As listed, two basic diesel engines are fitted, i.e., a four-cylinder engine (2.8 litres) and a five-cylinder engine (2.5 litres). The latter can operate with or without turbocharger. When dealing with the engine, make sure to identify the engine code letters, as changes in engine performance can only be identified by the engine code:

Four-Cylinder Engines:

Engine code AGK	From April 1997 to December 1998 125 kW (92 kW)
Engine code ATA	From January 1999 131 BHP (96 kW)

Five-Cylinder Engines:

Engine code AGX	From July 1997, 75 BHP (55 kW)
Engine code AHD	From January 1996, 102 BHP (75 kW)
Engine code ANJ	From April 1999, 109 BHP (80 kW)
Engine code APA	From April 1999, 90 BHP (66 kW)
Engine code BBE	From May 2001, 83 BHP (61 kW)
Engine code BBF	From April 1999, 95 BHP (70 kW)

The main difference between the two engines is the drive of the camshaft, injection pump and other accessories. In the case of the 2.5-litre engine, a toothed belt is used to drive the components mentioned above. In the case of a 2.8-litre engine, gearwheels and various idler gears are used to drive the camshaft, the power steering pump, the water pump and the injection pump. The drive gear assembly is located at the flywheel end of the engine.

The LT models covered in this manual have been developed together with Mercedes-Benz. The various models are available with different wheelbases and are at the time of writing as LT28, LT35 and LT46. LT46, however, is in many respects different, when compared with the two other models, as rear suspension, steering, brake system, transmission and other component groups have been specially designed for the LT46. Engine and fuel injection system are also different.

The front wheels are sprung with leave spring, fitted across the front axle. At the bottom of the suspension, you will find a suspension arm, attached with a ball joint to the steering knuckle. A telescopic shock absorber is fitted to the steering knuckle at the lower end and the body at the upper end. A stabiliser bar, connected to the two suspension arms, completes the front suspension.

A rigid rear axle with leaf spring suspension and hydraulic telescopic shock absorbers are fitted to the rear wheels.

All wheels are fitted with disc brakes that are different for the LT28/LT35 and LT46 models. The vacuum for operation of the brake servo unit is produced by a separate vacuum (exhauster) pump. ABS is either fitted as standard or available as optional extra. Vehicles can also be fitted with automatic brake differential, known as "ABD". Vehicles with the more powerful engines have an ABS/ABD system.

All vehicles are fitted with a power-assisted rack-and-pinion steering.

Important Notes for Repairs: To gain access to many parts of the engine, it will be necessary to remove the lock bonnet carrier, the maintenance cover, the front bumper, the air intake grille and in some cases also the front wing. The necessary operations are described in the "Body Parts" section.

With regard to the range of models covered in the manual, it must be pointed out that not all of these will be available in any particular country and that the manufacturer reserves the right to make modifications to standard equipment, optional extras and trim as may be necessary for any territory.

Since the variations between certain models are only in fittings and trim, it will be unnecessary in some parts of this manual to refer to individual model designations. If your vehicle is, however, an LT46, refer to the instructions for this model.

0.1. Identification

With the vehicle, you will have received various data cards that contain important vehicle details, for example, the chassis number, numbers of other components, details of optional extra equipment, etc. Further details will be in your Operator's Handbook.

You will know the location of the vehicle data plate, which gives you the total permissible weight, the permissible load on the front and rear axle and other important information.

The engine number is located at a different position, depending on the engine:

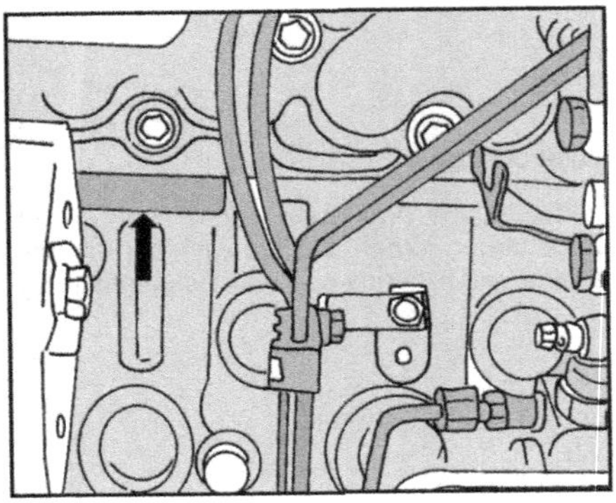

Fig. 0.1 – The location of the engine number in the case of the four-cylinder engine. The lower view shows where No. 1 cylinder is located.

Four-Cylinder Engine

The engine number, i.e., the three engine code letters and the serial number of the engine, is stamped into the L.H. side of the cylinder block, next to the injection pump, as can be seen in Fig. 0.1.

The number can also be seen on the vehicle data carrier (vehicle data plate). Note the No. 1 cylinder of these engines is located on the flywheel side (lower view of Fig. 0.1). The cylinder numbering commences from this cylinder.

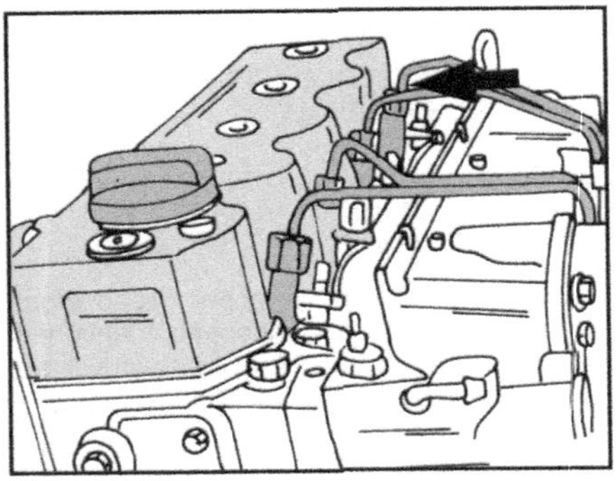

Five-Cylinder Engine

The engine number, i.e., the three engine code letters and the serial number of the engine, is stamped into the L.H. side of the cylinder block, next to the injection pump, at the location shown in Fig. 0.2. A sticker with the engine code and the engine number is additionally attached to the upper timing belt cover. The number is also shown on the data carrier.

These numbers are the essential information needed whenever spares or replacement parts are required. The vehicle manufacturer operates a policy of constant improvement and updating and it is only by quoting the vehicle and engine numbers that your dealer can provide you with the correct spares and replacements.

Note: If ever in the life of your LT, a replacement cylinder block is fitted, make sure that the original engine number has been transferred to the new cylinder block. This is a) a legal requirement and b) will avoid complications whenever you want to sell, part exchange or use the vehicle abroad.

The transmission and the automatic transmission are also identified by a number. As not all transmissions, i.e., gearboxes have the same reduction ratio for the final drive, it will be necessary to refer to these identification numbers (further details in Section 3) when, for example, a second-hand transmission is fitted. Each transmission has a code and is allocated to a particular engine within the vehicle range.

0.2. Dimensions and Weights

Wheelbase:
- LT28 .... 3000 or 3500 mm
- LT35 .... 3000/3350/4025 mm
- LT46 .... 3550/4025 mm

Front Track:
- LT28 and LT35, ET 82 wheel rims .... 1625 mm
- LT46, ET 108 wheel rims .... 1640 mm

Rear Track:
- LT28 and LT35, ET 83 wheel rims .... 1630 mm
- LT46, ET 108 wheel rims .... 1525 mm

Turning Circle between kerbs:
- 3000-mm wheelbase .... 11.20 metres
- 3350-mm wheelbase .... 12.80 metres
- 4025-mm wheelbase .... 14.30 metres

Max. Total Weight: .... Refer to Owner's Handbook

It is fairly impossible to provide information for every available model. Your operator's manual will give you the necessary information.

0.3. Filling Capacities

Engine (total capacity):
- Four-cylinder, incl. oil filter: .... 8.5 litres
- Four-cylinder, without oil filter: .... 8.0 litres
- Five-cylinder, with oil filter cooler: .... 7.8 litres
- Five-cylinder, without oil filter: .... 7.3 litres

Cooling System:
- 2.5-litre engine (approx.): .... 11.5 litres
- 2.8-litre engine: .... 11.5 litres

Manual Transmission: .... 3.4 litres
- Transmission type 000.1: .... 2.2 litres
- Transmission type 000.2: .... 1.6 litres

Rear Axle: .... 1.5 or 1.8 litres, depending on axle type

0.4. General Servicing Notes

The servicing and overhaul instructions in this workshop manual are laid out in an easy-to-follow step-by-step fashion and no difficulty should be encountered, if the text and diagrams are followed carefully and methodically. The "Technical Data" sections form an important part of the repair procedures and should always be referred to during work on the vehicle.

In order that we can include as much data as possible, you will find that we do not generally repeat in the text the values already given under the technical data headings. Again, to make the best use of the space available, we do not repeat at each operation the more obvious steps necessary – we feel it to be far more helpful to concentrate on the difficult or awkward procedures in greater detail. However, we summarise below a few of the more important procedures and draw your attention to various points of general interest that apply to all operations.

Always use the torque settings given in the various main sections of the manual. These are in most cases grouped together in separate sub-sections for convenient reference.

Bolts and nuts should be assembled in a clean and very lightly oiled condition and faces and threads should always be inspected to make sure that they are free from damage burrs or scoring. DO NOT degrease bolts or nuts.

All joint washers, gaskets, tabs and lock washers, split pins and "O" rings must be replaced on assembly. Seals will, in the majority of cases, also need to be replaced, if the shaft and seal have been separated. Always lubricate the lip of the seal before assembly and take care that the seal lip is facing the correct direction.

References to the left-hand and right-hand sides are always to be taken as if the observer is at the rear of the vehicle, facing forwards, unless otherwise stated.

Always make sure that the vehicle is adequately supported, and on firm ground, before commencing any work on the underside of the car. A small jack or a makeshift prop can be highly dangerous and proper axle stands are an essential requirement for your own safety.

Dirt, grease and mineral oil will rapidly destroy the seals of the hydraulic system and even the smallest amounts must be prevented from entering the system or coming into contact with the components. Use clean brake fluid or one of the proprietary cleaners to wash the hydraulic system parts. An acceptable alternative cleaner is methylated spirit, but if this is used, it should not be allowed to remain in contact with the rubber parts for longer than necessary. It is also important that all traces of the fluid should be removed from the system before final assembly.

Always use genuine manufacturer's spares and replacements for the best results.

Since the manufacturer uses metric units when building the cars, it is recommended that these are used for all precise units. Inch conversions are given in most cases, but these are not necessarily precise conversions, being rounded off for the unimportant values.

Removal and installation instructions, in this workshop manual, cover the steps to take away or put back the unit or part in question. Other instructions, usually headed "Servicing", will cover the dismantling and repair of the unit once it has been stripped from the vehicle. It is pointed out that the major instructions cover a complete overhaul of all parts, but, obviously, this will not always be either necessary and should not be carried out needlessly.

There are a number of variations in unit parts on the range of vehicles covered in this workshop manual. We strongly recommend that you take care to identify the precise model, and the year of manufacture, before obtaining any spares or replacement parts.

Std.: To indicate sizes and limits of components as supplied by the manufacturer. Also to indicate the production tolerances of new unused parts.

0/S Parts supplied as oversize or undersize or recommended limits for such parts, to enable them to be used with worn or re-machined mating parts.

U/S O/S indicates a part that is larger than Std. size. U/S may indicate a bore of a bushing or female part that is smaller than Std.

Max.: Where given against a clearance or dimension indicates the maximum allowable if, in excess of the value given, it is recommended that the appropriate part is fitted.

TIR: Indicate the total indicator reading as shown by a dial indicator (dial gauge).

TDC: Top dead centre (No. 1 piston on firing stroke).

MP: Multipurpose grease.

0.5. Jacking up the Vehicle

Remember that the vehicle is heavy, i.e., the jack used to lift the front of rear end must be strong enough to carry the weight. Always make sure that no additional loads are in the cargo room.

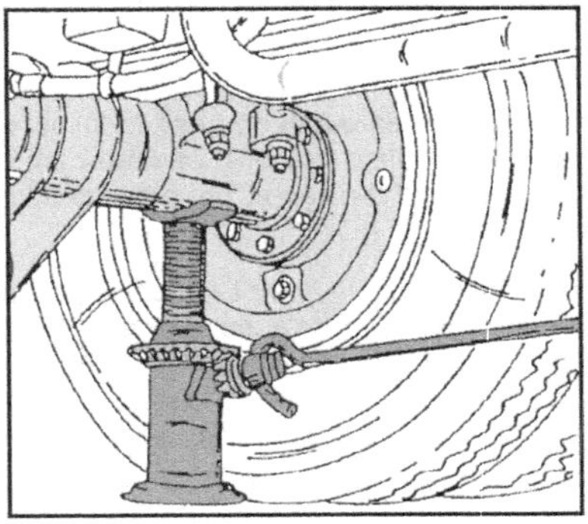

Fig. 0.3 – Jacking up the front end.

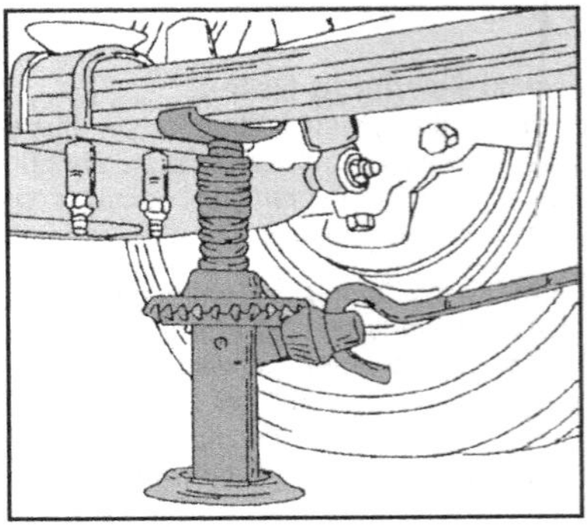

Fig. 0.4 – Jacking up the rear end.

When the front and rear ends of the vehicle must be placed on chassis stands, first jack up the front of the vehicle. Place chocks underneath the rear wheels or failing this, and apply the handbrake. A mobile jack is placed underneath the centre of the front end. If the hydraulic jack, supplied with the vehicle is used, apply it as shown in Fig. 0.3. To jack up the rear end of the vehicle, place the jack underneath the position shown in Fig.0.4. In this case, always chock the front wheels. Figs. 0.5–0.8 show further examples of positions where the jack can be applied. Depending on the work to be carried out, you will have the choice. Also note the jack application in Fig. 0.8 when twin rear wheels are fitted. The leaf spring mounting brackets are another good place to jack up the vehicle.

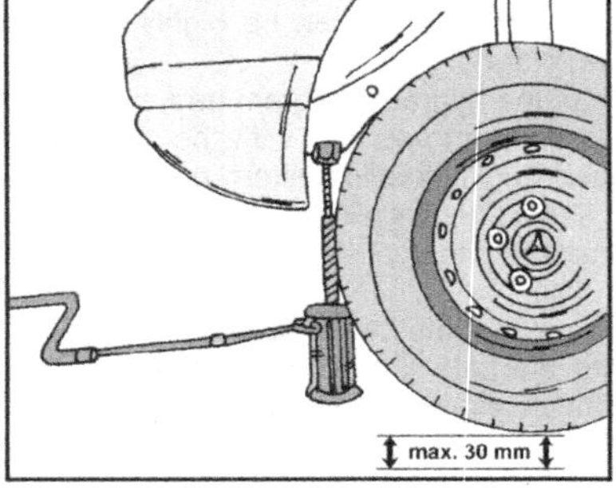

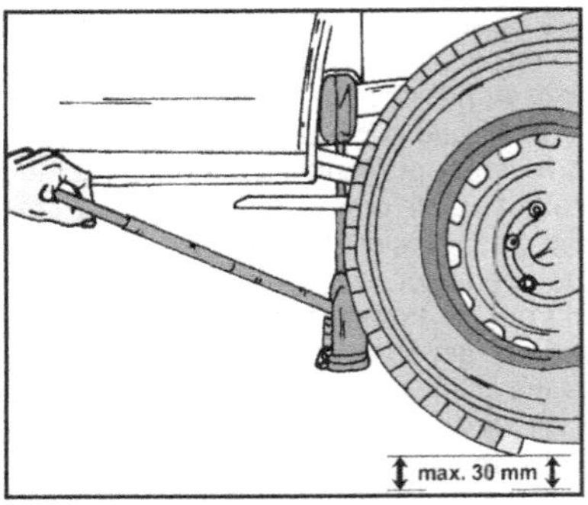

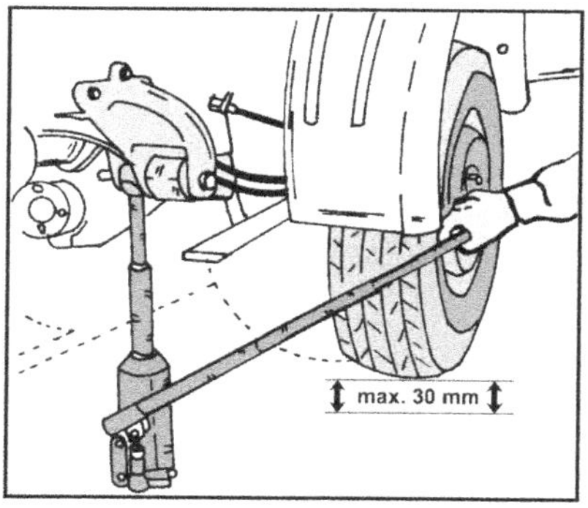

Fig. 0.7 – Jacking up the rear end.

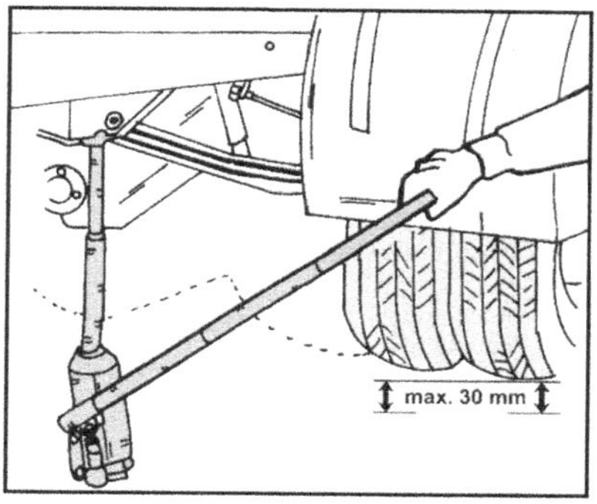

Fig. 0.8 – Jacking up the rear end.

Strong chassis stands must be placed underneath the front leaf spring or the sides of the rear axle tube. The bolt used for the height adjustment of the stands must be strong enough to carry the weight of the vehicle – remember this – makeshift bolts of small diameter will not do underneath the vehicle. Stands can be placed underneath the

It is always difficult to raise a vehicle first on one side and then on the other take care that the vehicle cannot tip-over when the first side is lifted. Ask a helper to *support the vehicle from the other side.*

1. ENGINES

1.0. General Data

Engine Identification (all listed):
- Four-cylinder engines: .... AGK to Dec. 1998, ATA after

Five-cylinder engines:
- From July 1996: .... AGX
- From January 1996: .... AHD
- From April 1999: .... ANJ
- From April 1999: .... APA
- From May 2001: .... BBE
- From April 1999: .... BBF

Number and Arrangement of Cylinders: .... 4 or 5, in line

Engine Capacity:
- Four-cylinder engines: .... 2.8 litres
- Five-cylinder engines: .... 2.5 litres

Cylinder Bore:
- Four-cylinder engine: .... 93.00 mm
- Five-cylinder engine: .... 81.00 mm

Piston Stroke:
- Four-cylinder engine: .... 103.00 mm
- Five-cylinder engine: .... 95.50 mm

Compression Ratio:
- Four-cylinder engine: 19.0:1
- Five-cylinder engine AGX to February 1999, AHD: 19.5:1
- Five-cylinder engine, all others and AGX from March 1999: 19.0:5

Injection Order:
- Four-cylinder: 1-3-4-2
- Five-cylinder: 1-2-4-5-3

Max. Performance:
- Four-cylinder – AGK: .... 125 BHP (92 kW) at 3500 rpm
- Four-cylinder – ATA: .... 131 BHP (96 kW) at 3500 rpm
- Five-cylinder – AGX: .... 75 BHP (55 kW) at 3800 rpm
- Five-cylinder – AHD: .... 102 BHP (75 kW) at 3500 rpm
- Five-cylinder – ANJ: .... 109 BHP (80 kW) at 3500 rpm
- Five-cylinder – APA: .... 90 BHP (66 kW) at 3500 rpm
- Five-cylinder – BBE: .... 83 BHP (61 kW) at 3500 rpm
- Five-cylinder – BBF: .... 95 BHP (70 kW) at 3500 rpm

Max. Torque:
- Four-cylinder – AGK: .... 28.0 kgm at 2000–2500 rpm
- Four-cylinder – ATA: .... 30.0 kgm at 2000–2500 rpm
- Five-cylinder – AGX: .... 16.0 kgm at 2000–2400 rpm
- Five-cylinder – AHD: .... 25.0 kgm at 2400 rpm
- Five-cylinder – ANJ: .... 28.0 kgm at 1900–2500 rpm
- Five-cylinder – APA: .... 22.0 kgm at 1800–2500 rpm
- Five-cylinder – BBE: .... 20.0 kgm at 1900–2400 rpm
- Five-cylinder – BBF: .... 16.0 kgm at 1900–2400 rpm

Valve Clearances (warm):
- 2.5-litre engine: .... Not applicable, hydraulic tappets
- 2.8-litre engine:
- Inlet valves: .... 0.20 mm (0.008 in.)
- Exhaust valves: .... 0.30 mm (0.012 in.)

Oil sump capacity: .... See Section 0.3

Valve Timing	2.5-Litre Engine	2.8-Litre Engine
Inlet valve opens	8° ATDC	--------
Inlet valve closes	28° ATDC	
Inlet valve 1 opens	------	5.3° ATDC
Inlet valve 1 closes	------	32° ABDC
Inlet valve 2 opens	------	3.7° ATDC
Inlet valve 2 closes	------	31.7° ABDC
Exhaust valve opens	37° BBDC	55.6° BBDC
Exhaust valve closes	10° ATDC	5.5° ATDC

1.1. Engine – Removal and Installation

A hoist or a hand crane is required to lift the engine out of the engine mountings. As the engine is lifted out towards the front, you will need a hoist with a beam. Remember that the power unit is heavy. The 2.8-litre engine can be removed with or without the gearbox; the 2.5-litre engine is always removed together with the gearbox, i.e., note the additional weight for the lifting capacity of the lifting tackle. The two engines are covered under separate headings. Also note the following when a 2.8-litre engine (4-cyl.) is removed.

- If the lifting tackle has a capacity of 500 kg, you can remove the engine together with the gearbox.
- If the lifting tackle has a capacity of only 300 kg, remove the engine without the gearbox.

The following instructions must be noted during the removal of all engine types.

- If a radio with anti-theft code is fitted, have the code number ready before the battery is disconnected.
- All engines are lifted towards the front out of the engine compartment.
- All cable binders that have been cut during the removal of the engine must be replaced and refitted to their original position. Therefore, mark the cable binders/clamps accordingly before cutting them.
- Coolant hoses are secured with spring clamps, which are opened with special pliers. Otherwise, cut the old clamps and use ordinary screw-type clamps during installation. Spring clamps, however, are preferred.

Remove the engine and gearbox (where applicable) as follows.

2.8-Litre Engine (4-cyl.)

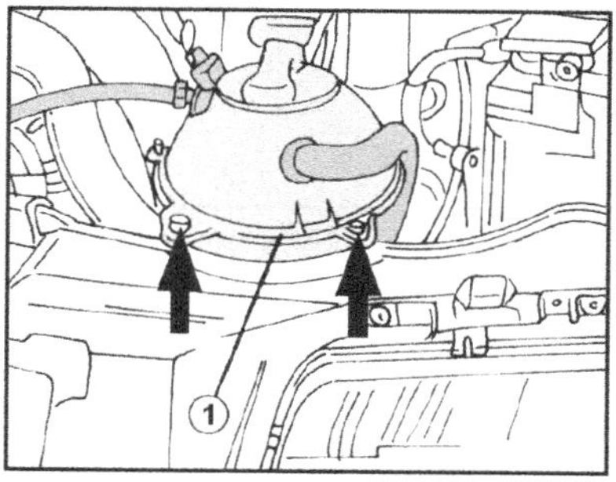

Operations connected with parts of the body that must be removed are covered in a separate section at the end of the manual.

- Switch off the ignition and disconnect the battery earth cable. Push the cable to one side to prevent it from "snapping" back to the battery terminal.
- Remove the maintenance cover inside the cab. This operation will be familiar to you (otherwise, refer to the body section).
- Drain the cooling system.
- Remove the front flap.

Fig. 1.1 – Remove the two "Torx" head screws to take out the coolant expansion tank and the spacer (1).

- Refer to Fig. 1.1 for the next operations. Remove the two torx-head screws securing the coolant expansion tank. Lift out the tank together with the spacer (1) and place the assembly to one side.
- Remove the L.H. and R.H. steps (bolted in position).

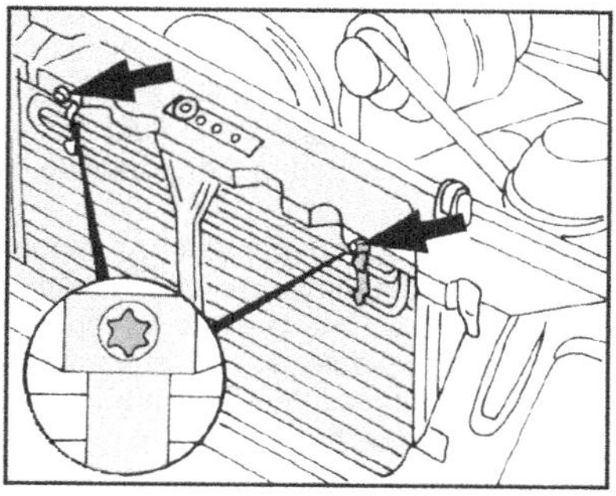

Fig. 1.2 – The cooling pipe for the power steering system is secured at the positions shown.

- Remove the front bumper and the lock carrier (body section).
- Remove the screws securing the cooling pipe for the power-assisted steering and place the pipe to one side (Fig.1.2).
- Remove the radiator as described in the section "Cooling System".
- Remove the hose clamps, shown with (1) in Fig.1.3, and withdraw the hose (arrow) from the oil separator. Remove the remaining hoses.
- Disconnect the hoses shown in Fig. 1.4 from the water pump side and the engine side. The coolant expansion tank and the spacer plate can now be removed from the engine compartment.
- Disconnect the throttle operating cable from the injection pump and place it on one side.

Fig. 1.3 – Remove the hose clamps (1) and withdraw the hoses. The hose with the arrow is connected to the oil separator.

- Remove the steering pump (two bolts) and remove it without disconnecting the fluid hoses.

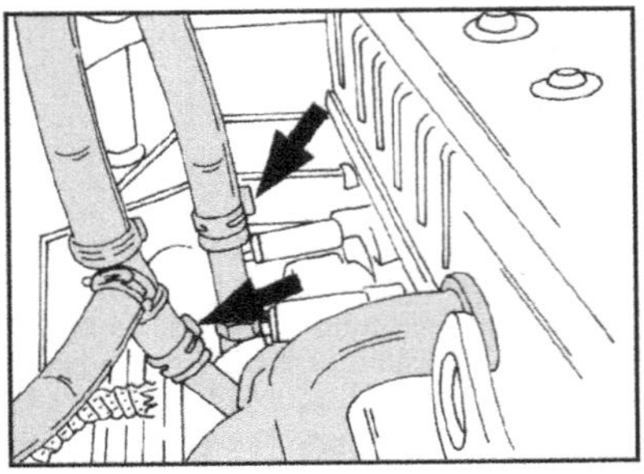

Fig. 1.4 – Disconnect the hoses from the water pump.

- The next operations are carried out by referring to Fig. 1.5. Disconnect the vacuum hose from the vacuum pump (L.H. view). Disconnect the fuel feed and return pipes from the injection pump and all electrical cables from the engine, starter motor, alternator, etc.

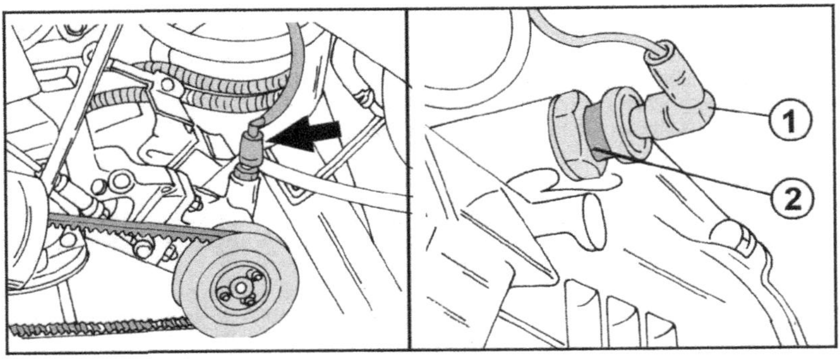

Fig. 1.5 – Disconnecting the vacuum hose from the exhauster pump and the cable from the reversing light switch.

The engine can now be removed with or without gearbox. The vehicle must be supported on chassis stands as required.

Removal of Engine and Gearbox

- Unscrew the gearchange lever knob, remove the rubber cover and remove the gearchange lever after unscrewing the clamp bolt (also refer to the section "Gearbox").
- Withdraw the cable connector plug (2) in Fig. 1.5 (R.H. view) from the reversing light switch. Also on the gearbox, remove a screw and take out the speedometer sensor. An earth cable must also be disconnected.
- Disconnect the exhaust pipe from the exhaust manifold and release the exhaust pipe from the retainer on the gearbox.
- Unscrew the clutch slave cylinder from the side of the gearbox, and without disconnecting the fluid hose, attach it with a piece of wire or cord to the underside of the chassis frame/body.
- Disconnect the propeller shaft from the gearbox flange.
- Position the lifting tackle with the beam over the engine and use chains or ropes to the lifting eyes on the cylinder head. Also place a mobile jack underneath the engine/gearbox unit to support the assembly from below. Before removing any of the engine and gearbox mountings, check once more that the power unit is securely suspended/supported.

Fig. 1.6 – Removal of the mounting crossmember and the centre gearbox mounting.

* Remove the mounting crossmember on the L.H. and R.H. sides and remove the centre mounting. Fig. 1.6 shows where the mountings are located.
* Unscrew the engine mountings from below (both sides) and lift out the assembly. The engine must be rotated slightly to guide the gearchange lever. Lifting tackle and jack must be operated at the same time until the assembly can be withdrawn towards the front.

Removal of the Engine without Gearbox

* Place a mobile jack underneath the gearbox until it is under slight tension and remove the bolts between engine and gearbox.
* Position the lifting tackle with the beam over the engine and use chains or ropes to the lifting eyes on the cylinder head. Lift the engine until the two engine mountings on the left and right are under slight tension. Unscrew the engine mountings from below (both sides). Check once more that the engine is securely supported on the ropes/chains.
* Separate the engine from the gearbox and lift it out towards the front. A helper will be required for this operation to guide the engine past other parts in the engine compartment.
* If repairs on the engine are required, place it immediately onto a workbench or wherever you intend to carry out the repairs.

2.5-Litre Engine (5-cyl.)

Some differences exist within the various engines which will be referred to.

* Carry out the operations described for the four-cylinder engine until the lock carrier is removed.
* Remove the radiator and the intercooler as described in the section "Cooling System" and remove the air intake hose from the air cleaner.
* Remove the following items by referring to Fig. 1.7. Referring to the L.H. view, withdraw the vacuum hose (1) from the air cleaner, the cable plug (2) from the air mass meter and the connecting hose (3) between the pipe and the air mass meter. In the case of engines with code **AHD, ANJ, APA, BBE and BBF**, disconnect the breather hose (4) from the pressure regulating valve and remove the air charge pipe (6) with the bracket. In the case of all engines, disconnect the cooling hose (5) from the cylinder head.
* If an engine with code **AHD, ANJ, APA, BBE or BBF** is fitted, refer to the R.H. illustration. Remove the securing screw (arrow) for the charge air pipe (1) and the pressure pipe (2) on the charge air pipe. In the case of all engines, withdraw the cable plug (3) from the intake air temperature sensor (**except engine ANJ**) and disconnect the cooling hoses (4) from the engine. The coolant expansion tank with hoses can now be removed.

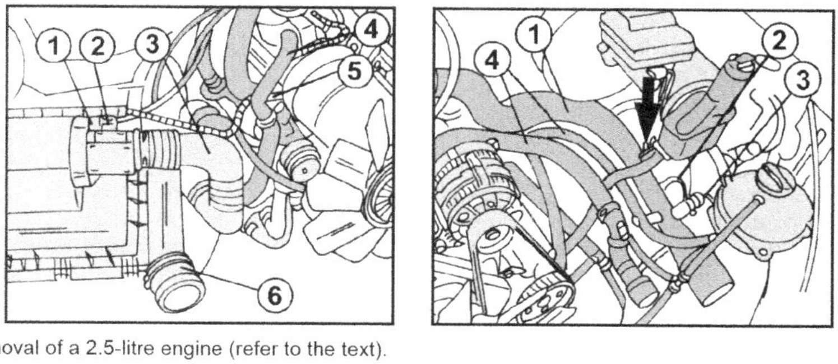

Fig. 1.7 – Removal of a 2.5-litre engine (refer to the text).

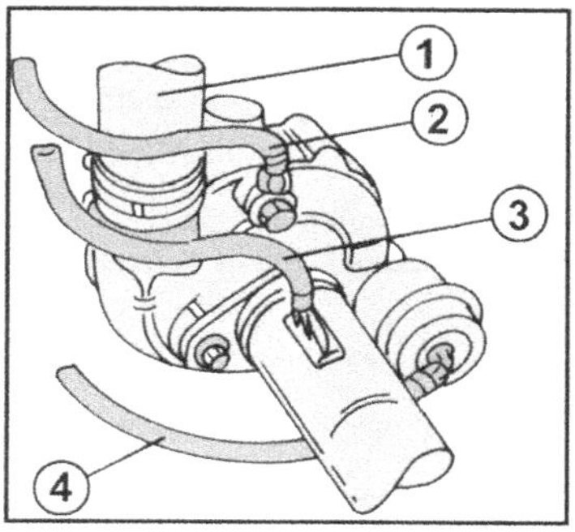

Fig. 1.8 – Removal of a 2.5-litre engine (only with certain engine codes, refer to the text).

* If an engine with code **AHD, ANJ, APA. BBE or BBF** is fitted, refer to Fig. 1.8. Remove the blue vacuum hose (2) from the turbocharger (1) – except engine code AJN. The yellow vacuum hose (3) from the connection pipe (except engine AJN), the red control pipe (4) from the charge pressure regulating valve and the bracket for the EGR valve (electrical identification N18) and the solenoid valve for the charge pressure limitation (N75).

The following instructions are valid for all engines.

* Remove the drive belt at the front of the engine (see separate heading in this section).
* Remove the steering pump without disconnecting the hoses and place it on one side.

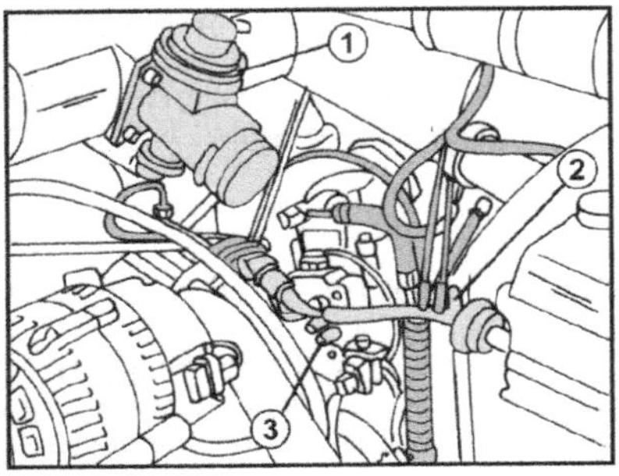

Fig. 1.9 – Disconnect/remove the items shown when removing a five-cylinder engine (refer to the text).

- Disconnect the connecting hose and the vacuum hose from the mechanical EGR valve (1) in Fig. 1.9 and remove the valve, and disconnect the vacuum hose (2) and the vacuum hose (3) for the brake servo unit from the exhauster pump (vacuum pump).
- Disconnect the small hose from the injector in cylinder No. 5 and unscrew the fuel feed and return pipes from the fuel pump.
- Remove the oil dipstick out of the dipstick guide tube and disconnect a coolant hose from the heat exchanger on the engine. Remove a second coolant hose from the heat exchanger and free it from the bracket.
- Disconnect the plus cable for the starter motor from the battery, open the protective sleeve and withdraw the cable plug from the starter motor.
- Withdraw the cable plug from the reversing light switch (Fig. 1.5, R.H. illustration). The sensor for the speedometer is removed after unscrewing the single securing screw. Also remove an earth cable from the transmission.
- Disconnect the exhaust pipe from the exhaust manifold and remove the exhaust pipe bracket from the side of the gearbox.
- Disconnect the fuel feed and return pipes from the injection pump and all other cables or leads connected to the engine, i.e., alternator, oil pressure switch, etc. Mark all cable connections if not sure where they were connected.
- Unscrew the clutch slave cylinder from the side of the gearbox and without disconnecting the fluid hose attach it with a piece of wire or cord to the underside of the chassis frame/body.
- Disconnect the propeller shaft from the gearbox flange.
- Position the lifting tackle with the beam over the engine and use chains or ropes to the lifting eyes on the cylinder head. Also place a mobile jack underneath the engine/gearbox unit to support the assembly from below. Before removing any of the engine and gearbox mountings, check once more that the power unit is securely suspended/supported.
- Remove the mounting crossmember on the L.H. and R.H. sides and remove the centre mounting. Fig. 1.6 shows where the mountings are located.
- Unscrew the engine mountings from below (both sides) and lift out the assembly. The engine must be rotated slightly to guide the gearchange lever. Lifting tackle and jack must be operated at the same time until the assembly can be withdrawn towards the front.
- If repairs on the engine are required, place it immediately onto a workbench or wherever you intend to carry out the repairs.

1.2. Engine – Installation

The installation of the engine is basically a reversal of the removal procedure. The following points should be noted before installation.

- As the room in the engine compartment is fairly restricted, make sure that cables, leads, pipes, hoses, etc. are well pushed to one side to avoid damaging them. Check oil and fuel pipes and the engine mountings for damage and/or wear and replace them if necessary.
- Check the clutch release bearing and replace if necessary.
- Check that the dowel sleeves are in the cylinder block before the gearbox is connected to the engine, mainly if the engine has been removed on its own.
- Do not connect any disconnected items before the engine mountings and/or transmission mountings have been refitted and the engine rests under its own weight on the mountings. Shake the engine in its mountings to settle the mountings before tightening them to 8.5 kgm (61 ft.lb.).
- Reconnect the propeller shaft (7.0 kgm/50 ft.lb.).
- Check the oil level in engine and gearbox before starting the engine and driving the vehicle.
- Before filling the cooling system, check that all drain points have been closed. Make sure that the correct anti-freeze strength has been pre-mixed.

Note: There is no need to remove the engine and/or gearbox to replace the mountings at the front or rear.

1.3. Engine Dismantling

Before commencing dismantling of the engine, all exterior surfaces should be cleaned, as far as possible, to remove dirt or grease. Plug the engine openings with clean cloth first to prevent any foreign matter from entering the cavities and openings. Detailed information on engine dismantling and assembly is given in the sections dealing with servicing and overhaul (sections commencing at 1.4.) and these should be followed for each of the sub-assemblies or units to be dealt with.

Dismantling must be carried out in an orderly fashion to ensure that parts, such as valves, pistons, bearing caps, shells, tappets and so on, are replaced in the same positions as they occupied originally. Mark them clearly, but take care not to scratch or stamp on any rotating or bearing surfaces. A good way to keep the valves in order is by piercing them through an upside-down cardbox and writing the number against each valve (refer to the previous page). Segregate together the tappets, the springs and retainers with collets for each valve, if possible in small plastic bags for each individual valve.

If a proper engine dismantling stand is not available, it will be useful to make up wooden support blocks to allow access to both the top and bottom faces of the engine. The cylinder head, once removed from the block, should be supported by a metal strap, screwed to the manifold face and secured by two nuts onto the manifold studs.

1.4. Overhaul of the Engine

1.4.1. CYLINDER HEAD

1.4.1.0. Technical Data

Material: .... Light alloy with pressed in valve guides and valve seat inserts
Max. distortion of cylinder head surface:
- 2.5-litre engine: .... Less than 0.20 mm (0.008 in.)
- 2.8-litre engine: .... Less than 0.10 mm (0.004 in.)
Max. re-facing of cylinder head face: .... Not permitted

Valves	***Four-Cylinder***	***Five-Cylinder***
Valve Seat Angle:	30° (inl.).45° (ex.)	45°, all valves
Valve Length:		
- Inlet valves:	Not available	96.85 mm
- Exhaust valves:	Not available	96.85 mm
Valve Stem Diameters:		
- Inlet valves:	6.952–6.970 mm	7.97 mm
- Exhaust valves:	6.952–6.900 mm	7.95 mm

Valve Stem Running Clearance in Guides:
- Inlet valves: .... 0.02–0.04 mm
- Exhaust valves: .... 0.10 mm

Valve Seat Width (only 2.5 litre available):
- Inlet valves: .... 2.7 mm
- Exhaust valves: .... 2.05 mm

Thickness of Valve Head Edge (margin):
- Inlet/exhaust valves – Alle engines.: .... 0.5 mm
Valve clearance: **..** See Section 1.0

Camshaft:
Camshaft identification: .... Different, depending on engine
Camshaft end float: .... max. 0.16 mm (4-cyl.), 0.15 mm (5-cyl.)
Bearing running clearance: .... max. 0.16 mm (4-cyl.), 0.11 mm (5-cyl.)

The following instructions should be followed when the cylinder head is replaced or refitted.

- When fitting the cylinder head with the camshaft in the head, coat the contact areas between the rocker arms and the cam surfaces (2.8 litres) or between the bucket tappets and the cam surfaces (2.5 litres) with engine oil.
- The valves of a new cylinder head are covered with plastic caps. Remove them just before the head is fitted.
- Replace the coolant if a new cylinder head is fitted.
- The tooth flank clearance of the intermediate shaft drive gear must be measured before the cylinder head is fitted (refer to respective heading).

1.4.1.1. Cylinder Head – Removal and Installation

Only remove the cylinder head when the engine temperature is less than 40 °C. The two engines are covered under separate headings.

2.8-Litre Engine (4-cyl.)

Fig. 1.10 shows the parts to be removed during the removal of the cylinder head. Also note the special notes given below.

Attention: Before the removal of the cylinder head, remove the valve rocker levers as described during the dismantling of the cylinder head. The rocker levers must be fitted after the head has been bolted down.

Remove the parts as described during the removal of the four-cylinder engine until access can be gained to all parts. The timing belt must be removed as described later on.

- The cylinder head cover (1) is removed after unscrewing screw (4). Underneath, you will find bushes (5) and sealing washers (3). Always replace the washers, if they have been damaged during removal of the cover. The same applies to the cover gasket (8). Replace only if damaged.
- Slacken the cylinder head bolts (7) in the order shown in Fig. 1.11. Slacken the bolts in several stages in the correct order until they can be removed by hand. *Bolts must always be replaced.*
- Remove the injection pipe bundle (8) as shown in the illustration without bending or kinking them. The union nuts are tightened to 3.0 kgm (22 ft.lb.) to the injectors and 2.7 kgm (19 ft.lb.) to the injection pump.

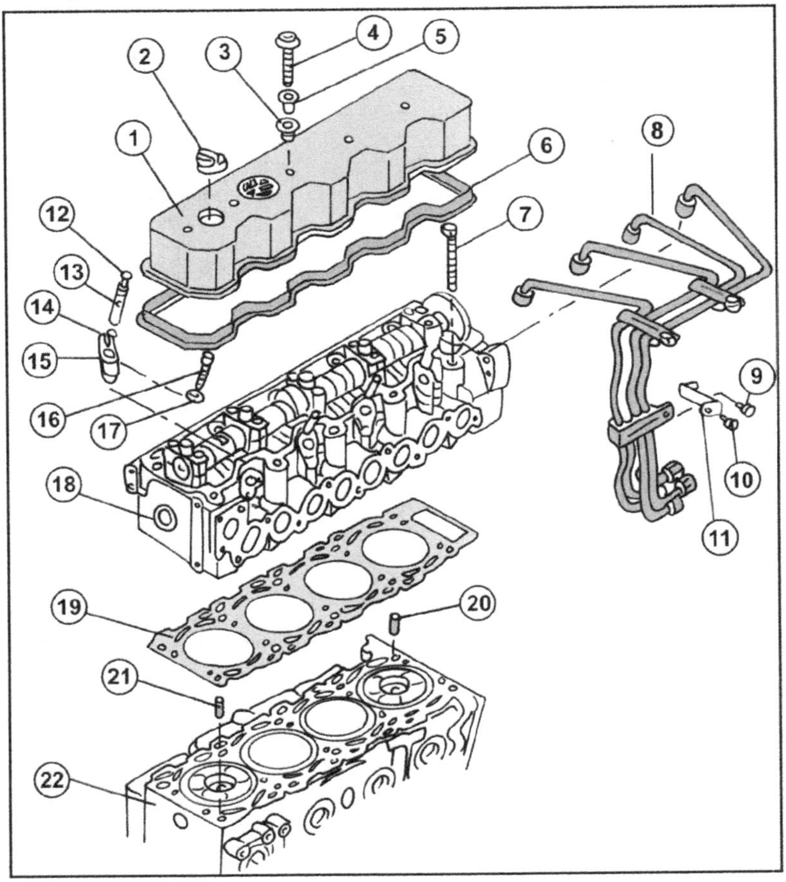

Fig. 1.10 – Details for the removal and installation of the cylinder head (2.8 litres). The items are referred to in the description.

- The bolts (9) are tightened to 2.0 kgm (14.5 ft.lb.), and the bolts (10) are tightened to 1.0 kgm (7.2 ft.lb.). The bolts secure the mounting bracket (11).
- Remove the injectors (13) as described in the section "Diesel Injection System". Note the fitted position of the rubber bush (12). The clamp bracket (15) secures the assembly. The heat protection ring (14) is not the same on all engines. As the ring must be replaced during installation, observe a thickness of 1 mm (engine AGK) or 3 mm (engine ATA).
- Tighten the bolt (16) to 4.0 kgm (29 ft.lb.). If the engine has been manufactured after December 1997, you will find the washer (17).

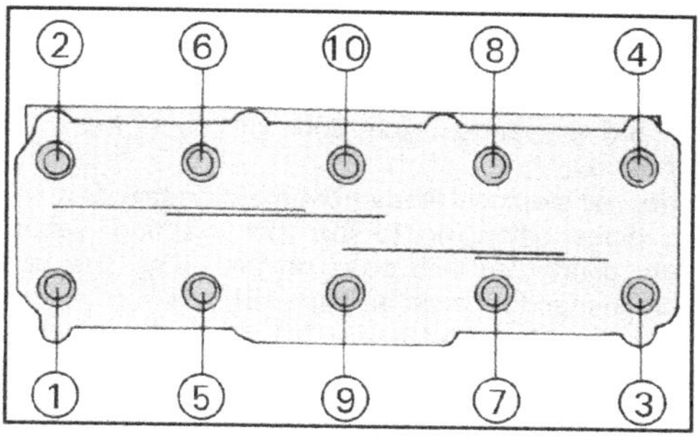

Fig. 1.11 – Slacken the cylinder head bolts in the order shown (2.8-litre engines).

- Check the cylinder head (18) after removal. To lift off the head, you can attach a hoist or small crane with ropes to the lifting eyes. Immediately, clean cylinder and cylinder block faces. If only the cylinder head gasket must be replaced, follow the instructions below. Otherwise, you can overhaul and/or repair the head as described in the following section.
- The cylinder head gasket (18) must always be replaced. The new gasket is wrapped in plastic and must only be unpacked before installation. Do not damage the gasket in any way in order not to harm the silicon coating.
- The spring sleeves (20) and (21) must be fitted as shown into the cylinder block (21) if they have come away from the cylinder block.

Note the following instructions during the installation of the cylinder head. Some of the points to be observed have already been mentioned above.

- Before fitting the head, check the end of the crankshaft to make sure that the alignment shown in Fig. 1.12 can be seen. Otherwise, rotate the crankshaft slowly (apply a socket to the bolt inside the pulley).

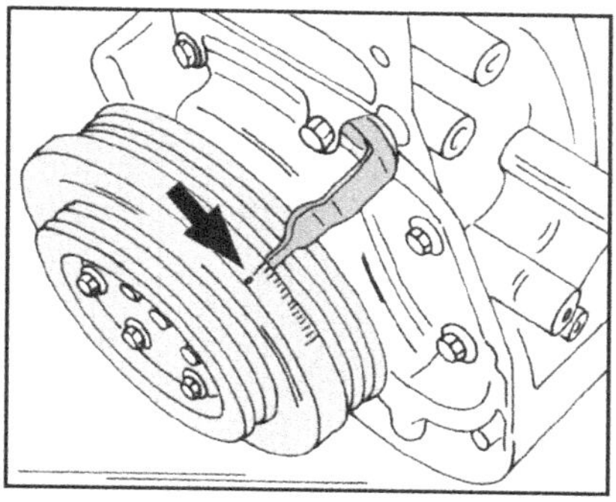

Fig. 1.12 – The crankshaft is at TDC when the mark in the vibration damper and the pointer are opposite to each other (arrow).

- Refer to Fig. 1.13 and check that the cams of No. 1 cylinder (flywheel side) are at the top dead centre position. If necessary, slowly rotate the camshaft until the alignment is obtained. The camshaft must now be locked against rotation. The workshop uses a special tool; you will have to make sure that the camshaft cannot move.
- Fit the new cylinder head gasket in position and fit the head over the gasket and the cylinder block. Make sure that the dowel sleeves engage. Both are on the same side of the block.
- Coat the cylinder bolt threads and the bolt contacting faces with engine oil. Remember NEW bolts. Screw the bolts hand-tight into the block and then tighten them in the reverse order of Fig. 1.11 to 3.0 kgm (22 ft.lb.).

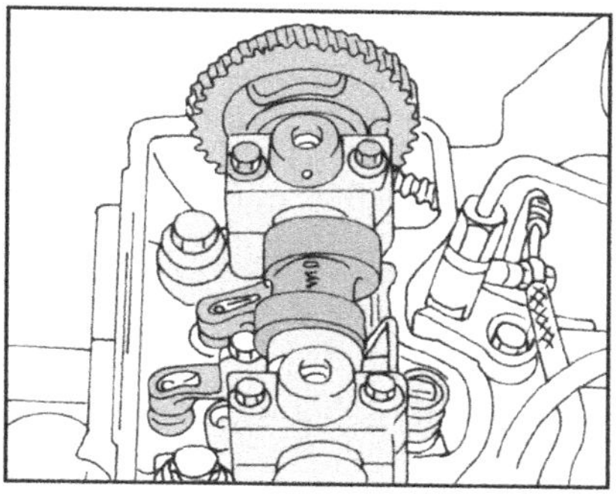

Fig. 1.13 – The camshaft must be aligned with the marking indicating TDC position for the first cylinder (flywheel side).

- Again, in reverse order of Fig. 1.11, tighten the bolts to 6.0 kgm (43 ft.lb.) and then again to 11.0 kgm (80 ft.lb.).
- Insert the bolt socket into the centre bolt (marked with "10" in Fig. 1.11 and without torque wrench tighten the bolt by 90° (quarter of a turn). The best method is to apply the tommy bar in the longitudinal direction and then tighten the bolt until the bar is across the cylinder head. Tighten all other bolts in reverse order of 1.11, i.e., 9 becomes 2, 8 becomes 3, etc.
- Repeat the above operation a second time, again tightening all bolts by a further 90° (quarter of a turn). *There is no need to re-tighten the cylinder head bolts after a certain mileage.*
- Refit the rocker levers and the cylinder head cover.
- The remaining operations are carried out in reverse order.

2.5-Litre Engine (Five-Cylinder)

The removal is carried out as follows. Fig. 1.15 shows the parts to be removed. Proceed as described for the four-cylinder engine until the cylinder head is accessible. Then, proceed as follows:

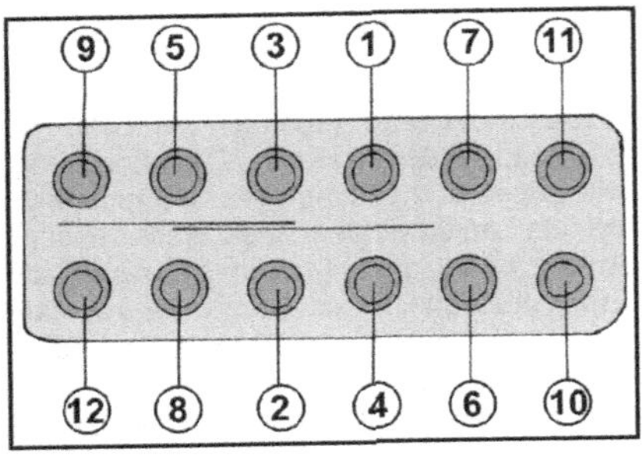

Fig. 1.14 – Tightening sequence for the cylinder head bolts of a five-cylinder engine. Slacken the bolts in reverse order (12 = 1, 11 = 2, etc.)

- Disconnect the battery earth cable, drain the cooling system and disconnect the cooling hoses from the cylinder head.
- Disconnect all electrical leads from the cylinder head and also all pipes belonging to the fuel injection system which are connected with parts on the body. The inlet and exhaust manifold must be removed.
- Unscrew the cylinder head cover (1). To do this, remove the bolts (2) with the sealing washers (3). The gasket is bonded to the cover. Take care not to damage it.
- Remove the timing belt as described for these engines later on. Also remove the camshaft gearwheel and on the other side of the engine the drive belt for the injection pump. These operations are described in conjunction with the injection pump.

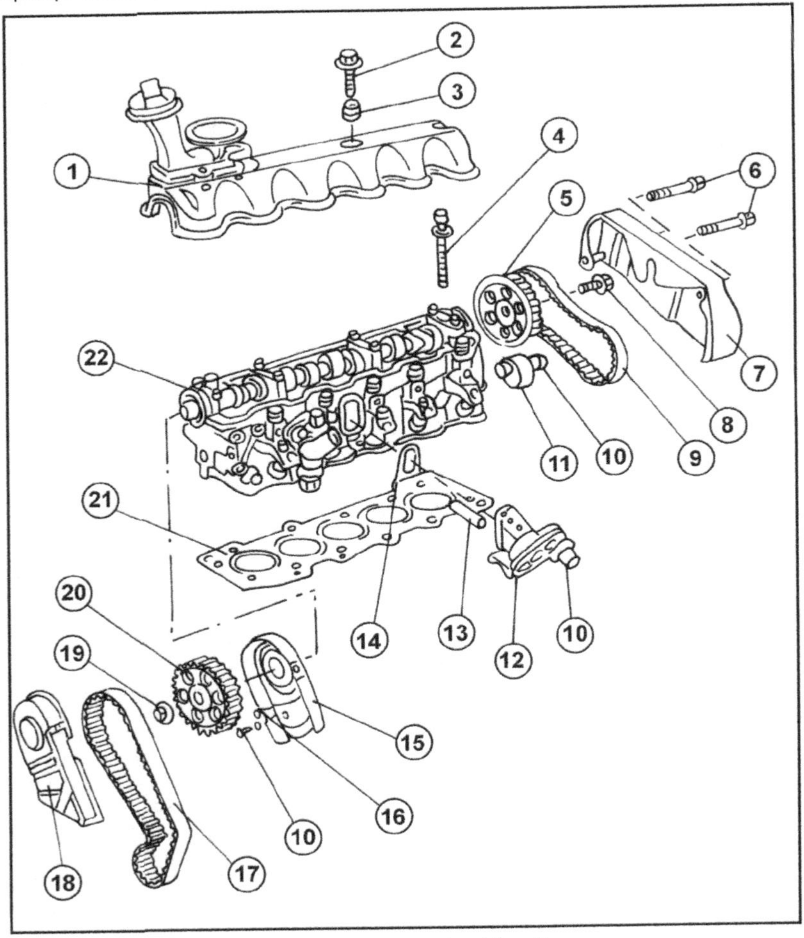

Fig. 1.15 – Details for the removal and installation of the cylinder head (5-cyl.).

1 Cylinder head cover
2 Bolt, 1.0 kgm
3 Sealing washer
4 Cylinder head bolt, replace
5 Injection pump gearwheel
6 Bolt, 2.0 kgm
7 Cover, injection pump belt
8 Bolt, 10.0 kgm
9 Injection pump belt
10 Nut. 2.0 kgm
11 Idler pulley
12 Vacuum pump
13 Operating push rod
14 "O" sealing ring
15 Inner belt cover
16 Bolt, 1.0 kg, fit with sealing compound
17 Timing belt
18 Upper belt cover
19 Bolt, note torque
20 Camshaft timing gear
21 Cylinder head gasket
22 Cylinder head

- Using a suitable socket, unscrew the cylinder head bolts in the reverse order shown in Fig. 1.14, i.e., start on the outside. Slacken the bolts in several stages. Remove the cylinder head and the cylinder head gasket.

Thoroughly clean the cylinder block and cylinder head surfaces. The following points must be noted before installation.

- When ordering the cylinder head gasket, the engine type and number must be quoted. Cylinder head gaskets are available in different thicknesses. The thickness of the gasket depends on the protrusion of the piston above the cylinder block face. When purchasing a new gasket, take the old gasket with you as it will have the identification marked at the position shown in Fig. 1.16. Either holes or notches are marked into the gasket. The arrow "A" shows where the part number can be found.

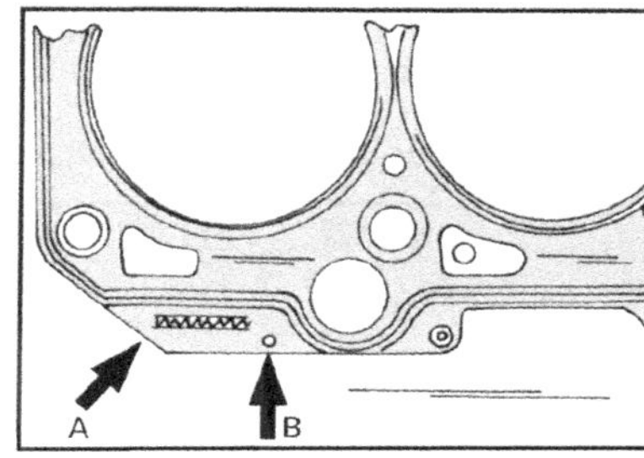

Fig. 1.16 – View of the cylinder head gasket. The arrow "A" points to the Part No. The arrow "B" shows where the holes or notches are located.

- To establish the thickness of the cylinder head gasket, set one of the pistons to T.D.C. position and with a steel ruler and a depth gauge, or with a dial gauge, as shown in Fig. 1.17, measure by how much the crown of the piston is above the cylinder block face. From the obtained dimension and the table below, select the correct cylinder head gasket. Remember to buy new cylinder head bolts when purchasing the gasket.

Piston Projection	*No. of grooves*
To 0.96 mm	1
0.97 – 1.01 mm	2
more than 1.01 mm	3

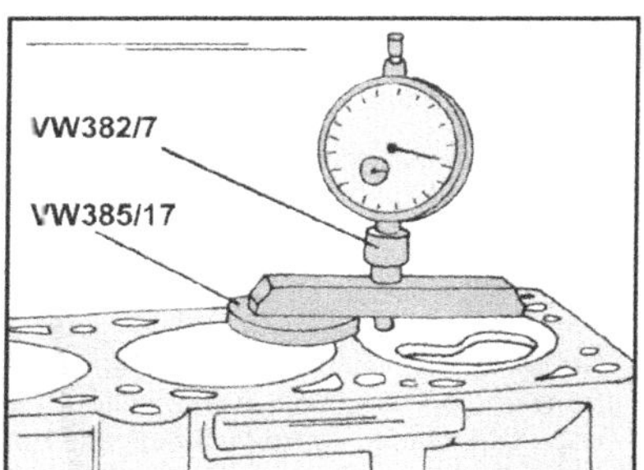

Fig. 1.17. – Measuring the projection of a piston above the cylinder block face.

- Special guide studs are available for the fitting of the cylinder head. These are fitted to the positions shown in Fig. 1.18 on the side of the exhaust manifold. Alternatively, studs can be prepared to screw in position. Studs must have a slot to unscrew them later on with a screwdriver.

Fit the cylinder head as described on the following pages.

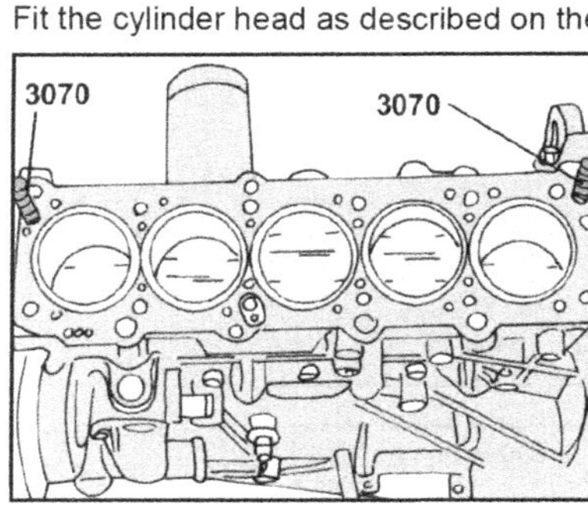

Fig. 1.18 – The two guide pins (1) fitted to the cylinder block of a five-cylinder engine to assist the fitting of the cylinder head gasket.

- Thoroughly clean the cylinder head and cylinder block faces. Place the cylinder head gasket over the engine block, the correct way round.
- Place the cylinder head over the block, engaging it with the two guide studs.
- Insert new cylinder head bolts and tighten them finger-tight. Remove the two studs and screw the cylinder head bolts in their place.
- Refer to Fig. 1.14 and tighten all bolts in the order shown to a torque reading of 4.0 kgm (29 ft.lb.). A special spline bit is used to tighten the bolts, together with a torque wrench. Your VW dealer or other suppliers may have these in stock and no attempt should be made to tighten the bolts with makeshift tools.
- Tighten all cylinder head bolts in the order shown to a torque reading of 6.0 kgm (45 ft.lb.)
- Insert the spline bit into each bolt, in the order shown in the two illustrations, make a note of the position of the torque wrench. Tighten each bolt by exactly one half of a turn, but in two stages (2 × a quarter of a turn, i.e., 90°).

- There is no need to re-tighten the cylinder head bolts after the engine has reached its operating temperature. Earlier instructions to re-tighten the bolts after the engine has had its first run are no longer applicable.
- All other operations are carried out in reverse order to the removal procedure. The installation of the timing parts and the adjustment of the timing are described under a separate heading.
- Replace the cylinder head cover gasket and the half-round rubber seal in the camshaft bearing cap.

1.4.1.2. CYLINDER HEAD – DISMANTLING AND OVERHAUL

Five-Cylinder Engines *(see Fig. 1.19)*

- Remove all auxiliary parts from the cylinder head. Unscrew the five injectors from the cylinder head and put them away safely where they cannot suffer damage during the work on the cylinder head. Make up a bracket that can be attached to the studs for the exhaust manifold and clamp the bracket into a vice.
- Remove the camshaft as described later on.

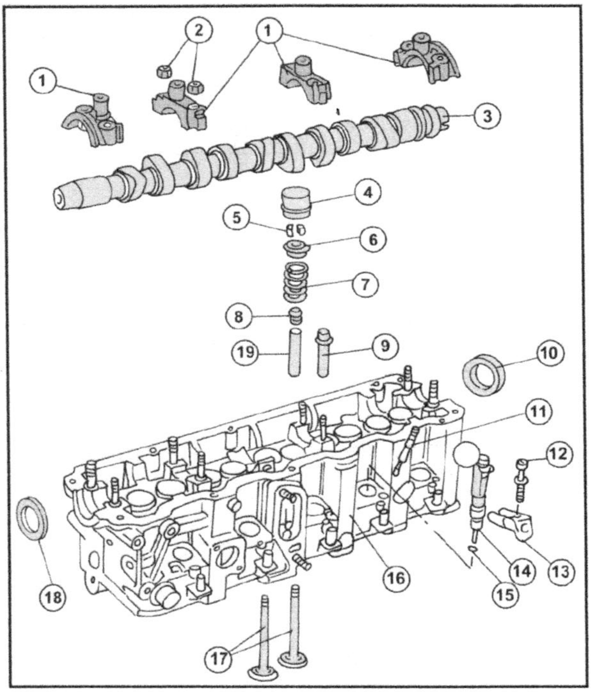

Fig. 1.19 – Exploded view of the cylinder head of a five-cylinder engine.

1 Camshaft bearing caps
2 Nuts, 2.0 kgm
3 Camshaft
4 Hydraulic tappet
5 Valve cotter halves
6 Upper valve cup
7 Outer valve spring
8 Valve guide seal
9 Valve guide
10 Oil seal
11 Glow plug
12 Bolt, 3.0 kgm
13 Injector clamp
14 Injector
15 Thermal protection washer
16 Cylinder head
17 Valves
18 Camshaft oil seal
19 Valve guide without shoulder

- Remove the half-round sealing insert from the bearing cap, if it is still in the cap. Otherwise, remove the seal from the cylinder head.
- Carefully lift the camshaft out. Using soft-metal jaws, clamp the camshaft into a vice and remove the bolt securing the camshaft gearwheel. Remove the gearwheel and slide the camshaft oil seal from the end of the shaft.

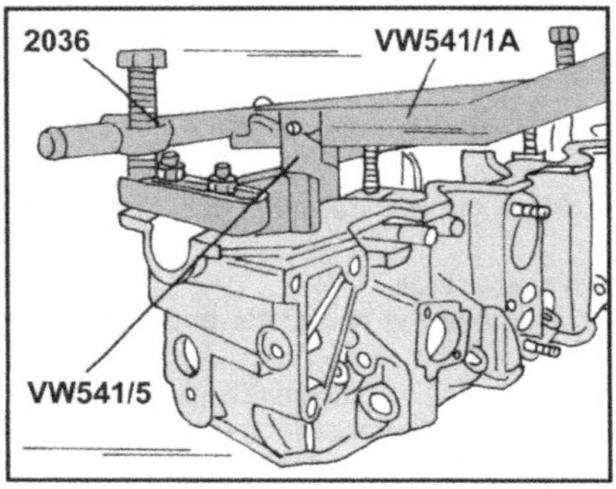

Fig. 1.20 – Removal of the valve springs and valves with a valve lifter. Shown is the special tool, but a similar device can be used.

- Remove the tappets. Keep the adjusting disc with its tappets and lay each of the tappets out in order (on a sheet of paper) so that they can be replaced in the same position.
- A valve lifter is necessary for the removal of the valves and should be shown in Fig. 1.20 or working with a similar principle. Press down the retainer to allow the collets to be extracted with a pair of thin-nosed pliers. An alternative method is to place a piece of tube over the valve spring retainer and to tap the tube with a short blow of a hammer. The valve cotter halves will be freed and collected in the inside of the tube. If this method is used, support the valve head from the other side. The valve lifter should, however, be available for the installation.
- Immediately, clean all removed parts and inspect them as described on the following pages. We recommend that a major overhaul of the cylinder head is carried out by an engine shop.

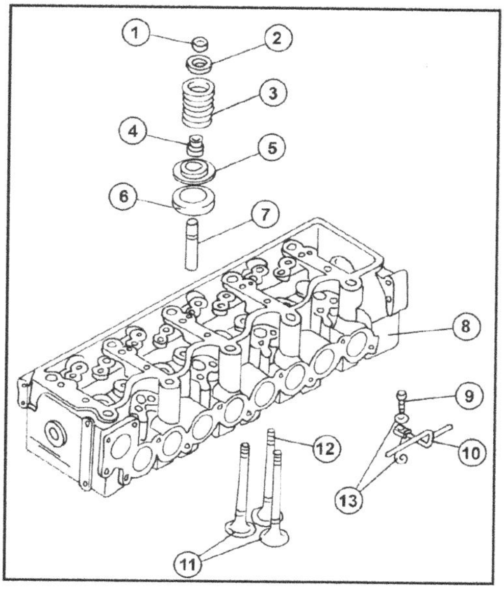

Fig. 1.21 – Cylinder head and valves of the four-cylinder engine.

1 Valve cotter halves
2 Upper valve spring cup
3 Valve spring
4 Valve stem seal
5 Lower valve spring cup
6 Spacer ring (if fitted)
7 Valve guide
8 Cylinder head
9 Banjo bolt. 1.0 kgm
10 Oil splash jet
11 Inlet valves
12 Exhaust valve

Four-Cylinder Engines

Figs. 1.21 and 1.22 show the component parts of the cylinder head together with the camshaft. Fig. 1.21 is followed when the valves are removed.

As already mentioned, rocker levers must be removed before the removal of the cylinder head and must be refitted after the head has been bolted to the cylinder block.

Removal and Installation of Rocker Levers

- Remove the cylinder head cover.
- Remove the rubber cap inside the vibration damper at the front end of the crankshaft and rotate the crankshaft with a ring spanner, applied to the bolt inside the crankshaft pulley/damper, until the tips of the cylinder in question are pointing towards the top.
- Using a wooden or plastic wedge, push the valve spring in the direction of the arrow in Fig. 1.23 until the rocker lever is free and can be removed. As can be seen in Fig. 1.22, different rocker levers are used for the inlet and exhaust valves. Do not interchange the two inlet valve rocker levers, if both are removed. Mark the accordingly, for example, L.H. and R.H. side and cylinder number.

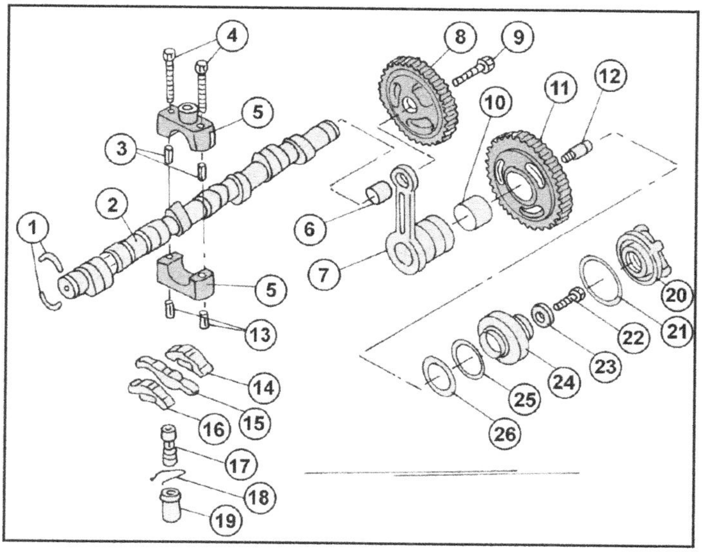

Fig. 1.22 – Component parts of the valve operating drive of a four-cylinder engine.

1	Thrust washer	10	Bearing bush	19	Threaded insert
2	Camshaft	11	Intermediate gear	20	Closing cover
3	Spring pins	12	Bolt, 8.0 kgm	21	Sealing ring
4	Bolt, 3.0 kgm	13	Spring pins	22	Bolt, 6.5 kgm
5	Upper bearing cap	14	Rocker lever, exh.	23	Washer
6	Bearing bush	15	Rocker lever, inlet	24	Threaded sleeve
7	Adjusting bearing	16	Rocker lever, inlet	25	Retaining ring
8	Camshaft gearwheel	17	Valve adjusting screw	26	Shim
9	Camshaft gearwheel bolt	18	Retaining clip		

Fig. 1.23 – Push the valve spring downwards with a "soft" wedge until the rocker lever is free and can be removed (4-cyl.).

- The installation is a reversal of the removal procedure, but the markings must be observed when the original parts are fitted. Oil the contact areas of the rocker levers and cams. The valve clearances must be adjusted as described later on.
- Fit the cylinder head cover and refit the rubber cap inside the crankshaft vibration damper (1.0 kgm/7 ft.lb.).

Removal of Valves

Remove the valves in the manner described for the five-cylinder engines, with the difference that each cylinder has three valves. Mark the valves for refitting, if required.

Camshaft Removal

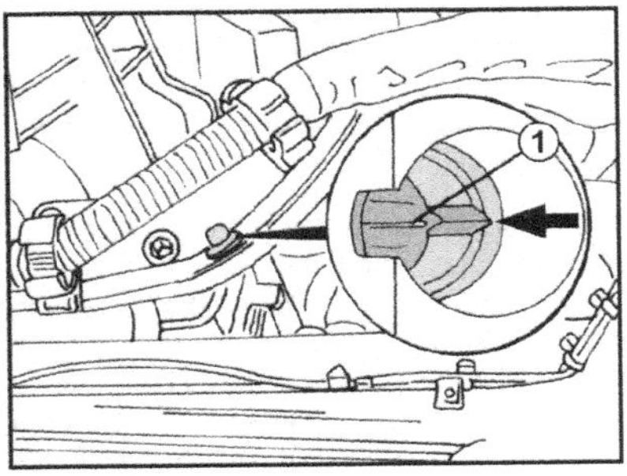

The camshaft is shown in Fig. 1.22. As can be seen from the illustration, upper and lower bearing caps are used. As all bearing caps are the same, make sure that you mark them with their number after removal. Remove the camshaft as follows.

- Remove the cylinder head cover.
- Remove the rubber cap in the centre of the crankshaft vibration damper and rotate the crankshaft (ring spanner, bolt inside the pulley) until the TDC mark in the flywheel is in line with the notch in the intermediate plate as shown in Fig. 1.24.

Fig. 1.24 – Rotate the crankshaft until the TDC mark (1) on the flywheel is in line with the notch on the intermediate plate.

- Prevent the crankshaft from rotating. Try to engage a gear or use a ring spanner as shown in Fig. 1.25 (a helper is required in this case), and slacken and unscrew the camshaft gearwheel bolt. Remove the gearwheel.

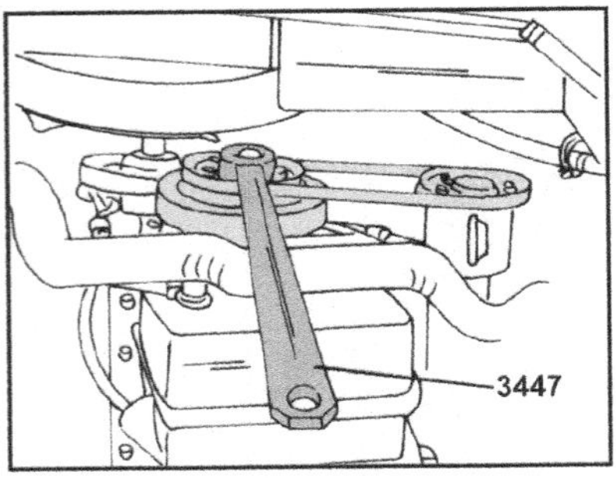

Fig. 1.25 – Rotating the crankshaft. The illustration shows the special wrench, which is basically a ring spanner.

- Slacken the camshaft bearing cap bolts from the centre towards the outside in several stages until all are slack and can be removed. Remove and mark the caps as you take them off. No. 1 is located on the flywheel side. The camshaft can now be lifted out.

Remaining Parts in/on the Cylinder Head

To remove the intermediate gearwheel, shown in Fig. 1.22, remove the camshaft in the manner described above. Before the gearwheel is refitted, you will have to check the tooth flank clearance and the end clearance as described during the assembly of the cylinder head. The bolt (12) for the intermediate gearwheel is tightened to 8.0 kgm. Remove the bearing bush (10) from the gearwheel. The bush must be lubricated with oil during installation. Remove the adjusting bearing (7) with the bush (6).

Remove the closing cover (20) and the sealing ring (21) and remove the bolt (22) underneath. Tighten to 6.5 kgm during installation. Take off the washer (23) and the threaded sleeve (24). The retaining ring (25) and the shim (26) are in the given order. The shim (26) is available in several thicknesses and is used to adjust the end clearance.

The installation of the various items should be carried out by referring to Fig. 1.22.

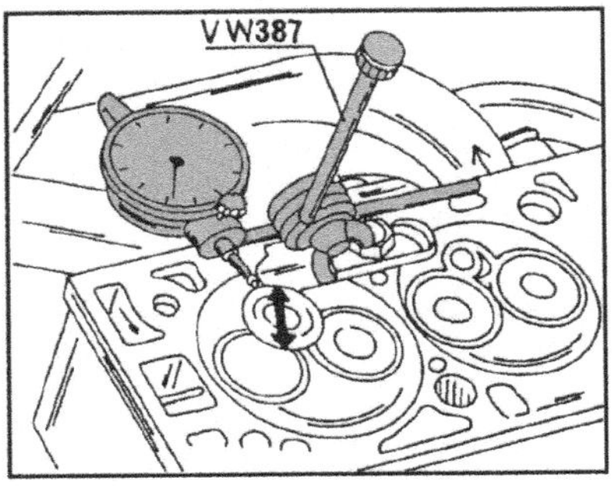

Fig.1.26a – Checking the valve stems for wear with the special tool on a five-cylinder engine. Rock the valve to and fro and read off the dial gauge.

Valve Guides: Valve guides and valve stems should first be inspected for visible wear. Clean the inside of the valve guides with a rag moistened in petrol, inserting the rag into the guides, and move it to and fro to clean the inside of the bore. Valve stems are best cleaned by clamping them into an electric drill and using a wire brush to clean off any carbon deposits.

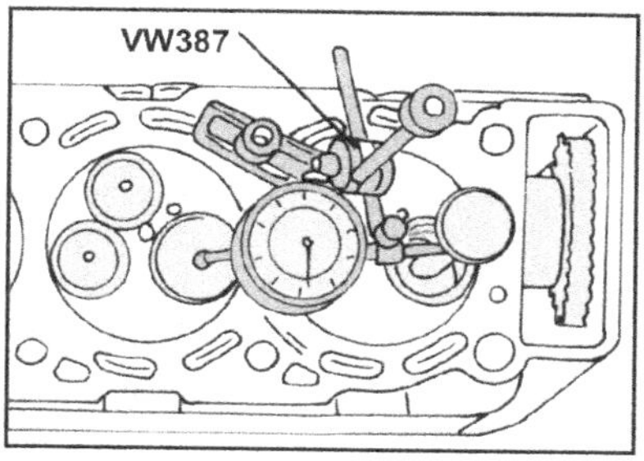

Fig.1.26b – Checking the valve stems for wear with the special tool on a four-cylinder engine. Rock the valve to and fro and read off the dial gauge.

Valve guides can be checked for wear with the special VW tool VW 387, as shown in Fig. 1.26a for a 5-cyl. or Fig. 1.26b for a 4-cyl., which is used in conjunction with a dial gauge. To use this tool with success, the following instructions should be observed.

- Attach the bracket to the side of the cylinder head as shown and lift the valve in question from its guide until the dial gauge feeler can be applied against the valve head edge.
- Push the valve to one side and set the dial gauge to zero. Move the valve to the opposite side and read the indication of the dial gauge.
- The max. permissible reading is 1.3 mm for both inlet and exhaust valves (all engines). If this value is exceeded, either the valves or the valve guides, or both must be replaced.

To replace the valve guides, press out the old guides from the cylinder head (from the camshaft side), using a suitable mandrel or carefully drive them out with a drift. Press in the new valve guides from the camshaft side of the cylinder head until the shoulder of the guide touches the head. The pressure must not exceed 1 ton, and otherwise, the guides can break off. The cylinder head must be cold and the guides must be coated with engine oil before they are pressed in.

After all guides have been pressed in position, use an 8-mm reamer (five-cylinder) or 7-mm reamer (four-cylinder) to open them up to the correct diameter, observing the running clearance of the guides. **Valve seats must be re-faced after valve guides have been replaced.**

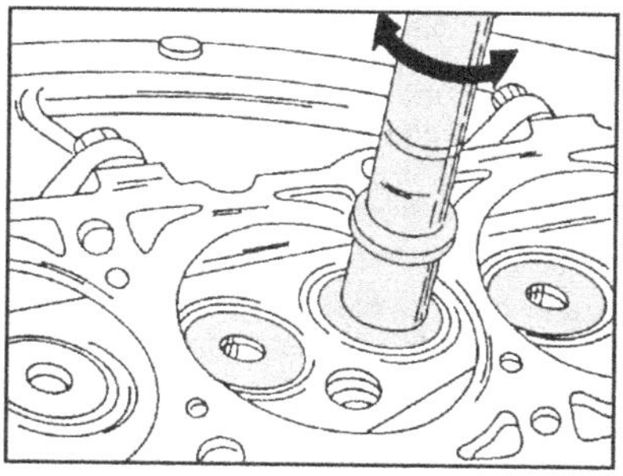

Fig. 1.27. – Grinding-in of valves with a suction tool. Move the handle in the direction of the arrows and lift from time to time, rotate by 90° and continue grinding.

Valve Seats: Valve seats can be re-machined, but we recommend to take the cylinder head to an engine shop to have the valves renovated.

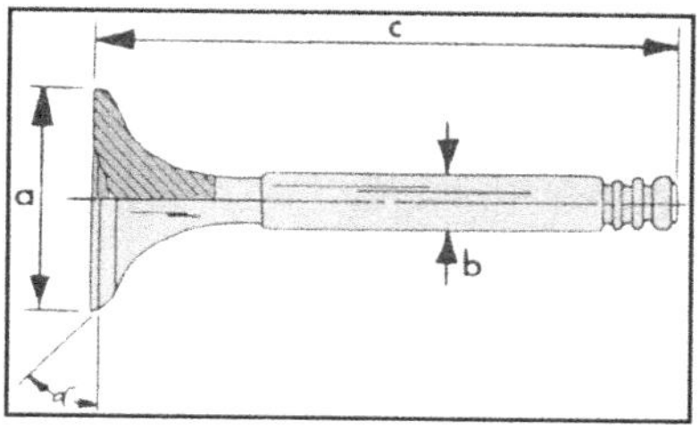

Fig. 1.28 – The principal dimensions of a valve.
a Valve head diameter
b Valve stem diameter
c Valve length

Valves: Valves with bent or pitted stems should be replaced. Grinding or straightening of valve stems is not permissible. Valves must not be re-machined. Only grinding-in or lapping with lapping paste is allowed. Use the grinding-in tool, as shown in Fig. 1.27. Some exhaust valves are sodium-filled and must be taken to a VW workshop to have them scrapped. We recommend to take all exhaust valves to a dealer, irrespective if they are sodium-filled or not. Fig. 1.28 shows the principle valve dimensions. Refer to Section 4.1.1.0 (page 15).

Valve Springs: Springs should be compared with new springs and rejected if weak or cracked. Lay each spring on a surface plate and check that the ends of the springs are square. Slide the originally fitted spring and a new spring together over a long bolt (with large washers under bolt head and nut) and screw a nut onto the end of the bolt. Clamp the bolt head into a vice and slowly tighten the nut. If the coils of the original spring close before the coils of the new spring, replace the springs as a set, as the other springs will have similar defects.

Camshaft: If a camshaft is replaced, always quote the engine number to obtain the correct camshaft. The camshaft for the five-cylinder engines requires special attention. Fig. 1.29 shows where the camshaft identification can be found. Shafts for all engines are marked at "A" with the letters "WZO" near the cams for cylinder No. 1 and with the identification "046/74" or "074" near the cams for No. 2 cylinder (at "B"). At "C", you will find a number that gives the engine to which the camshaft is fitted.

In the case of a camshaft for the four-cylinder engine, quote the engine type and engine number.

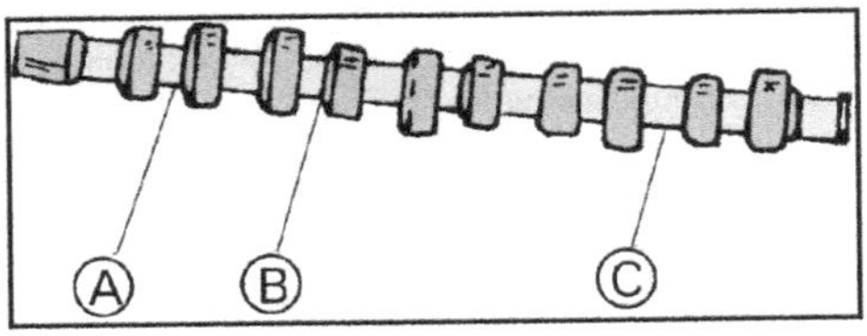

Fig. 1.29 – Camshaft identification in the case of a five-cylinder engine. Letters and numbers are stamped into the shaft as shown.

Fig. 1.30 – Checking a camshaft for run-out.

Place the camshaft with the two end journals into V-blocks or clamp the shaft between the points of a lathe, as shown in Fig. 1.30, and apply a dial gauge to the centre journal. Slowly rotate the camshaft and read off the dial gauge. If the indication is more than 0.01 mm, replace the shaft, as it is bent.

Check all bearing journals for visible wear. To check the camshaft end float, place the shaft into the cylinder head without fitting the tappets (2.5 litres only) in order to avoid the pressure of the valve springs onto the camshaft. Fit the front and rear bearing caps.

Arrange a dial gauge as shown in Fig. 1.31, with the stylus of the gauge resting against the end of the camshaft. Move the camshaft to and fro and read off the dial gauge. The dial gauge reading must not exceed 0.15 mm (5-cyl.) or 0.16 mm (4-cyl.). If this is the case, check the bearing cap end face and the camshaft. These are the positions where the camshaft thrust is taken.

Cylinder Head: Thoroughly clean the surfaces of cylinder head and cylinder block. The cylinder head surface must be checked for distortion. To do this, place a straight edge over the gasket surface as shown in Fig. 1.33 and insert a feeler gauge between the steel ruler and the cylinder head face.

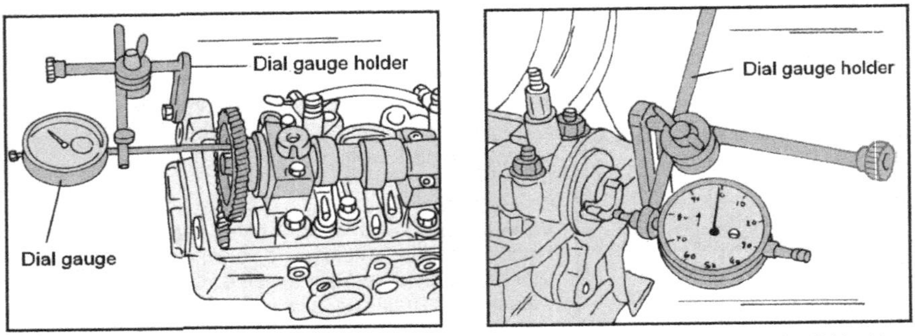

Fig. 1.32 – Checking the camshaft end float. On the L.H. side on a five-cylinder engine and on the R.H. side on a four-cylinder engine.

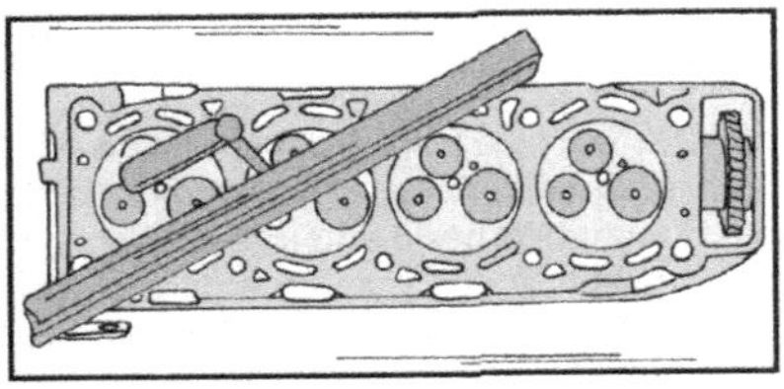

Fig. 1.33 – Checking the cylinder head surface for distortion, using a steel ruler and a feeler gauge.

Checks must be carried out in longitudinal, diagonal and transverse directions. If a feeler gauge of more than 0.10 mm (0.004 in.) in thickness in the case of a four-cylinder engine or 0.20 mm (0.008 in.) in the case of a five-cylinder engine can be inserted, replace the cylinder head, as it cannot be refaced.

1.4.1.3. Cylinder Head Assembly

Re-assemble the cylinder head components in accordance with the notes below, following the dismantling procedure in the reverse sequence. Fig. 1.19 shows an exploded view of the cylinder head for a five-cylinder engine. Figs. 1.21 and 1.22 show the construction of the cylinder head and associated parts of a four-cylinder engine.

All Engines

- Lubricate the valve stems and guides before installation of the valves. If valves have been lapped, make sure that each valve is fitted to its correct seat.

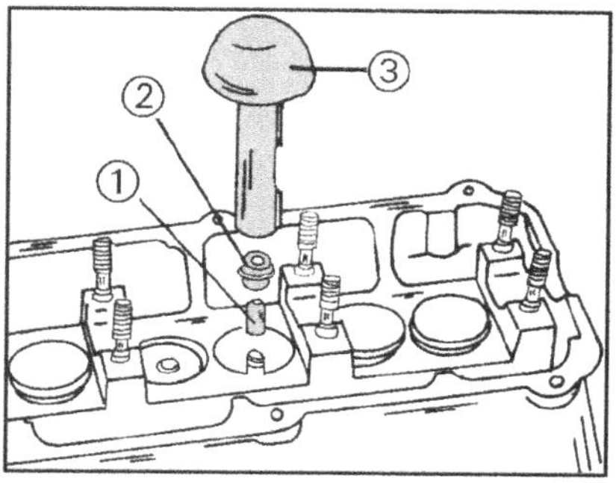

- Fit new valve stem seals. A special tool is available for this purpose. A plastic sleeve is pushed over the valve guide and the well-oiled seal is placed over the sleeve. The tool is shown in Fig. 1.34 and should be used as follows.
- Place the plastic sleeve (1) over the protruding valve guide and press firmly in position.

Fig. 1.34 – Fitting the valve stem oil seals (refer to the text).

- Lubricate the sealing ring (2) and slide it over the sleeve (1).
- Place the tool (3) over the valve stem seal and carefully push it over the valve guide. We would like to point out that a special tool or a similar arrangement must be used in order to avoid damaging the valve stem seals.
- Check the end float of the camshaft as described earlier on.
- Fit the valve springs, place the upper spring retainer over the valves and using the valve compressor and compress the springs until the valve cotter halves can be inserted by means of a thin-nosed pair of pliers. Remove the valve compressor and check that the cotters have engaged in their groove by tapping the ends of the valve stems lightly with a hammer.

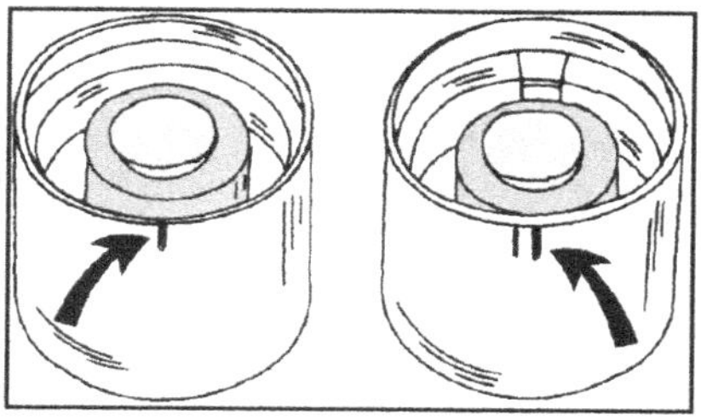

Fig. 1.35 – Tappets should have been marked with their installation position in the cylinder head.

Five-Cylinder Engine

- Insert the tappets into their bores (well oiled). If the same tappets are used, insert them into their original bores (Fig. 1.35).

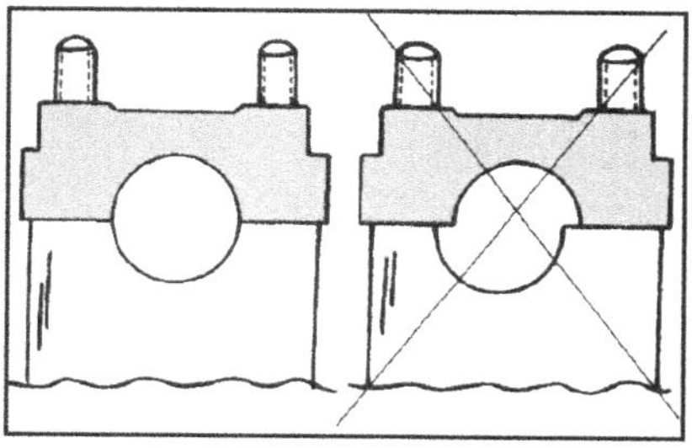

Fig. 1.36 – Camshaft bearing caps must be fitted as shown in the L.H. side.

- Refit the camshaft. Generously lubricate the camshaft journals and the bearing bores with engine oil. Place the camshaft in position. The two cams for the valves of the first cylinder must be facing upwards. Install the camshaft as follows.
- Fit the camshaft bearing caps Nos. 2 and 4 and evenly tighten the four nuts. Refer to Fig.1.36 to make sure that the offset is shown. Coat the No.4 bearing cap face with sealing compound (AMV 174 004 01 is specified). Fit the remaining bearing caps Nos. 1 and 3 in the same manner. In this case, coat the No. 1 bearing cap face with sealing compound.
- Using a piece of tube of suitable diameter, fit the camshaft oil seal over the shaft and into the bore. The spring inside the oil seal must be facing towards the cylinder head. Lubricate the oil seal lip and the running area on the camshaft with engine oil before fitting. A second oil seal is fitted to the head of a five-cylinder engine on the other side of the cylinder head. The camshaft oil seals(s) can also be fitted as described later on under "Camshaft and Camshaft Oil Seal".
- Slide the camshaft timing gear over the camshaft (do not forget the Woodruff key) and fit the bolt with the washer. Hold the timing gear against rotation and tighten the bolt to 8.5 kgm (62 ft.lb.) or 10.0 kgm (72 ft.lb.), but first read the note below.

Note: If the camshaft timing gear bolt is marked "8.8", tighten the gearwheel to 8.5 kgm. If the bolt head is marked "10.9", the tightening torque is 10.0 kgm (72 ft.lb.). Check the marking of the bolt before applying the torque wrench.

- Refit the inlet and exhaust manifolds to the cylinder head (new gaskets) and refit the injectors and the glow plugs. Refit the remaining parts to the cylinder in reverse order to the removal procedure.

Four-Cylinder Engine

Many of the operations required for the assembly of the cylinder head have already been mentioned during the dismantling of the head. The following information concerns mainly the installation of the camshaft and the adjustment of the valve.

Camshaft Installation

- Place the lower bearing caps in position (make sure that the dowel sleeves are fitted).
- Coat the camshaft bearing journals with engine oil and place the shaft into the bearing bores. Turn the camshaft until the cams of No. 1 cylinder (flywheel end) are pointing upwards.
- Place the upper bearing caps in position, fit the bolts and tighten them from the centre towards the outside evenly to 3.0 kgm (22 ft.lb.).
- Rotate the camshaft until the cams of No. 1 cylinder are in the TDC position, i.e., the cams for the inlet valves and the cam for the exhaust valve must be facing upwards. Prevent the rotation of the camshaft until the head is fitted.
- Fit the camshaft gearwheel and the bolt and tighten the bolt finger-tight.
- Block the crankshaft against rotation by applying a ring spanner to the centre bolt inside the crankshaft pulley (see Fig. 1.25) and tighten the camshaft gearwheel bolt to 10.0 kgm (72 ft.lb.).
- Refit the dust cover to the crankshaft vibration damper and tighten the screws to 1.0 kgm (7 ft.lb.).
- Refit the cylinder head cover.

Adjusting the Valve Clearances

A ring spanner to turn the crankshaft (Fig. 1.25), a feeler gauge to measure the valve clearances and a screwdriver to adjust the clearances are required to check and/or adjust the valves. Note the following before any adjustment.

- The valve clearance must only be adjusted on a cold engine. If the engine is fitted and has been running allow at least 2–3 hours before any check.
- The gearbox must be in neutral.
- The valve adjusting screws (see Fig. 1.22) have a locking device that will engage every half turn, indicated by an audible "click". The clearance must, therefore, only be measured after the "click" has been heard. If the valve adjusting screw is only "at halfway" and the adjusting screw is turned, the spring retaining clip will come away and the screw advances to the next locking position, with the resulting incorrect valve adjustment. *Apart from this, serious engine damage can result.*

Fig. 1.37 – Checking the valve clearances on a four-cylinder engine.

Check and/or adjust the valve clearances as follows.

- Remove the cylinder head cover and then the rubber cap from the crankshaft vibration damper.
- Rotate the crankshaft in the manner shown in Fig. 1.25 (ring spanner or special wrench shown) until the cams of the cylinder to be dealt with are pointing upwards. For reference, remember that No. 1 cylinder is on the flywheel side. The adjustment can be carried out in the firing order or the engine, i.e., first set No. 1 cylinder to TDC and then turn the crankshaft accordingly.
- Check the valve clearances of the cylinder in question. Adjust the two inlet valves to 0.20 mm (0.008 in.) and the exhaust valve to 0.30 mm (0.012 in.). Check with a feeler gauge, as shown in Fig. 1.37. The feeler gauge must slide in with little resistance. A good sign is if the gauge can just be inserted and then slides in with a little pressure.
- Adjustments are carried out on the adjusting screw, *but remember, only adjust the screw by a half a turn at a time.*
- Refit the cylinder head cover and then the rubber dust cover to the crankshaft vibration damper (1.0 kgm).

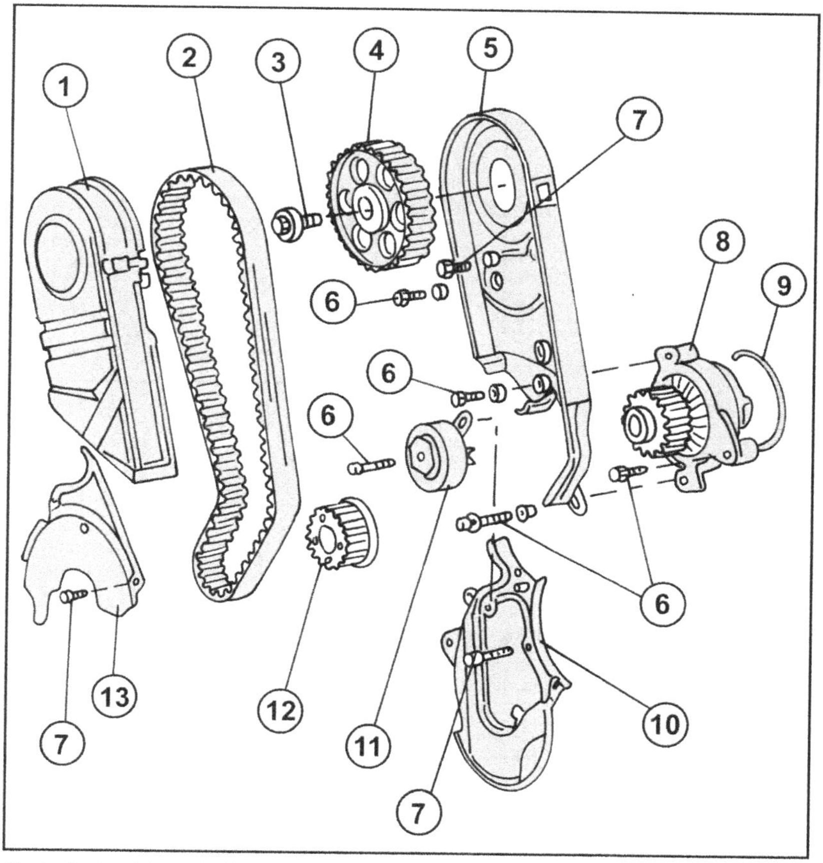

Fig. 1.38 – The toothed belt drive of a five-cylinder engine.

1 Upper timing belt cover
2 Toothed belt
3 Camshaft gearwheel bolt
4 Camshaft gearwheel
5 Inner timing belt cover
6 Screw (with sealer)
7 Screw (sealer)
8 Coolant pump
9 "O" sealing ring
10 Lower inner belt cover
11 Belt tensioner
12 Crankshaft gear
13 Lower belt cover

1.4.2. Timing Drive – 2.5-Litre Engine

The component parts of the timing drive are shown in Fig. 1.38. Study the illustration before any work is carried out. Follow each instruction step by step to avoid mistakes, as wrong installation of the timing belt can lead to immediate damage to the engine. Camshaft and crankshaft must not be rotated when the timing belt is removed, as the valves may hit the pistons. The following description assumes that the necessary special tools or other means to counterhold the camshaft timing gear, locking the injection pump gear, etc. are available or can be obtained. If this is not the case, do not attempt the replacement of the timing belt.

The drive of the overhead camshaft is by means of a toothed belt, running on toothed sprockets on the crankshaft, camshaft, water pump and the tensioning roller. The drive gear for the injector pump is located on the opposite side of the timing drive. The injection pump is driven by a separate toothed belt. Operations on this belt drive are described under a separate heading.

Timing Belt Replacement

Figs. 1.38 should be referred to during the following operations. The front of the vehicle must be removed as described during the removal of the engine to gain access to the individual parts. Proceed as follows.

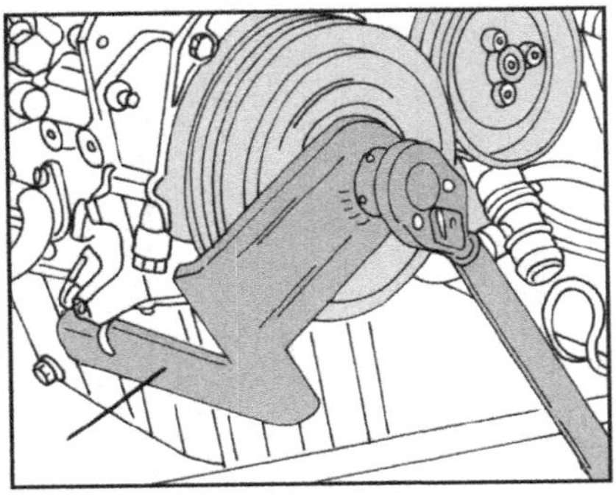

Fig. 1.39 – Preventing the vibration damper from rotation when slackening the bolt in the centre of the damper.

- Remove the ribbed V-belt as described later on in this section.
- Remove the upper timing belt cover (1).

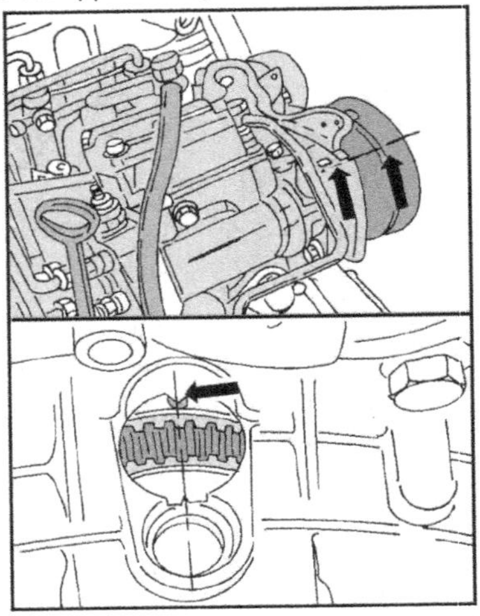

Fig. 1.40 – Aligning the timing marks on the injection pump drive gear and the mounting console (upper view) and the flywheel and the clutch housing (lower view).

- Remove the visco clutch and the air guide from the radiator (see the section "Cooling System") and remove the visco clutch pulley.
- Remove the maintenance cover inside the cab.
- Remove both toothed belt guards (timing belt and injection pump belt) and the cylinder head cover. Note that the cylinder head cover is secured with bolts, screwed into the camshaft bearing caps.
- Prevent the crankshaft from rotating similarly as shown in Fig. 1.39 and slacken the centre bolt of the vibration damper/crankshaft pulley.
- Rotate the engine until the piston of the No.1 cylinder is at the top dead position. This is the case when the marks in Fig. 1.40 are aligned as shown. Note that this position can only be obtained every second revolution. Also check on the flywheel to make sure that the two marks are aligned.
- Remove the drive belt for the injection pump as described later on.
- Completely remove the bolt from the centre of the vibration damper/crankshaft timing gear and unscrew the four bolts securing the damper to the timing gear. The vibration damper can now be removed.
- Mark the running direction of the toothed belt. Use a felt pen and mark an arrow into the outside of the belt.
- Slacken the securing bolt of the belt tensioner (11) in Fig. 1.38 and use a ring spanner to release the belt tension until the belt can be removed. Remove the toothed belt.
- Without rotating the camshaft, lock the shaft with the setting bar shown in Fig. 1.41 at the rear of the cylinder head. If this bar is not available, make sure that the camshaft cannot rotate during the following operations.

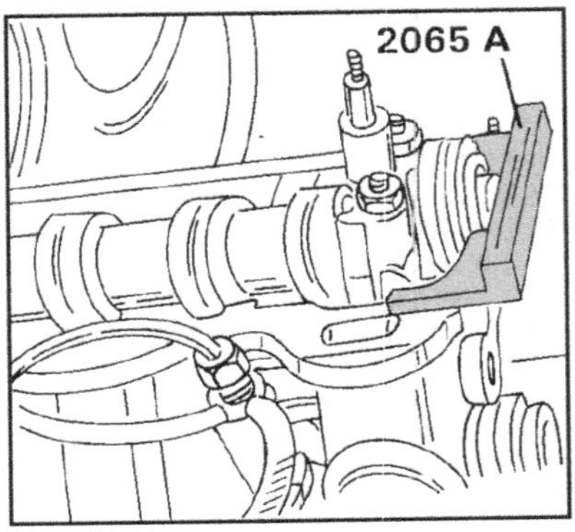

Fig. 1.41 – Locking the camshaft in position with the special tool before removal of the timing belt.

Before the new belt is fitted, on a removed engine, align the flywheel and the injection pump, as shown in Fig. 1.39. The marks should still be in the set position if the engine is fitted. Remove the injection pump drive sprocket and check that the camshaft is locked with the setting tool, as shown in Fig. 1.41. If the special tool is used, it will have to be centred as follows.

- Turn the camshaft until one end of the setting bar touches the cylinder head. Measure the gap at the other end with feeler gauges.
- Take half of the thickness of the feeler gauges and insert this thickness between the setting bar and the cylinder head.

- Now, turn the camshaft until the bar rests on the feeler gauges and insert the second feeler gauge of the same thickness between the other end of the bar and the head.

The timing belt is now fitted as follows.

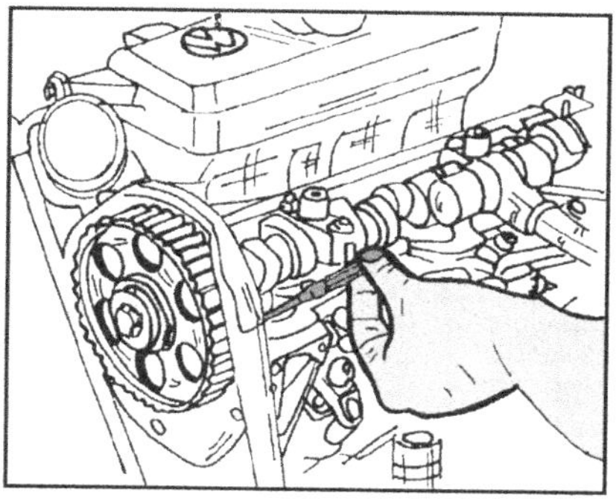

Fig. 1.42 – To remove the camshaft timing gear from the taper of the camshaft, insert a drift at the position shown and hit it with short blow of a hammer.

- Slacken the camshaft sprocket securing bolt by half a turn and release the camshaft sprocket from the camshaft taper by tapping with a hammer and a drift through the rear toothed belt guard opening. This can be seen in Fig. 1.42. Make sure that the gearwheel cannot rotate if the setting gauge is not used.

For the next operation, refer to Fig.1.43.

- Fit the tensioning roller (1) so that the roller tab seats in the lower toothed belt guard moving pin. This is indicated by (A) in the illustration. In this position, fit the camshaft timing belt and tighten the roller securing bolt finger-tight.

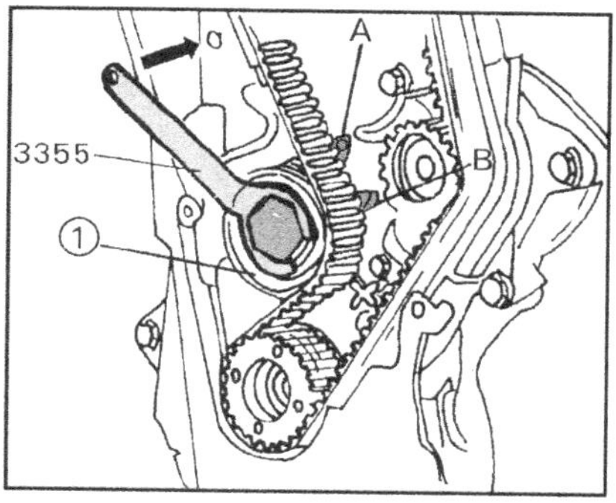

Fig. 1.43 – Details for the installation of the timing belt. The numbers and letters are referred to in the text.

- Turn the tensioning roller clockwise in the direction of the arrow.

Note: A ring spanner can be used in place of the special tool shown in the illustration.

- Turn the special tool or ring spanner shown in the illustration, as far as the stop and from this position turn the tensioning roller backwards until bolt indicators align at position (B) in the illustration. Retain this position and tighten the bolt in the centre of the roller to 2.0 kgm.
- Check once more that the crankshaft is still at the top dead centre position (see Fig. 1.40). Otherwise, correct until the marks are aligned.
- Hold the camshaft timing gear against rotation and tighten the bolt. If the bolt head marking is "8.8", tighten the bolt to 8.5 kgm (61.5 ft.lb.), if the bolt head is marked "10.9", tighten the bolt to 10.0 kgm (72 ft.lb.). Make absolutely sure that the correct torque is applied.
- Remove the setting bar (Fig. 1.41) from the camshaft.
- Fit the crankshaft timing gear and the vibration damper and tighten the bolt. Hold the damper against rotation, as shown in Fig. 1.39, and tighten the centre bolt to 16.0 kgm (115 ft.lb.). From the final position, tighten the bolt a further 180° (half a turn), if necessary in several stages. Fig.1.39 shows how the vibration damper is locked against rotation in the workshop. A similar device must be made-up to lock the shaft. You may be able to engage a gear to lock the crankshaft, but due to the elasticity of the drive, it may be difficult to tighten the bolt without having the crankshaft locked solid.
- Fit the vibration damper to the crankshaft timing gear and tighten the four M8 bolts to 2.0 kgm.
- Rotate the crankshaft by two complete turns and check once more that the tensioning roller is in the correct position. If the R.H. edges of the pointers "A" and "B" are not aligned, re-adjust the belt.
- Install the toothed belt for the injection pump drive as described below.
- Install the two toothed belt guards and the cylinder head cover.
- Install the maintenance cover.
- Refit the ripped drive belt as described later on in this section. Finally, refit the protective cover underneath the vehicle.

Replacing the Injection Pump Drive Belt

A tensioning roller is fitted to tension the belt for the injection pump of the 2.5-litre engines fitted to the LT. Various special tools are recommended and we advise you to read through the text before you attempt the work.

Fig. 1.44 shows the parts on the side of the injection pump. Removal and installation takes place by referring to the illustration. Note that two rollers are fitted: one is the tensioning roller and the other one is a guide roller.

The commencement point of injection must be checked and/or adjusted and the job must be carried out by a workshop. The following instructions only deal with the removal and installation and the tensioning of the drive belt.

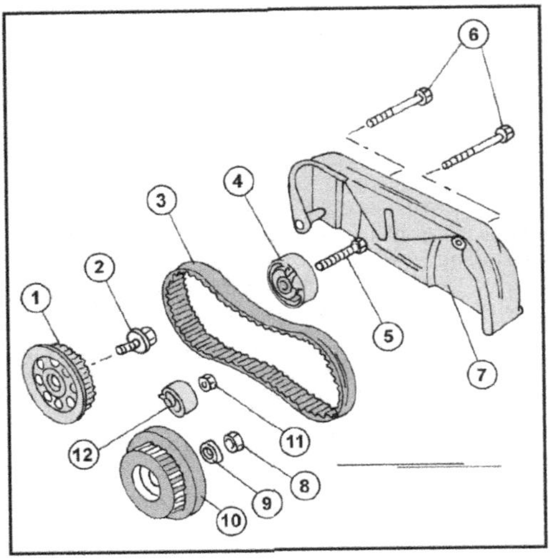

Fig. 1.44 – The component parts of the injection pump drive belt.

1 Injection pump drive gear
2 Bolt, 10.0 kgm
3 Pump drive belt
4 Tensioning roller
5 Bolt, 1.5 kgm
6 Bolts. 2.0 kgm
7 Belt cover
8 Nut, 9.0 kgm
9 Plain washer
10 Pump drive gear
11 Nut, 2.0 kgm
12 Idler roller

- Remove the maintenance cover inside the cab.

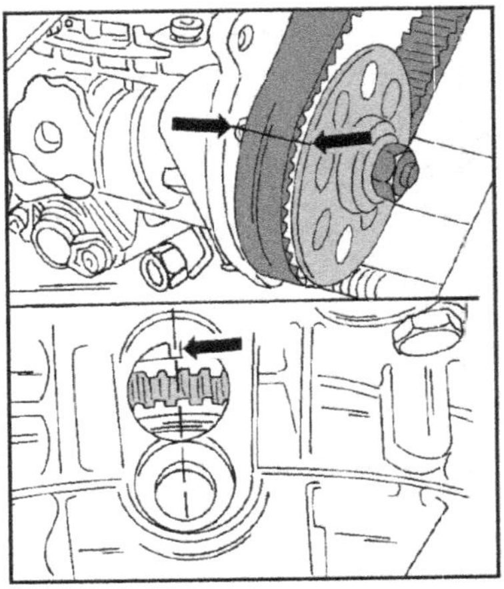

Fig. 1.45 – Alignment of the timing marks on the injection pump drive gear and the pump mounting console and the flywheel and the clutch housing opening in the lower view.

- Remove the belt cover (7) after removal of screws (8).
- Rotate the crankshaft in the direction of rotation until the piston of No. 1 cylinder is at the top dead centre position. This can be checked by referring to Fig. 1.45. This position is only possible every 2nd TDC position. Also check that the marks shown in Fig. 1.46 on the crankshaft vibration damper and the lower timing belt cover are aligned as shown.

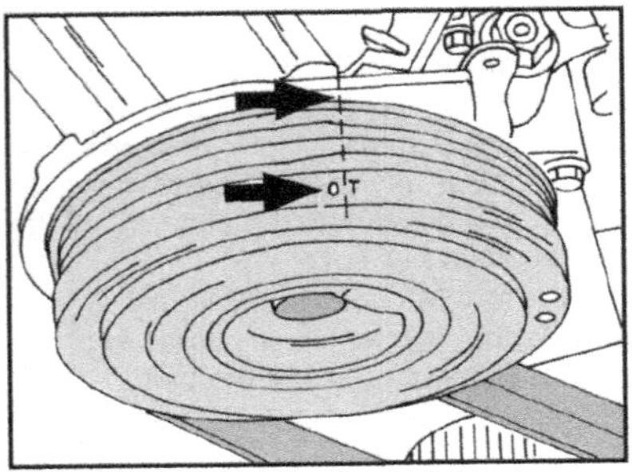

Fig. 1.46 – Alignment of the timing marks on the crankshaft vibration damper and the lower timing belt cover. Note the inscription "OT" (TDC).

- Mark the direction of running of the drive belt into the outside of the belt, using a felt-tip pen, if the same belt is to be refitted.

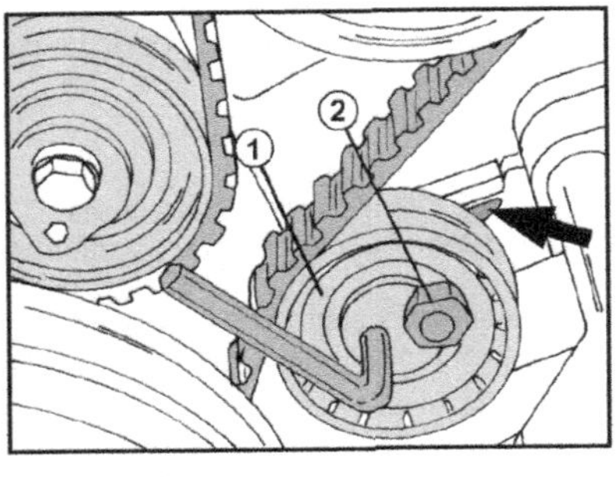

Fig. 1.47. – Align the idler roller for the injection pump drive belt in accordance with the description in the text.

- Counterhold the injection pump drive gear and remove the bolt in the centre. Take off the drive gearwheel and the belt (3). Remove the tensioning roller (4) after removal of the bolt (5), if required. The idler roller (12) can also be removed. If necessary, unscrew the nut (11).

The injection timing must be checked after installation of the drive belt. The following description only deals with the installation and tensioning of the drive belt.

- Place the new belt over the injection pump drive gear and the gearwheel on the camshaft and tighten the securing bolt for the injection pump until it can just be rotated by hand. Observe the marking of the belt, if the original one is fitted.

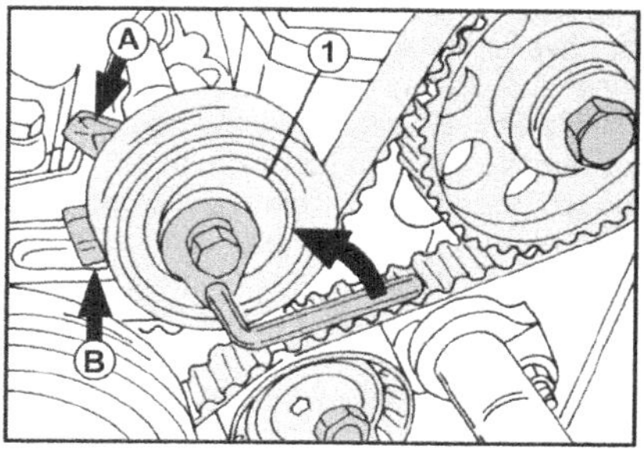

Fig. 1.48 – Details for the adjustment of the drive belt tension for the injection pump.

- Check the installation position of the idler roller (12) in accordance with Fig. 1.47. The idler roller indicator (1) must be in line with the cylinder head flange (arrow). If necessary, slacken the securing nut (2) and turn the roller until the alignment is obtained. Tighten the nut to 2.0 kgm (14.5 ft.lb.).
- Refit the tensioning roller (1) in Fig.1.48 and adjust it. A tab (A) on the roller must be positioned in the cut out of the console, shown in the illustration by arrow (B). Tighten the bolt in the centre of the tensioning roller hand-tight when in the position described.
- Turn the tensioning roller in Fig. 1.48 in an anti-clockwise direction as far as the inner stop and from this position turn it back until both pointers align as shown by (A). Tighten the securing bolt in the centre to a torque setting of 1.5 kgm (11 ft.lb.).
- Counterhold the injection pump drive gear and tighten the securing bolt to 16.0 kgm (115 ft.lb.). Install the toothed belt cover. All other operations are carried out in reverse to the removal procedure.

1.4.4. Camshaft

1.4.4.1. Camshaft – Removal and Installation

The camshaft on all engines can be removed with the cylinder head fitted. Prepare the vehicle to have access to the engine compartment. The removal and installation of the camshaft for the four-cylinder engine has already been described during the removal and/or dismantling of the cylinder head. First, remove the cylinder head cover and, of course, the timing belt and the injection pump timing belt. You must therefore refer to the previous sections.

The following instructions apply to the 2.5-litre engine.

- Rotate the engine until the piston of No.1 cylinder is at the top dead centre position. Refer to the instructions during the removal of the timing belt.
- Remove the timing belt and the injection pump drive belt as described in the last section. Section 1.4.2 (pages 27 and 30) should be referred to. Also remove the vacuum pump and the operating tappet.
- Slowly slacken the nuts of the bearing caps 1 and 3 in a diagonal manner. At this stage, do not slacken the nuts of bearing caps Nos. 2 and 4.
- Now, unscrew the nuts of bearing caps No. 2 and No. 4 in a diagonal fashion until the spring pressure is released and the two caps are fully free. The pressure will be retained to the last moment, as the springs will push the camshaft upwards. Remove the half-round oil seals from the No.1 and No. 5 bearing cap, if it is still in the cap. Otherwise, remove the seals from the cylinder head.
- Carefully remove the camshaft and remove the camshaft timing gear from one end and the injection pump drive gear from the other end. The camshaft must be clamped into a vice, using soft-meal jaws.
- The tappets can either be removed or left in position (mark where they came from). Refer to the section on the tappets before refitting them. Tappets must be checked before they are refitted.

The installation is a reversal of the removal procedure, but the tightening sequence of the bearing caps is important.

- If the tappets have been removed, check them as described in the next section. Then, insert each tappet (well oiled) in its original bore (if the same tappets are used), in accordance with the markings.
- Generously lubricate the camshaft bearing journals and the bearing bores with engine oil. Place the camshaft in position. The two cams for the valves of the first cylinder must be facing upwards. Coat the contact faces of No. 1 and No. 4 with sealing compound (if possible AMV 004 01). Fit the bearing caps Nos. 2 and 4. The caps are slightly off-centre and must be fitted the correct way round. Fig. 1.36 shows the correct and incorrect installation of the bearing caps and should be referred to.
- Tighten the nuts of the fitted bearing caps diagonally by a few turns until the caps are down on the cylinder head and then to a final torque of 2.0 kgm (14.5 ft. lb.).
- Fit the remaining bearing caps (No. 1 and No. 3) and tighten the nuts in a diagonal manner to 2.0 kgm (14.5 ft.lb.).
- Fit the camshaft oil seal or the two seals as described below.

1.4.4.2. Replacing the Camshaft Oil Seal(s) – 2.5 Litres

The oil seals at both ends of the camshaft can be replaced with the engine fitted. VW-Workshops use a special tool to remove the oil seal, but it is also possible to remove the bearing cap to remove and install the seal(s). The oil seal is located behind the timing wheel, the second oil seal is located behind the drive wheel for the injection pump, i.e., the wheel (and drive belt) will have to be removed as described in the section applicable to the engine fitted. Proceed as follows to remove the seal(s).

- Slacken the alternator mountings, push the unit towards the inside and remove the ribbed belt.
- Unscrew the three screws securing the timing belt guard. Also remove the cylinder head cover.
- Rotate the engine until the piston of the first cylinder is at the top dead centre. This is indicated when both valves of this cylinder are closed. Turn the engine by applying a socket or ring spanner to the crankshaft pulley centre bolt.
- Slacken the timing belt, as described, and lift off the timing belt. Tie the belt in position to prevent it from slipping off the other timing wheels.
- Counterhold the camshaft timing wheel and slacken the camshaft timing gear bolt by one turn.
- Insert a small drift into the bore shown in Fig. 1.43 and hit the end of the drift with a short blow of a hammer. The cone seating of the gearwheel will be split and the gearwheel can be removed after fully removing the bolt and the washer. Using a pair of side cutters, remove the Woodruff key.
- Unscrew the rear toothed belt guard from the cylinder head. The bearing cap can now be unscrewed (or both bearing caps) and the oil seal replaced. The arrow in the oil seal on the injection pump side must face in the direction of camshaft rotation.

The installation of the oil seals is a reversal of the removal procedure. Clean the camshaft journal and the inside of the bearing cap and lubricate the oil seal lip and the journal with engine oil. Orientate the spring side of the seal towards the camshaft and place the seal over the shaft and into the bearing cap. Use a suitable piece of tube and carefully knock the oil seal in position until flush.

Fit the Woodruff key, slide the timing wheel over the shaft and the key and fit the bolt and washer. Counterhold the timing wheel and tighten the bolt to the correct torque, noting that on some engines, the bolt head is marked with a number that regulates the applicable torque.

Carry out all other operations in reverse order to the removal procedure.

1.4.4.3. Camshaft – Removal and Installation (four-cylinder)

The removal is described on page 22 , and the installation and the subsequent adjustment of the valves are described on page 26.

1.4.4.4. Hydraulic Tappets

The following points must be observed at all times in order to prevent damage to the engine by the tappets, due to incorrect treatment.

- Always store hydraulic tappets in the upright position, with the contact area at the bottom.
- Always replace tappets as complete units as they cannot be adjusted or repaired.

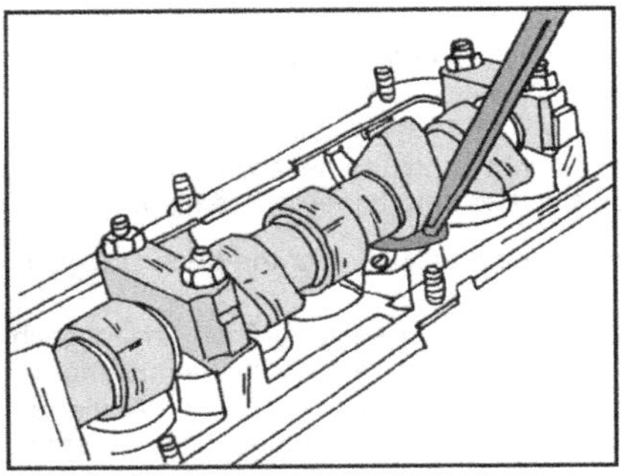

Fig. 1.49. – Checking the tappets (refer to text).

- Valve noises when the engine is started are normal, but excessive noise can be due to the tappets. In this case, check the tappets one after the other as follows.
- Start the engine and let it run until the cooling fan cuts in.
- Increase the engine speed to 2500 rpm for approx. 2 minutes.

If the tappets are still noisy, stop the engine and remove the cylinder head cover. To find the noisy tappet, turn the engine over until the cam in question is pointing with the tip upwards and using a piece of wood or plastic, as shown in Fig. 1.57, and push the tappet downwards. If the play is more than 0.10 mm (0.04 in.) before the valve begins to open, fit a new tappet.

In this case, it may be best to replace the complete tappet set, as other tappets could be equally in bad condition.

Note: If new tappets are fitted, do not start the engine for about 30 minutes after assembly, as this may cause the valves to strike the pistons.

1.4.5. PISTONS AND CONNECTING RODS

1.4.5.0. Technical Data

Piston Diameter:
- Five-cylinder engine: .... 80.96 mm
- Four-cylinder engine: .... 92.96 mm

Oversizes: .... Refer to workshop

Running Clearance of Pistons:
- Standard: .... 0.03 mm
- Wear limit: .... 0.07 mm

Piston Rings:	***Four-Cylinder***	***Five-Cylinder***
Number of piston rings:	3	3
Side clearance of rings in grooves:		
- Std., upper rings:	0.085–0.090 mm	0.07–0.11 mm
- Wear limit:	0.25 mm	0.25 mm
- Std., centre rings:	0.070–0.105 mm	0.05–0.08 mm
- Wear limit:	0.25 mm	0.25 mm
- Std., oil control rings:	0.03–0.065 mm	0.03–0.06 mm
- Wear limit:	0.15 mm	0.15 mm
Piston ring gaps:		
- Std., compression rings:	0.30–0.55 mm	0.25–0.45 mm
- Std., oil control rings:	0.30–0.50 mm	0.25–0.50 mm
- Wear limit:	1.0 mm	1.0 mm

Piston Protrusion above Cylinder Block: See the main text. Regulated by thickness of cylinder head gasket. Five-cylinder engine only.

Connecting Rods:
- Bearing running clearance: .... 0.251 mm (4-cyl.), 0.40 mm (5-cyl.)
- Connecting rod bearings: .... Refer to "Crankshaft"
- Piston pin running clearance: .... approx. 0.01–0.02 mm

1.4.5.1. Servicing and Overhaul

The component parts of a piston and connecting rod assembly for a four-cylinder engine are shown in Fig. 1.50.

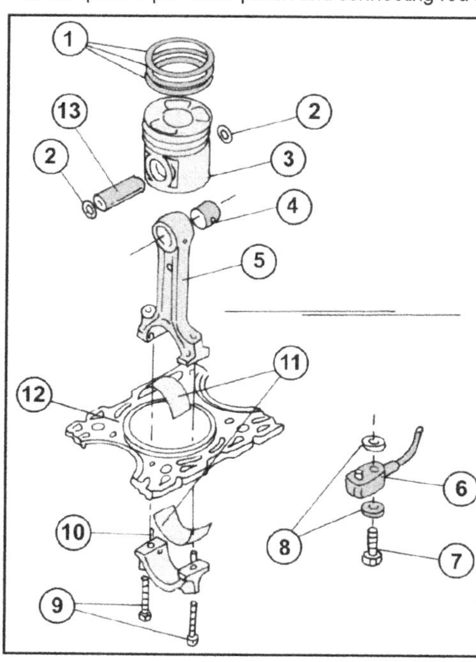

Fig. 1.50 – Piston and connecting rod of a four-cylinder (2.8 litres) engine.

1 Piston rings
2 Piston pin snap ring
3 Piston
4 Small end bush
5 Connecting rod
6 Oil splash jet
7 Banjo bolt, 1.0 kgm
8 Sealing washers
9 Connecting rod bolts
10 Spring dowel sleeve
11 Big end bearing shell
12 Cylinder block
13 Piston pin

Similar parts are used in the five-cylinder engine, with the difference of the shape of the oil splash jet. The four-cylinder engine has two splash jets. Also note the tightening torque for the bearing cap bolts (9) – different for four- and five-cycle engines. The piston crowns of the two engines are, however, not the same. The pistons of the four-cylinder engine have recesses in the piston crowns to accommodate the three valve heads. The following points must be observed in particular when a five-cylinder, 2.5-litre engine is dealt with.

- The cylinder head does not have combustion chambers, also known as swirl chambers. Instead, the combustion chambers are machined into the piston crowns.

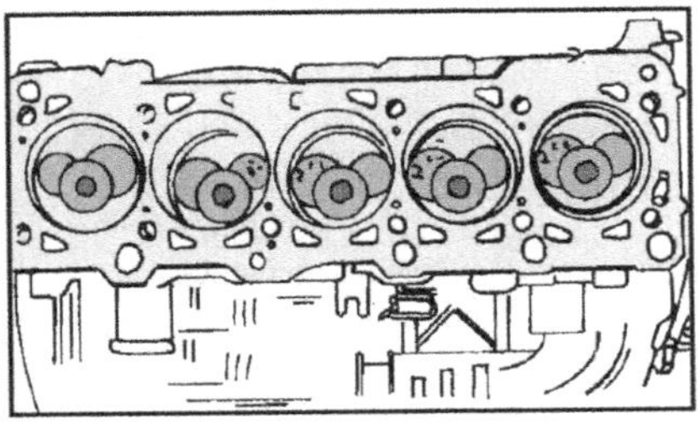

Fig. 1.51. – View of the cylinder block with the fitted pistons of a TDI engine. Note where the large recess for cylinders 1/2 and 3/4/5 are located.

- Pistons for the engine are different in shape. They have, what Volkswagen call an Omega recess which forms the combustion chamber.
- After removal of the cylinder head, you will see the pistons arranged, as shown in Fig. 1.51. Each piston has a larger recess for the inlet valves, but not all are facing towards the same side. The recess on a piston for cylinders Nos. 1 and 2 must be nearer the flywheel end, and the recess of a piston for cylinders 3, 4 and 5 must be near the pulley end. New pistons are either marked with 1/2 or 3/4/5 to indicate where they must be fitted. When pistons are removed, mark their fitted position, as wrong installation of pistons will lead to engine damage.

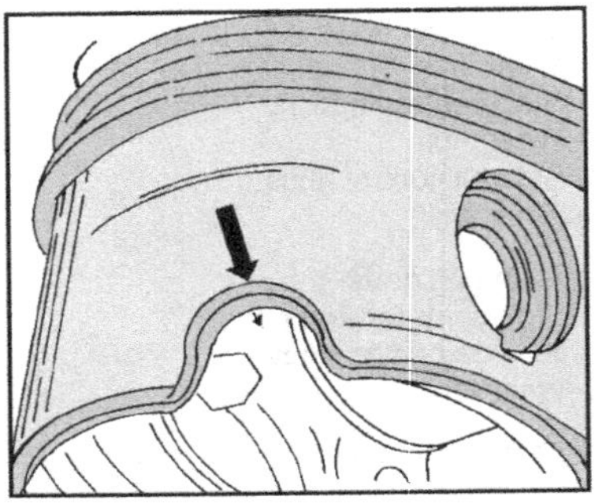

Fig. 1.52 – The pistons for a diesel engine are recessed on one side to prevent contact between the oil spray jets and pistons. The piston side with the recess must be on the same side as the oil jets in the cylinder block.

- The pistons of all engines have a recess for the oil spray jets on one side (see Fig. 1.52) to prevent contact of the jets and the pistons at the bottom top dead centre. These sides must be on the same side as the oil jets.

Removal of Pistons and Connecting Rods

The pistons and connecting rods are pushed out from the top of the cylinder block after removing the bearing caps and the lower bearing shells. Before carrying out these operations, note the instructions below regarding marking and identification. Note the references to the 2.5-litre engine given above.

- Mark each of the piston centres and connecting rods to the cylinder bore from which it is removed. This can be carried out by numbering the piston crowns with the numbers 1, 2, 3 and 4 (and 5), as shown in Fig. 1.53 and with an arrow each, which should be facing towards the crankshaft pulley. Remember that the arrow marked into the piston crown cannot be seen due to the carbon deposits.

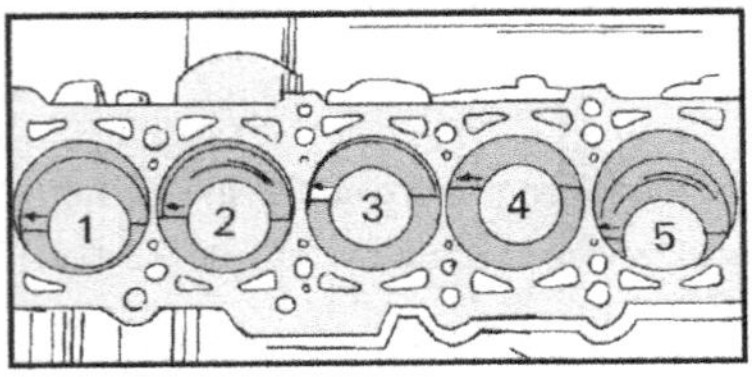

Fig. 1.53 – Piston identification. The top of the pistons should be marked with their cylinder number and arrows. Remember that cyl. No. 1 of a four-cylinder engine is on the flywheel side.

- Refer to Fig. 1.54 and note the arrows pointing to the cast lugs and their identification marks on the connecting rod and bearing cap. These lugs should face towards the crankshaft pulley.

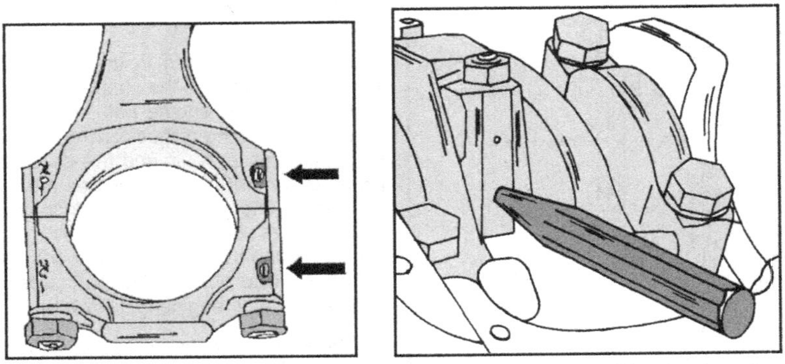

Fig. 1.54 – The two casting lugs (arrows, L.H. side) on the connecting rod and big end bearing cap – these must match when the bearing caps are refitted. Always use new connecting rod bolts and nuts. Connecting rods and bearing caps can be marked with a centre punch as shown on the R.H. side before the cap is removed. The illustration shows a cap with nuts. These engines use, of course, bolts to secure the bearing caps.

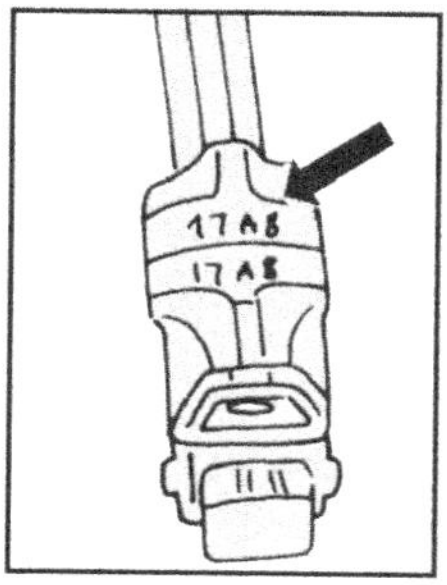

Fig. 1.55 – Connecting rods and bearing caps of a four-cylinder engine are marked at the position shown.

- Before removing the connecting rod bearing caps, mark each of the con rods and its cap with a centre punch to ensure assembly in the same position (Fig. 1.54, R.H. side). In the case of a four-cylinder engine, refer to Fig.1.55. The bearing cap and connecting rod are marked with letters and numbers, which must be opposite to each other when the parts are assembled.
- Mark the bearing shells and the rod to which they belong and mark the upper and the lower shells as soon as they have been removed.
- If pistons are replaced, make sure that they are of the correct diameter and the correct engine.
- Remove the bearing caps and shells to take out the piston assemblies, making sure that there is no ridge of carbon in the bore that might prevent removal.

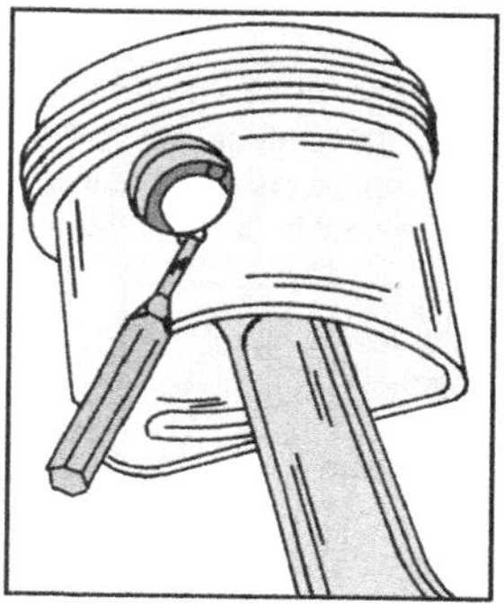

Fig. 1.56 – Piston pin securing clips can be removed with a pointed drift, inserted into the notch of the piston pin bore.

- Push out the piston pin after having removed the circlips, by inserting a pointed drift into the notches of the piston pin bore, as shown in Fig. 1.56. Use a suitable mandrel and press to extract the piston pin. If the pin has a tight seat, heat the piston in hot water (at least 60 °C/140 °F).
- Using a pair of piston ring pliers, as shown in Fig. 1.57, remove the piston rings carefully from the pistons (over the crown) and keep them matched if they are to be re-used. If no piston ring pliers are available, insert some metal strips under the piston ring to be removed, on opposite sides of the piston and lift the ring this way over the piston crown. Clean all parts thoroughly, avoiding abrasive or sharp tools that might scratch the pistons.

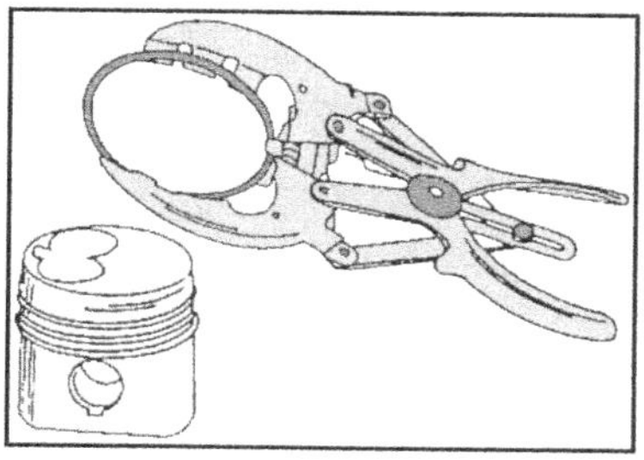

Fig.1.57 – Removal or installation of piston rings. Only the use of the expander made for this purpose will prevent breakage of the piston rings.

Carry out the inspections given below, all data being given in Section 1.4.5.0. Pistons must always be replaced in sets.

- Check the piston diameter 10 mm (0.4 in.) from the bottom edge of the piston skirt, as shown in Fig. 1.58, at right angles to the piston pin. The piston must be replaced if outside the limits given.

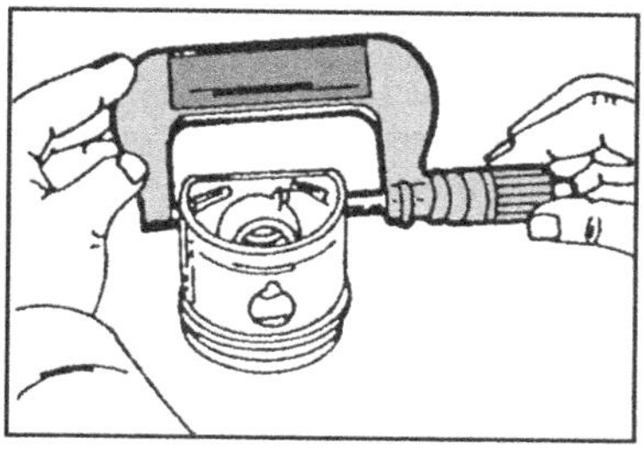

Fig. 1.58 – Measuring the piston diameter. Note where the jaws of the micrometre are applied.

- Check the clearance of the piston rings in their grooves as shown in Fig. 1.81. The grooves must be thoroughly cleaned before this check is carried out. Piston ring grooves can be cleaned with a broken-off piston ring, which should be ground or filed flat at the broken end. Go around the circumference of the groove, being careful not to lift off any metal.
- Check the ring gap by inserting the ring into the cylinder bore, as shown in Fig. 1.60. Make sure that the ring is inserted squarely – use the crown of a piston to press the ring down. Note that the gap is measured at the bottom end of the cylinder bore. Compare with the values given in Section 1.5.1.

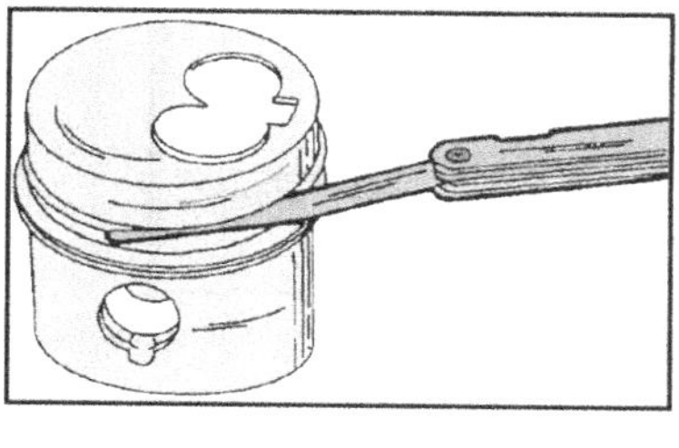

Fig. 1.59 – Checking the side clearance of the piston rings by means of a feeler gauge.

- Check the piston pins and the bores of the connecting rods for wear. Connecting rods must be replaced in complete sets.

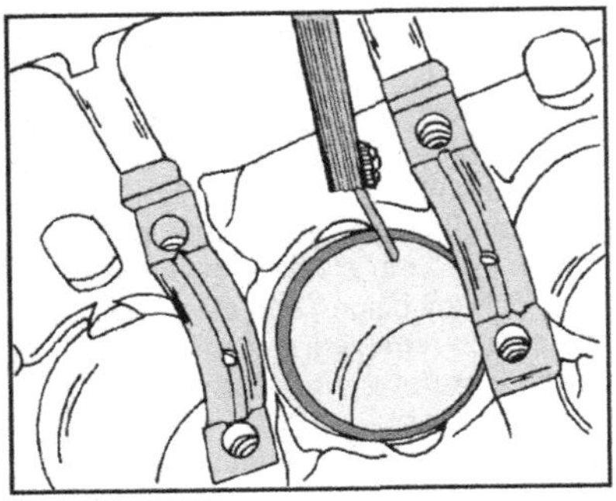

Fig. 1.60 – Checking the piston ring gaps, rings at bottom of cylinder bore.

- If possible have the connecting rods checked for distortion, especially if the wear pattern on the old pistons is more pronounced on one side.
- Replace the connecting rod bolts for wear. The bolts stretch with the tightening of the nuts, i.e., once tightened they are useless.

To install the piston pin into the piston and connecting rod, a proper mandrel must be used with a press, and the piston must be heated to at least 60 °C (140 °F). The piston can be inserted into hot water to heat it up. The disposition of the connecting rod must be as follows.

Five-Cylinder Engines

- Fit the pistons so that the cylinder numbers are arranged, as shown in Fig. 1.53. After installation, compare the fitted pistons with Fig. 1.51, to make sure that the piston bowls face into the correct direction.
- The cast bosses on the rod and cap (Fig. 1.54) must point towards the crankshaft pulley end of the engine.

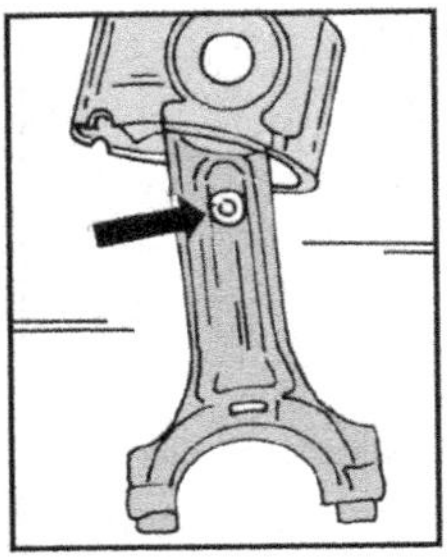

- The cylinder No. marking on the connecting rod and big end bearing cap (Fig. 1.54) must be opposite to each other after the cap has been bolted to the connecting rod.
- Push the piston pins into their bores with the use of a mandrel and fit the retaining rings to both sides. Allow the assemblies to cool off and check that the pistons can move easily on their piston pins.
- When installing the piston rings, note that the rings are marked on one side with "Top" or "Oben". This marking must be facing towards the piston crown when the rings have been fitted. Rings can easily break and we suggest to carry out this operation very slowly and with the greatest care.

Fig. 1.61 – Arrange the small round boss on each connecting rod so that this side faces towards the flywheel side of the engine.

Four-Cylinder Engine

The assembly of pistons and connecting rods is carried out similarly as described above, but an important difference must be noted. Each connecting rod has on each side a small round boss. This boss is important during the installation of the assemblies. Otherwise, assemble pistons and con rods in accordance with the markings, if the old parts are refitted.

Pistons and Connecting Rods – Installation

- If the oil splash jets for the cooling of the pistons have been removed, refit them and tighten the bolts to 1.0 kgm (7 ft.lb.). In the case of a five-cylinder engine, the bolts must be coated with sealing compound (AMV 188 100 02) and are fitted as shown in Fig. 1.62 at the bottom of the cylinder bores. In the case of a four-cylinder engine, the two jets are fitted as shown in Fig. 1.62 on the R.H. side. The two jets have a small centring pin (arrows), which also assures the correct installation. Banjo bolts without sealing compound are used (1.0 kgm/7 ft.lb.). Sealing washers are fitted to both sides of the banjo bolts (replace washers).

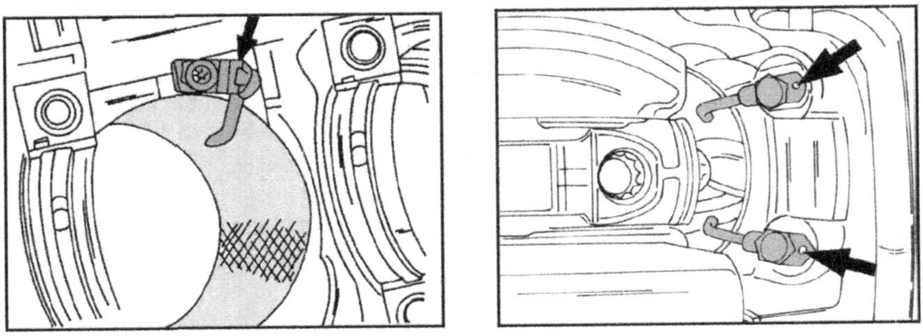

Fig. 1.62 – Location of the oil splash jets for the cooling of the pistons. In the L.H. view for the 2.5-litre engine and in the R.H. view for the 2.8-litre engine.

- Before fitting the pistons into the cylinder bores, arrange the bottom ring gap so that it faces the front of the engine, with the other two rings arranged at 120° to each other and facing the rear.
- Oil the cylinder bores before inserting the pistons into the cylinders. Use a ring compressor, as shown in Fig. 1.63 to avoid damage to the piston rings. If no ring compressor is available, make up a strip of thin sheet metal, lay it around the piston rings and squeeze together the ends of the strip to push the rings into their grooves. Install the pistons after checking that the piston pin circlips are in position on both sides, as it is very easy to forget one, after the first one has been fitted. The easiest method to install the pistons is to place the cylinder block on its side and turn the crankshaft until two crankpins are at the bottom. Fit these two pistons and connecting rods and then turn the crankshaft until the other pins are at the bottom. Fit the remaining assemblies. Check once more that the cylinder markings on connecting rods and big end bearing caps are on the same side of the engine and that the connecting rods are fitted the correct way. Refer to Fig. 1.61 when fitting a piston/connecting rod assembly of a four-cylinder engine.

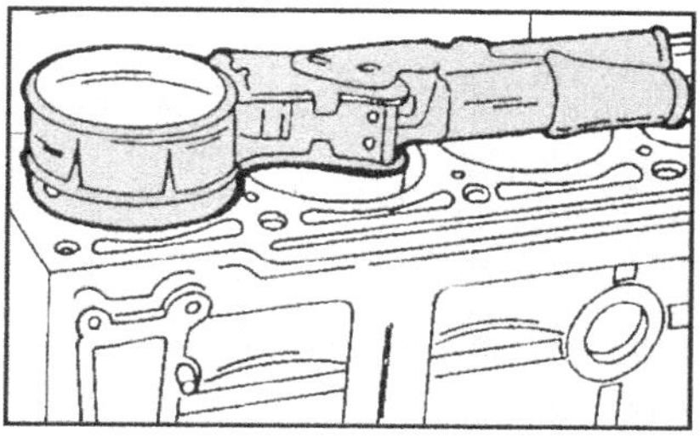

Fig. 1.63 – Pushing the piston rings into their grooves with the help of a proper piston ring compressor.

- Place the second bearing shell into the big end bearing cap, lubricate the bearing shell surface with engine oil and push the bearing cap against the big end of the con rod. Re-check that all markings are opposite each other, as shown in Figs. 1.54 or 1.55. Also make sure that the dowel pin is fitted into the face of the connecting rod big end. Otherwise, remove it from the bearing cap and knock into the connecting rod.
- The connecting rod nuts or bolts must always be replaced. Oil the nut contact faces and tighten the two bolts. In the case of a 2.5-litre engine, tighten the bolts in turn to 3.0 kgm (22 ft.lb.). From this position, tighten the nuts a further 90°. The bolts of a four-cylinder engine are tightened in the same manner, but the initial tightening torque is 2.0 kgm (14.5 ft.lb.).

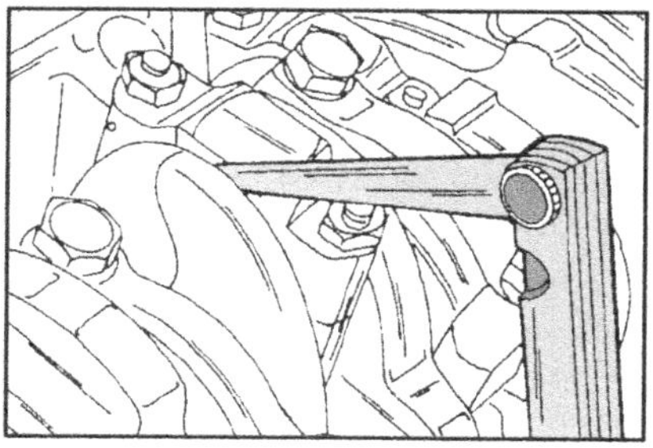

Fig. 1.64 – Checking the connecting rod end float by means of a feeler gauge. Insert the feeler gauge between the con rod thrust face and the crankshaft.

- After installation of the connecting rod bearings, check with a feeler gauge at the position shown in Fig. 1.64 to make sure that the connecting rod side play is less than 0.40 mm (0.016 in.) in the case of a 2.5-litre engine or 0.25 mm in the case of a four-cylinder engine.
- Temporarily fit the crankshaft pulley with the bolt, apply a socket to the bolt head and rotate the crankshaft assembly a few times. Any hard spots will be spotted immediately and can be rectified before the engine is further assembled. Do this after each assembly is fitted. After completed installation, refit the oil sump to the bottom of the crankcase as described in the relevant section. All other operations are carried out in reverse order to the removal procedure.

Oil Cooling Jets for Pistons

The pistons have a cut-out (recess) on one side, as shown in Fig. 1.52. During installation of the pistons, take care that the pistons are inserted so that they are on the same side as the spray jets. The jets must be removed before the piston ring gaps are measured as described earlier on. During installation of the jets at the position shown in Fig. 1.62, coat the threads of the screws with thread locking compound (VW uses compound with Part No. AMV 188 100 02). Tighten the bolts to 1.0 kgm (7.2 ft.lb.). Refit the jets for the four-cylinder engine as described on page 36.

1.4.6. Crankshaft and Flywheel

1.4.6.0. Technical Data

Number of main bearings:	5 (four-cylinder), 6 (five-cylinder)
Markings (main bearing caps):	1–5 or 1–6, commencing at the crankshaft pulley end of the engine
Axial play taken at:	By means of thrust half washers, at one of the bearings
Main/big end bearings:	Thin-wall shell type. Shells in block with oil drillings, shells in caps without drillings

Crankshaft End Clearance:
- Std. – 5-cyl.: .... 0.07–0.18 mm (0.0028–0.0069 in.)
- Std. – 4-cyl.: .... 0.08–0.251 mm (0.003–0.10 in.)
Wear limits: .... 0.25 mm (0.10 in.)

Main Bearing Running Clearance:
- Std.: .... 0.044–0.008 mm (4-cyl.), 0.016–0.075 mm (5 cyl.)
- Wear limits .... 16 mm (0.006 in.)

Big End Bearings:
- Std. running clearance: .... 0.02–0.076 mm (0.0008–0.003 in.)
- Max. clearance: .... 0.098 mm (4-cyl.), 0.08 mm (0.0035 in.) (5-cyl.)
- Std. axial clearance: .... 0.05–0.31 mm (0.002–0.0122 in.)
Max. clearance: .... 0.40 mm (0.016 in.)

Main Journal Diameters – 5-cyl.:*
- Basic: .... 58.00 mm (2.304 in.)
- Undersize 1: .... 57.75 mm (2.294 in.)
- Undersize 2: .... 57.50 mm (2.284 in.)
- Undersize 3: .... 57.25 mm (2.275 in.)

Main Journal Diameters – 4-cyl.:*
- Basic: .... 69.981–70.000 mm
- Undersize 1: .... 69.731–69.750 mm
- Undersize 2: .... 69.481–69.500 mm
- Undersize 3: .... 69.231–69.250 mm

Crankpin Journal Diameters – 5-cyl.:*
- Basic: .... 47.80 mm
- Undersize 1 .... 47.55 mm
- Undersize 2 .... 47.30 mm
- Undersize 3 .... 47.05 mm

Crankpin Journal Diameters – 4-cyl.:*
- Basic: .... 54.981–55.000 mm
- Undersize 1 .... 54.731–54.750 mm
- Undersize 2 .... 54.481–54.500 mm
- Undersize 3 .... 54.231–54.250 mm

* All diameters are toleranced at –0.022 to –0.042 mm (only 5-cyl.)

1.4.6.1. Crankshaft – Removal and Installation

The crankshaft of the four-cylinder engine has five main bearings, and the shaft of the five-cylinder engine has six bearings. The component parts together with the bearing shells are shown in Figs. 1.65 and 1.66. The crankshaft end float of the shaft must be measured before removal in the manner described below. The end float is different for the two engine types, as is the end float wear limit. New half-thrust washers must be used to correct the end float.

The oil pump must be removed from the front of the engine when the crankshaft of a **five-cylinder engine** is removed. The bearing caps are marked with numbers 1–6, commencing at the pulley end. The bearing shells have oil drilling when fitted in the cylinder block and no oil drillings when fitted into the bearing caps.

A driver for the oil pump is fitted into the end of the crankshaft but should only be removed if the crankshaft is replaced or the driver damaged. A new driver must be heated to 200 °C and driven over the crankshaft end.

Four-Cylinder Engine: The bearing cap No. 5 is located on the crankshaft pulley end of the engine. Bearing cap No. 3 has recesses to locate the thrust washers for the end float control. Bearing shells Nos. 1, 2, 4 and 5 in the main bearing caps have no lubrication grooves, the shells in the cylinder block have lubrication grooves. Bearing shell No. 3 has a lubrication groove when fitted to the cylinder block. The oil pump is driven from the front of the crankshaft.

A further difference will be found in the flywheel. Both engine types have a so-called dual-mass flywheel. The removal of this type is not straightforward and is described in detail later on.

Proceed as follows to remove the crankshaft. The operations are similar for four- and five-cylinder engines. In the case of the four-cylinder engine, remove the drive gear set and other applicable parts as described later on. Also read the special note given for the five-cylinder engines at the end of this section.

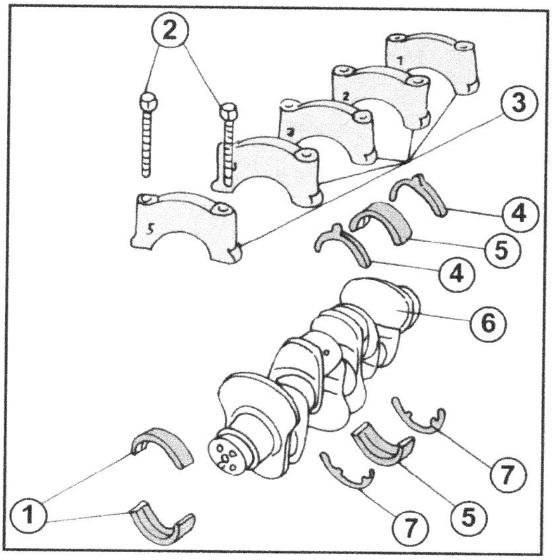

Fig. 1.65 – Crankshaft and bearings of a four-cylinder engine.

1 Main bearing shells
2 Bearing cap bolt
3 Main bearing caps
4 Thrust washers
5 Bearing shell No. 3
6 Crankshaft
7 Thrust washers

Measuring the Crankshaft End Float

It will be more convenient to measure the crankshaft end float before the crankshaft is removed, so proceed as follows.

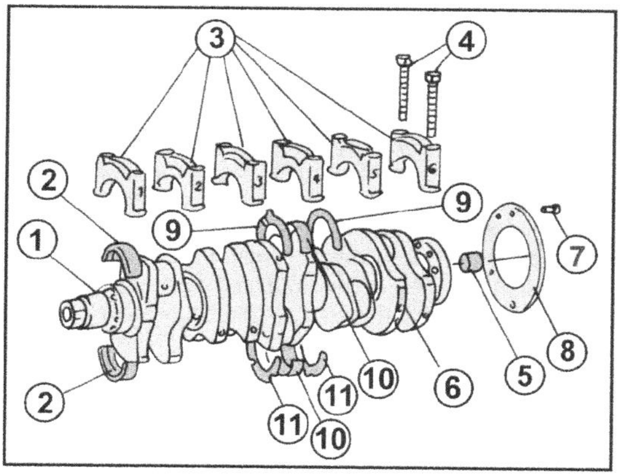

Fig. 1.66 – Crankshaft and bearings of a five-cylinder engine.

1 Oil pump driver
2 Lower bearing shells
3 Main bearing caps
4 Main bearing bolts
5 Needle roller bearing
6 Crankshaft
7 Screw, replace
8 Impulse sender wheel
9 Lower thrust washer
10 Bearing shell No. 4
11 Upper thrust washer

- Insert a feeler gauge at the centre bearing as shown in Fig. 1.67 or use a dial gauge as shown in Fig. 1.67 on the R.H. side and check that the end float is within the max. value given in Section 1.4.5.0.
- Make sure that all of the end float is taken up in each direction by applying a lever (preferably the wooden handle of a tool to avoid damage) and urging the crankshaft as far as it will go in each direction.

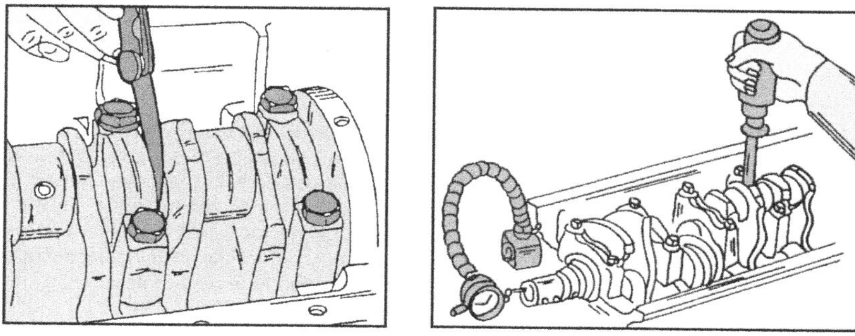

Fig. 1.67 – Checking the crankshaft end float with a feeler gauge (left) or with a dial gauge (right).

To remove the crankshaft, proceed as follows.

- Remove the engine.
- Remove the flywheel and the clutch as described under separate headings. Clutch and flywheel must be marked on opposite points to align them during installation. Remove the crankshaft vibration damper and the bolt securing the crankshaft timing gear, whilst the flywheel is locked against rotation.

- If an automatic transmission is fitted, remove the drive plate in a similar manner.
- In the case of a four-cylinder engine, remove the flywheel housing the gear drive assembly (see the description later on).
- Remove the injection pump.
- Remove the cylinder head together with the inlet manifold and the exhaust manifold.
- Remove the water pump pulley in the case of a 2.5-litre engine.
- Remove the toothed belt covers and mark the direction of rotation of the toothed timing belt, and remove the timing belt as already described (5-cyl. only).
- Remove the oil sump and the oil pump (see the section "Lubrication").
- If only the crankshaft requires removal, you can leave the pistons in the cylinder block. Otherwise, remove the pistons and connecting rods as already described. Make sure to mark the bearing shells, big end bearing caps and connecting rods accordingly.
- Remove the front cover from the cylinder block and take off the gasket.
- Remove the engine intermediate plate and unscrew the rear oil seal cover. The two oil seals must be removed from the flanges immediately, as they must be replaced.
- Remove the bearing caps from the crankshaft, noting that they are marked with their number 1–5 or 1–6. No.1 bearing cap of a five-cylinder engine is at the front of the engine, at the flywheel end in the case of a four-cylinder engine. Mark the shells and keep them matched with their caps. Remove the thrust half washers for the crankshaft end float from the centre bearing. It should be noted that all the bearing shells in the caps are without oil drillings.
- Remove the crankshaft carefully and place to one side. Now, look at the bearing shell halves remaining in the cylinder block. You will see that they have a lubrication groove and these halves must always go into the block (see Fig. 1.68). Also notice the nicks in the shells which must always align with the locators in the cylinder block and caps.

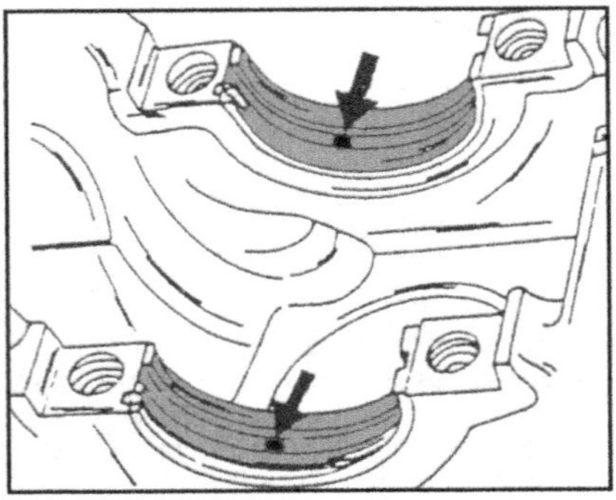

Fig. 1.68 – View of the bearing shells in the crankcase. These are the ones with the oil grooves.

Note for Five-Cylinder Engine: The oil pump must be removed from the front of the engine. Refer to the applicable description. Crankshaft bearing caps are marked from 1 to 6, No. 1 at the timing end. On the end of the crankshaft, you will find the driver for the oil pump. Only remove when a new crankshaft is fitted. The driver must be heated to 200 °C during installation and is driven over the crankshaft end with a piece of steel tube of suitable diameter.

Install the crankshaft in the reverse sequence to removal, noting the points already mentioned. The crankshaft and all other parts must be absolutely clean before installation and the journals should be well lubricated. Tighten the bearing caps gradually, finally tightening to the specified torque. Refer to the next section for further information on the fitting of the front and rear oil seals and to Section 1.4.5.3. regarding the crankshaft main bearings.

NOTE: Check the running clearance of the main and connecting rod bearings before the crankshaft is refitted and re-check the end float after completed installation.

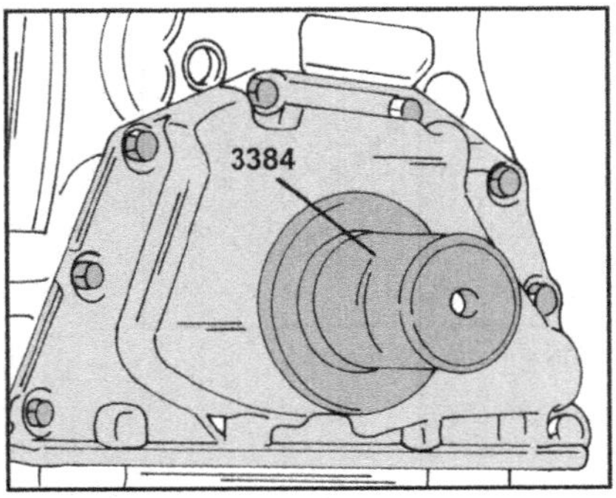

1.4.6.2. Crankshaft Oil Seals

Front and rear oil seals should always be replaced once the crankshaft oil seal flanges have been removed. The rear oil seal is replaced together with the cover flange. Both seals can be replaced with the engine fitted to the vehicle if necessary, but special tools are necessary, mainly for the front oil seal to carry out this operation successfully. Refer to the heading in question to replace a seal.

Front Crankshaft Oil Seal – Four-Cylinder Engine

- Remove the ripped drive belt (see later on, this section).

Fig. 1.69 – Installation of the front crankshaft oil seal of a four-cylinder engine, showing the special driver/mandrel.

- Remove the dust cap from the inside of the crankshaft vibration damper.

- Prevent the crankshaft from rotating (see Fig. 1.25) and remove the bolt in the centre. To slacken the crankshaft timing gear bolt, engage 5th gear and apply the handbrake to lock up the engine and undo the bolt with a socket.
- To remove the front oil seal, a special puller is normally used. Otherwise, screw two self-tapping screws into the seals and use screwdrivers and a pair of pliers to remove the seal.
- Coat the sealing lip of the new oil seal with engine oil and drive the seal into the oil pump housing and over the crankshaft end. The workshop uses the special tool shown in Fig. 1.69. The illustration shows a mandrel that must fit over the entire face of the oil seal.
- Fit the vibration damper over the crankshaft and tighten the bolts finger-tight.
- Prevent the crankshaft from rotating and tighten the centre bolt to 10.0 kgm (72 ft.lb.). Also tighten the vibration damper bolts.
- Fit the rubber cap into the damper and tighten the bolts to 1.0 kgm (7.2 ft.lb.).
- Finally, refit the drive belt.

Rear Crankshaft Oil Seal – Four-Cylinder Engine

Gearbox, clutch and driven plate and flywheel must be removed to gain access to the rear oil seal. Prevent the crankshaft from rotating as shown in Fig. 1.25 and then proceed as follows.

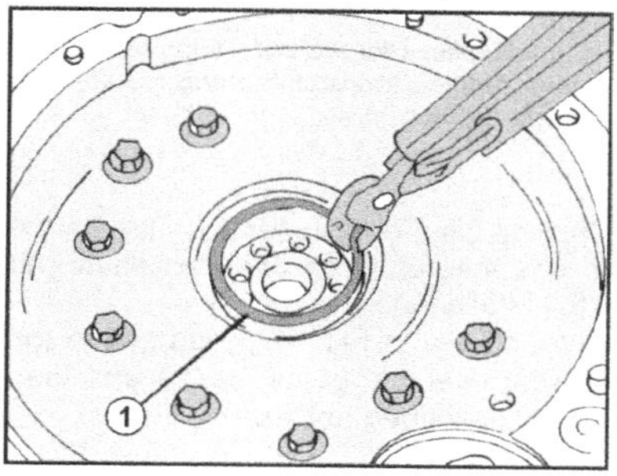

- Refer to Fig. 1.70 and remove the bush (1) with a pair of pliers as shown.
- Use a tool with a hooked end of screw two self-tapping screws into the oil seal as described above to withdraw the seal.
- Lubricate the oil seal lip with engine oil and carefully drive the seal in position. Take care not to damage the oil seal. Remember that the workshop will use a guide bush.

Fig. 1.70 – Removal of the bush (1) in the inside of the crankshaft bore, using a pair of pliers.

- Place the crankshaft bush in Fig. 1.70 in position and carefully drive it in with a plastic mallet.
- Refit the flywheel and tighten the bolts to 12.5 kgm. All other parts are refitted in reverse order to removal.

Front Crankshaft Oil Seal – Five-Cylinder Engine

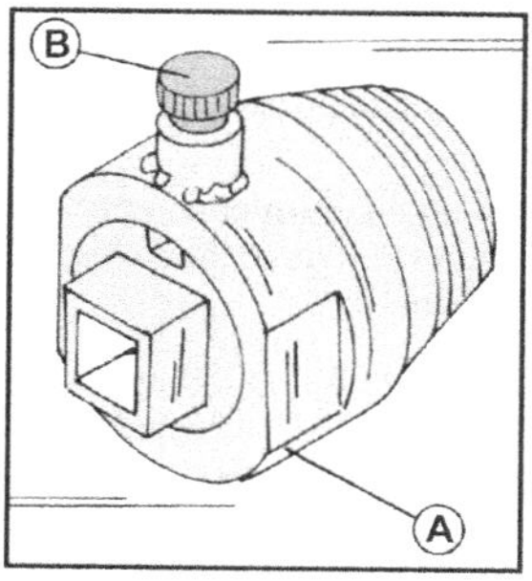

Ripped drive belt, the timing belt cover, timing belt and vibration damper, crankshaft timing gear and the hydraulic pump with the drive belt must be removed to gain access to the oil seal. Prevent the crankshaft from rotating to slacken the centre bolt in the timing gear (try to engage a gear, as above).

Fig. 1.71 – Puller (3203) for the removal of the front crankshaft oil seal (five-cylinder engine).

It should be mentioned that two different types of oil seals are used. Either a so-called PT-FE oil seal is used, which has no internal spring or a conventional oil seal.

The special puller, shown in Fig. 1.71, is normally used to remove the oil seal.

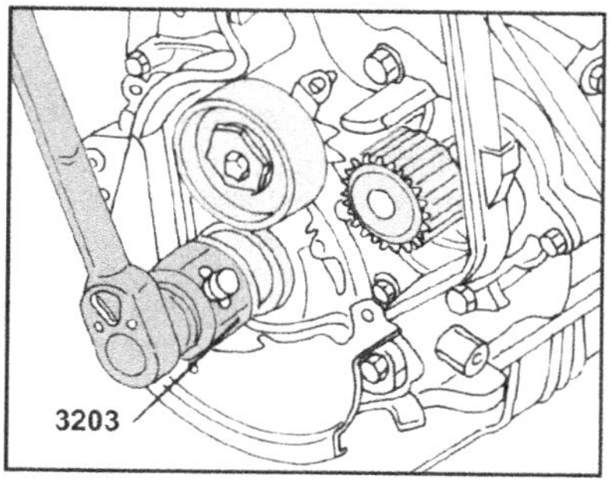

Fig. 1.72 – Removal of the front crankshaft oil seal of a 2.5-litre engine with the puller shown in Fig. 1.71.

Remove the seal as follows.

- Unscrew the inner part "A" of the special puller shown in Fig. 1.71 by two turns out of the outer part, and in this position, tighten the lock bolt "B".
- Lubricate the threaded end of the tool with engine oil and slide it over the crankshaft and into the oil seal and tighten the tool with a ratchet, as shown in Fig. 1.72. Tool and ratchet must be pressed inwards during the extraction of the seal.
- Unscrew the lock screw (B) in Fig. 1.71 and screw the inner part against the oil seal until it is free of the crankshaft end.
- Clamp the tool into a vice and remove the oil seal with a pair of pliers.

The installation is carried out as described below, depending on which type of oil seal is used. Otherwise, the installation is a reversal of the removal procedure.

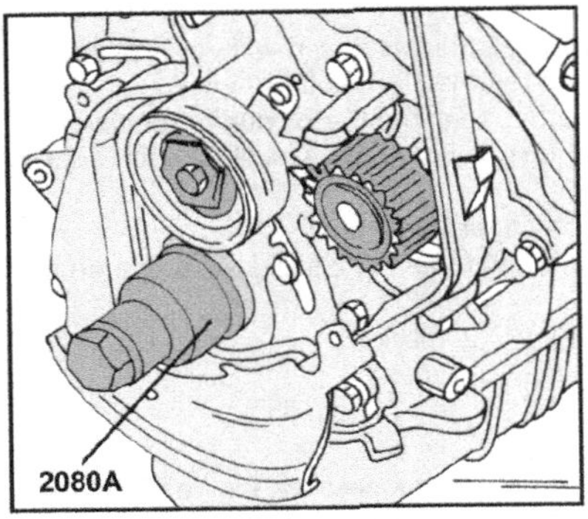

Fig. 1.73 – Installation of the front crankshaft oil seal of a 2.5-litre engine. Otherwise, you can follow the description in the text for an alternative method.

Normal Oil Seal

- A guide sleeve (2080A) and an installation tool are used in the workshop to refit the oil seal. The sleeve is a piece of tube of a suitable diameter which is placed over the outer face of the oil seal, as shown in Fig. 1.73. The centre bolt of the crankshaft timing gear is then screwed into the end of the crankshaft, as shown in the illustration. By tightening the bolt, the seal will be pulled in position. It is, of course, possible to drive the oil seal in position with a drift, but the danger of damage cannot be overstressed.

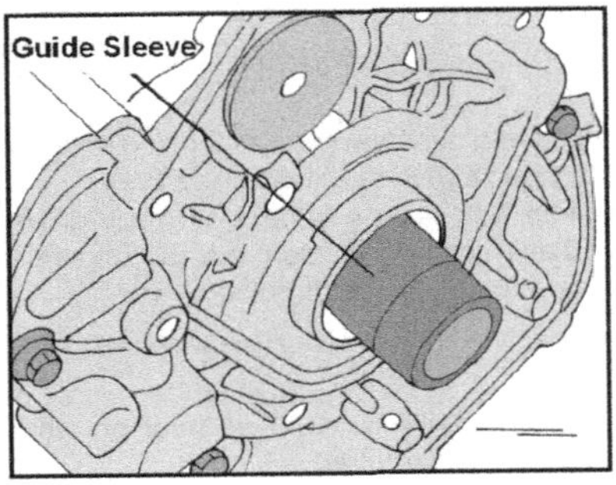

Fig. 1.74 – The special VW guide sleeve for the installation of the front PTFE oil seal. Using any other method to install the ring requires the greatest care to avoid damage to the ring.

PTFE-Seal

This type of seal must not be oiled or greased, i.e., it must be fitted as received. Any remnants of oil on the crankshaft end must be removed prior to installation.

A special guide sleeve, shown in Fig. 1.74, is used in the workshop. Slide the oil seal over the guide sleeve and then proceed in a similar manner described above.

1.4.6.3. Crankshaft Bearings

Data for the crankshaft main bearings and big end journals and bearings are given in Section 1.4.6.0. together with the allowable clearances. Clean the crankshaft thoroughly and inspect the journals for signs of wear or damage. Measure the journals at two positions around the circumference and along its length. Out-of-round and/or a taper of more than 0.03 mm (0.0012 in.) will mean that the crankshaft will have to be re-ground.

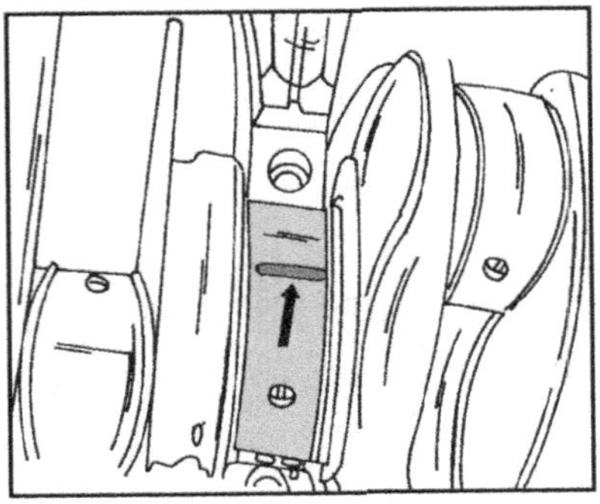

Fig. 1.75 – "Plastigage" strips are placed over the bearing journals as shown before the bearing cap is fitted.

Main bearing and crankpin bearing clearances can be assessed by the use of "Plastigage" as described below. Three types are available, with ranges of:

PG1 Green	0.025–0.075 mm (0.00098–0.0029 in.)
PR1 Red	0.050–0.150 mm (0.00196–0.0059 in.)
PB1 Blue	0.100–0.230 mm (0.0039–0.0090 in.).

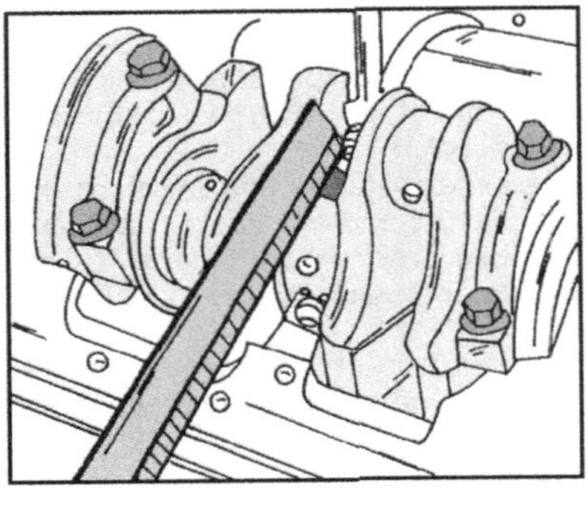

Fig. 1.76 – Measuring the width of the flattened "Plastigage" wire to determine the bearing running clearance. Main bearings and big end bearings are checked in the same manner.

To use the "Plastigage", first remove all dirt and grease from the journals and shells. Place a length of plastic wire along the complete length of the journal, in the direction shown in Fig. 1.75, and fit the shell and bearing cap. Tighten the cap bolts to a tightening torque of 6.5 kgm (47 ft.lb.) in the case of a 2.5-litre engine or 2.5 kgm + 1/3 of turn in the case of a 2.8-litre engine, when the main bearings are measured or to 3.0 kgm (22 ft.lb.) in the case of the big end bearing caps. Use the original connecting rod bolts for this check. Take care that the crankshaft cannot rotate once the bearing caps are fitted.

Now, remove the bearing cap in question, refer to Fig. 1.76 and compare the width of the compressed "Plastigage" wire with the scale provided in the kit. If the clearance is excessive, the crankshaft bearing journals must be re-ground and undersize bearing shells fitted. Your dealer will advise you about the availability of bearing shells. We recommend, however, that the crankshaft together with the block is taken to an engine shop to have the shaft re-ground and re-shelled.

1.4.6.4. Crankshaft – Installation

The following points must be noted in particular when the crankshaft is refitted.

- Refer to Figs. 1.65 and 1.66 for the location of parts.
- Thoroughly clean the bearing bores and place the bearing shells with the oil lubrication grooves into the "notches" of the bearing bores. These are the shells shown in Fig. 1.68. Generously lubricate the shells with engine oil. Place the two thrust washers in position left and right of the centre bearing.
- Slowly lower the crankshaft into the bearing shells. If the piston/connecting rod assemblies are still in the block, you will have to guide the big ends with the inserted shells over the journals.
- Place the lower bearing shells into the respective main bearing caps (lugs into "notches") and generously oil the shell's faces. Fit the two lower thrust washers against the centre bearing cap. The lubricating grooves are on the side of the crankshaft.

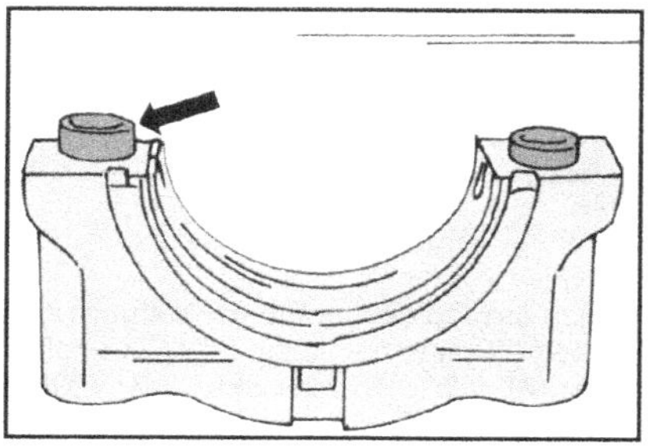

Fig. 1.77 – When fitting the main bearing caps of a four-cylinder engine, make sure that the dowel sleeve with the arrow is on the correct side.

Note for 2.8-Litre Engine: The crankshaft main bearing caps have two bores of different depths. The crankcase also has two similar bores, again of different depths. During installation of the bearing caps, make sure that the side with the longer centring dowel, shown in Fig. 1.77 with the arrow, is located on the side of the oil filter.

- Fit the bearing caps (noting the above note) and tap them in position with a rubber or plastic mallet.
- Tighten the main bearing cap bolts from the centre towards the outside to 6.5 kgm (47 ft.lb.) in the case of the five-cylinder engine. In the case of a four-cylinder engine, tighten the bolt to 2.5 kgm (18 ft.lb.) and from the final position by a further 1/3 of a turn (120°). Rotate the crankshaft a few turns after tightening the bearing caps to check for binding.
- Check the end float of the crankshaft once more as described earlier on (refer also to illustrations in question).
- Refit the piston and connecting rods as described on page 36.

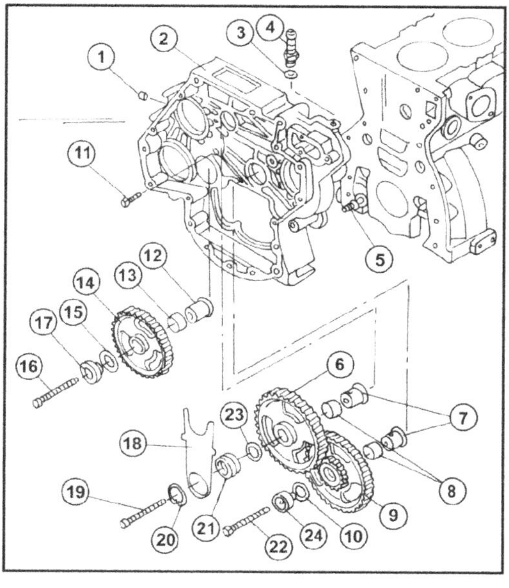

- Refit the two oil seals as described for the engine in question in Section 1.4.6.2. commencing on page 40.
- In the case of a five-cylinder engine, fit the crankshaft timing gear.
- Fit the timing belt (5-cyl.) or the drive gear assembly (4-cyl.). Fit the flywheel and clutch, if applicable.
- Refit the oil pump and the oil sump (refer to Section 1.6).
- All remaining operations are carried out in reverse order to the removal procedure.

1.4.7. Drive Gear Assembly (Four-Cylinder Engine)

Fig. 1.78 shows the parts of the various gearwheels at the rear end of the engine. Removal and installation can be carried out by referring to the exploded view, but various operations are necessary to gain access to the parts shown. Note the following points.

Fig. 1.78 – Details for the removal and installation of the drive gear set. The numbers are referred to in the text.

- A closing cap (1) is fitted into the flywheel housing (2). The sealing faces of the housing must be thoroughly cleaned and coated with Loctite 515 during installation. Note the correct location when fitting the housing.

- The coolant fitting (4) is fitted with a sealing washer (3) into the housing. The sealing washer must always be replaced.
- A closing plug (5) is fitted at the position shown.
- The intermediate gearwheel (6) drives the camshaft. Refer to Fig. 1.79 to find out where the gearwheel is located on the gear drive set. The end clearance of the gearwheel can be adjusted as described later on.
- Bearing (7) and bearing bush (8) are fitted to the gearwheel (6). The bearing bush can be replaced, but special tools are required, therefore, workshop job.
- The intermediate gearwheel (9) drives the coolant pump (water pump). Refer to Fig. 1.79 for its location. The end float of the gearwheel can be adjusted as described later on.
- The shim (10) is used to adjust the end float of the intermediate gearwheel for the coolant pump.
- The bolts (11) are tightened to 6.0 kgm (43 ft.lb.).
- The bearing (12) and the bearing bush (13) are fitted into the drive gearwheel (14) for the steering pump. The shim (15) in front of the gearwheel is used to adjust the gearwheel end clearance. The bolt (16) is tightened to 6.0 kgm (47 ft.lb.). A thrust washer (17) for the gearwheel is fitted under the bolt head.

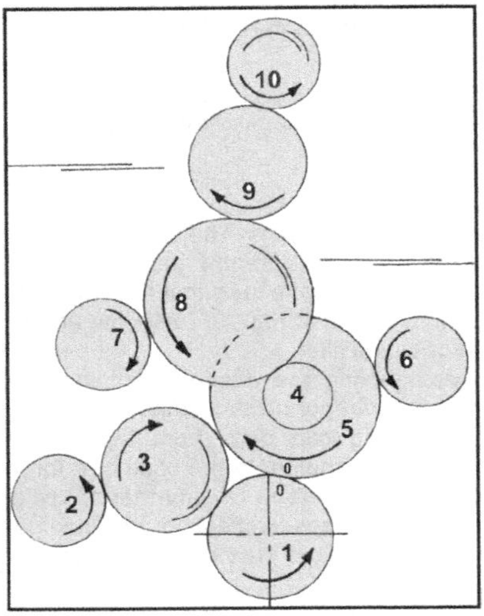

Fig. 1.79 – The arrangement of the drive gears and intermediate gearwheels of the four-cylinder engine.
1 Crankshaft gearwheel
2 Gearwheel for steering pump
3 Intermediate gearwheel
4 Intermediate gearwheel
5 Intermediate gearwheel
6 Coolant pump drive gear
7 Injection pump drive gear
8 Intermediate gearwheel
9 Intermediate gearwheel
10 Camshaft drive gear

- The spacer fork (18) is used to secure the parts shown and is secured with the bolt (19) and the circlip (20). The bolt is tightened to 6.0 kgm (47 ft.lb.). The thrust washer (23) is used to adjust the end clearance of the gearwheel (6). The thrust washer (21) is placed against the gearwheel.
- The bolt (21) is tightened to 6.0 kgm (47 ft.lb.). The bolt secures the coolant pump drive gear and has a thrust washer (24) for the gearwheel.

The following operations on the gearwheel drive set should be mentioned in particular: the end clearance of gearwheels 3, 4, 5, 8 and 9 in Fig. 1.79 should be checked and if necessary adjusted, if parts have been replaced.

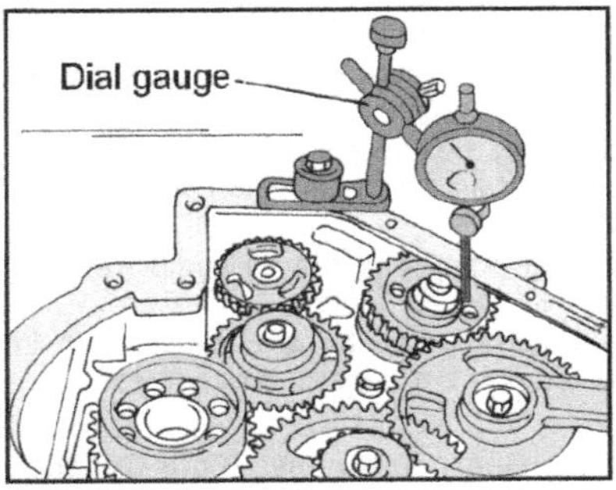

Fig. 1.80 – Gearwheel end floats are checked as shown.

Checking Clearances in the Gearwheel Drive System

- Check the end clearance of the camshaft gearwheel (10) in Fig.1.79, as shown in Fig. 1.32.
- The end clearances of gearwheels 3, 4, 5 and 8 are checked, as shown in Fig. 1.80. To do this, place the stylus of the dial gauge onto the gearwheel in question and move the wheel up and down. The reading on the dial gauge is the end float, which should be between 0.06 and 0.12 mm. Shims of different thicknesses (2.45–2.49 mm, 2.50–2.54 mm and 2.56–2.60 mm). These are the shims (10), (15) and (23) in Fig. 1.78.

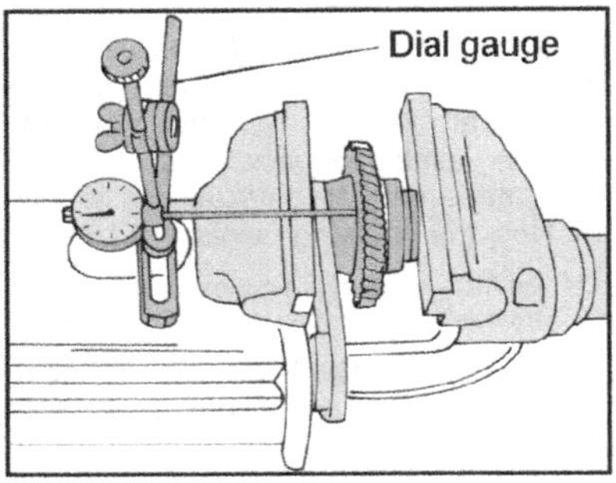

Fig. 1.81 – The end float of the intermediate gearwheel for the camshaft (9 in Fig. 1.79) is measured in the manner shown.

- Check the end float of the intermediate gearwheel for the camshaft (6, Fig. 1.78). The end float is the same as given above, but the check is carried out, as shown in Fig. 1.81. If corrections are necessary, use the shim thicknesses given above.
- The tooth flank clearance of the intermediate gearwheel (9) and the camshaft drive gearwheel (10) in Fig. 1.79 can be adjusted. The operations are, however, rather complicated.

As the clearance must only be measured when parts are replaced, we recommend to take the assembled engine to an engine shop, dealing with this type of engine, and have the clearance checked and if necessary adjusted.

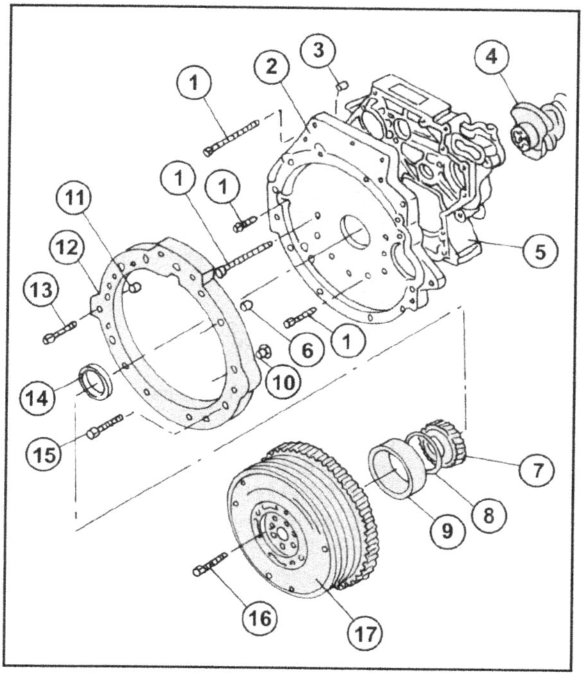

Fig.1.82 – Details for the removal and installation of the flywheel and other parts on the timing gear drive of a four-cylinder engine. The various numbers are referred to in the text.

1.4.8. Dual-Mass Flywheel – Four-Cylinder

The parts shown in Fig. 1.82 must be removed to gain access to the component parts of the gearwheel drive. The drive gear on the crankshaft is also hidden. This is also the reason that No. 1 cylinder is on the flywheel side of the engine. The removal and installation of the parts can be carried out by referring to the illustration.

Note the following points.

- The bolts (1) secure the flywheel housing (2). Before installation, clean the faces of the housing and coat them with "Loctite 515". Carry out the installation carefully, as the dowel sleeve (3) must engage properly.
- The crankshaft (4) carries the drive gear (7) for the gearwheel set (1, in Fig. 1.79), an "O" sealing ring (8) and a bush (9). The sealing ring must always be replaced.
- The dowel sleeve (6) will centre the parts during assembly. A threaded insert (10) is fitted at the position shown.
- The dowel sleeve (11) locates the intermediate flange (12). The bolts (13) and (15) are tightened to 5.0 kgm (36 ft.lb.).
- The crankshaft oil seal (4) can be replaced as already described earlier on.
- The dual-mass flywheel (17) is fitted with the ring gear for the starter motor. The bolts (16) are tightened to 12.5 kgm (90 ft.lb.). The vibration damper on the opposite side must be held against rotation when the bolts are tightened (see, for example, Fig. 1.25).

The following operations can be carried out on the gearwheel drive set and associated parts.

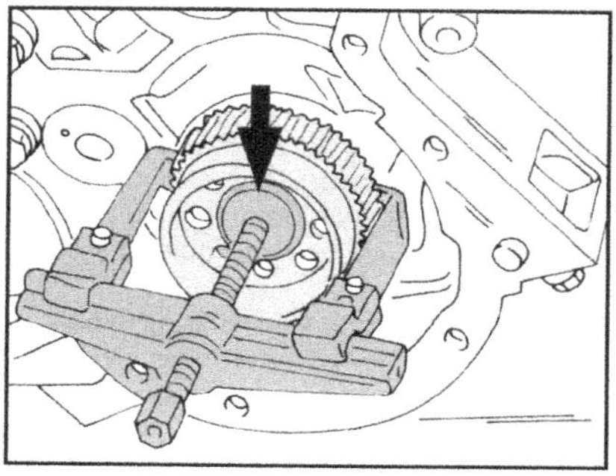

Fig. 1.83 – Removal of the crankshaft gearwheel (4-cyl.).

Removal and Installation of Crankshaft Drive Gear

- Remove the transmission (see the relevant section).
- Remove the rubber cap in the centre of the vibration damper and prevent the crankshaft from rotating as, for example, as shown in Fig. 1.25.
- Remove the clutch and the clutch-driven plate.
- Remove the flywheel.
- Remove the crankshaft gearwheel as shown in Fig. 1.83, using a conventional two-arm puller. A suitable pressure pad must be placed inside the crankshaft bore (arrow) to prevent damage. The crankshaft bush will come away with the gearwheel.
- The gearwheel must be heated to a temperature of 60 °C before installation (hot plate) and then quickly driven over the crankshaft end. Make sure that the gearwheel is fitted in its correct position. Allow the gearwheel to cool off and then insert the bush, using a plastic mallet.
- Refit the flywheel (12.5 kgm/90 ft.lb.). All other operations are carried out in reverse order.

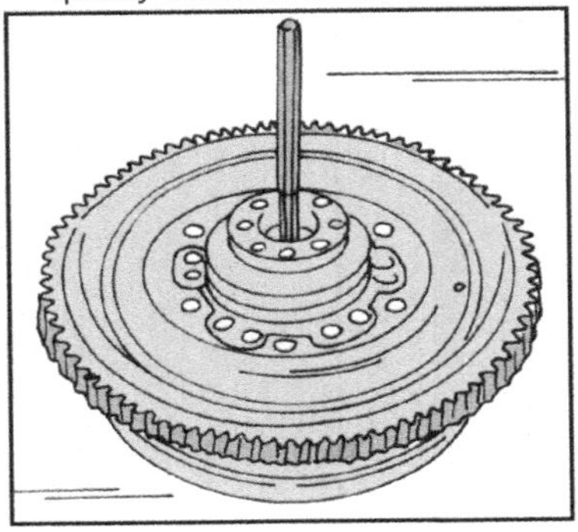

Replacing the Needle Roller Bearing Inside the Flywheel

The flywheel must be removed. The needle roller bearing is then driven out from the rear of the flywheel, as shown in Fig. 1.84. The new bearing is driven into position from the opposite side of the flywheel. Use a mandrel of the outside diameter of the bearing to prevent damage.

Fig. 1.84 – Removal of the needle roller bearing from the flywheel. Install from the other side.

- Refit the flywheel to the engine and carry out all other operations in reverse order to the removal procedure.

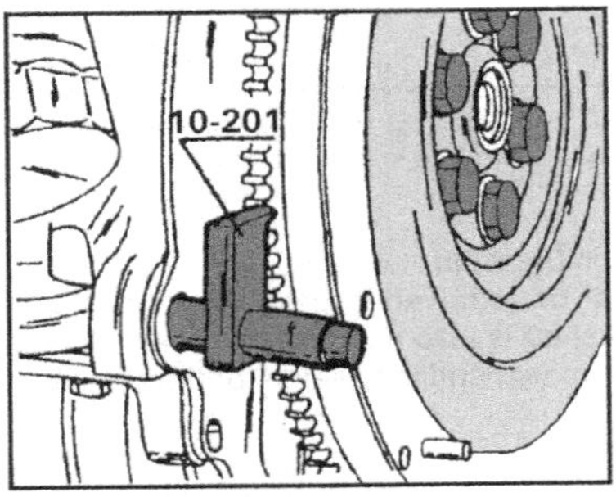

Fig. 1.85 – A flywheel can be locked against rotation with a so-called flywheel brake. Shown is the special tool, but a similar device can be used without problems. Otherwise, refer to Fig. 1.86.

1.4.8. Flywheel (2.5 Litres)

A dual-mass flywheel is fitted to the 2.5-litre engine. To remove the flywheel (after removal of the clutch and clutch-driven plate), counterhold it suitably. If the oil sump (pan) has been removed, insert a wooden block between the crankshaft and the side of the crankcase.

Fig. 1.86 – Bolt a steel bar to the flywheel when slackening or tightening the flywheel bolts. This arrangement shows the bar when the bolts are slackened.

Otherwise, insert a strong screwdriver into the teeth of the flywheel. Fig. 1.85 shows how the special flywheel lock is inserted into the flywheel teeth to lock the flywheel. A similar tool can be used. Otherwise, refer to Fig.1.86 and bolt a strong steel bar to the flywheel as shown (using the clutch bolts). Arrange the bar as applicable to slacken or tighten the flywheel bolts.

The flywheel bolts cannot be used a second time. Tighten the bolts to a torque reading of 6.0 kgm (43.5 ft.lb.). The flywheel must be locked against rotation as described during the removal. From this final position, turn each bolt evenly across by a further quarter of a turn (90°).

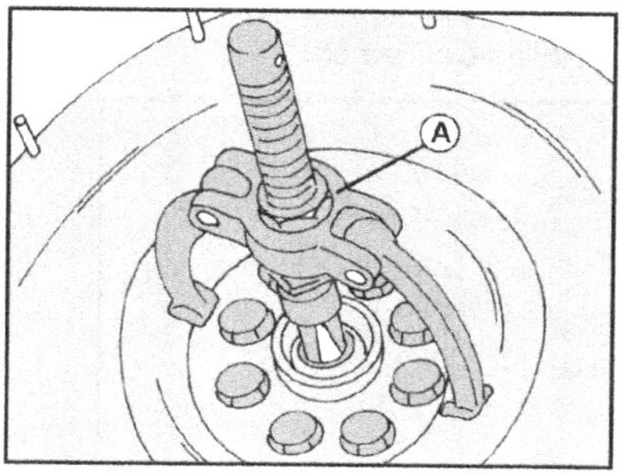

The needle roller bearing can be replaced, but the flywheel must be fitted to extract the bearing. A suitable puller, as shown in Fig. 1.87, must then be used to withdraw the bearing. The claws of the puller must be located underneath the bearing.

Fig. 1.87 – Removal of the crankshaft needle roller bearing (2.5-litre engine).

Lubricate the new bearing with engine oil and drive it in position with a suitable mandrel.

1.4.9. Cylinder Block

1.4.9.0. Technical Data

Cylinder bores: Cylinder bores are 0.03 mm (0.0012 in.) larger in diameter than the piston belonging to any one cylinder bore. Full details are given in Section 1.4.5.0.

Max. cylinder bore wear: .... 0.04 mm (0.0016 in.) from the specified size

Measuring points for cylinder bores: .... 10 mm (0.4 in.) from top edge of bore, 10 mm (0.4) from lower edge of bore and in centre of bore. One measurement to be in the axial direction and one in the transverse direction.

Cylinder bore diameter:
- 2.8-litre engines .... 93.00–93.022 mm
- 2.5-litre engine: .... 81.01 mm

1st Oversize:
- 2.5-litre engine only: .... 81.26 mm

1.4.9.1. Servicing

Special attention should be given to the cylinder block at each major overhaul of the engine, irrespective of whether the bores are to be re-machined or not. Thoroughly clean all cavities and passages and remove all traces of foreign matter from the joint faces. If any machining or honing of the bores has taken place, it is essential that all swarf is removed before the assembly of the engine takes place.

Measurement of the cylinder bores should be left to an engine shop. We recommend not to attempt this operation yourself. If the bores are outside the wear limit, the block requires re-boring and new pistons must be fitted.

1.4.10. Ripped V-belt Drive

Four-Cylinder Engine

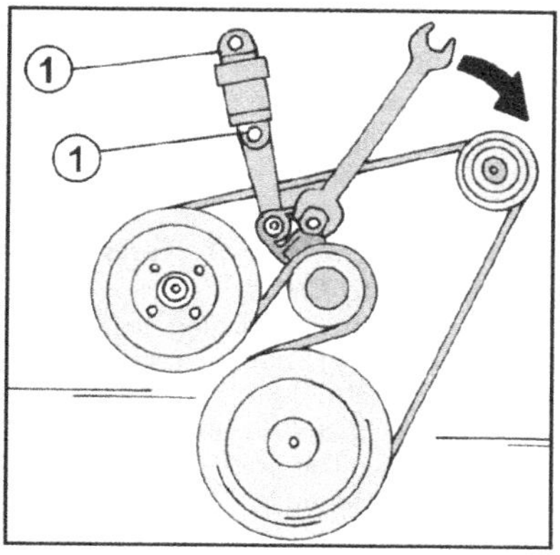

The drive belt must be properly tensioned when alternator, crankshaft pulley, viscous clutch or, if fitted, the compressor for an air conditioning system has been removed. A second alternator can be fitted to these engines. A single, endless ripped V-belt is fitted, which is automatically tensioned by means of a special tensioning device. Depending on the layout of the belt drive, an idler roller can be fitted (with A/C system). A separate drive belt is used to drive the vacuum pump.

Fig. 1.88 – Releasing the drive belt tension on a four-cylinder engine.
1 Securing bolts of tensioning element

Removal and Installation of Drive Belt

- Carry out the necessary operations to gain access to the belt drive.
- Mark the running direction of the belt into the outside, using a felt pen, if it is intended to refit the belt.

- Apply a 27-mm A/F spanner to the bolt shown in Fig. 1.88 and move the spanner in the direction shown by the arrow. The belt tension will be released and the belt can be lifted off.

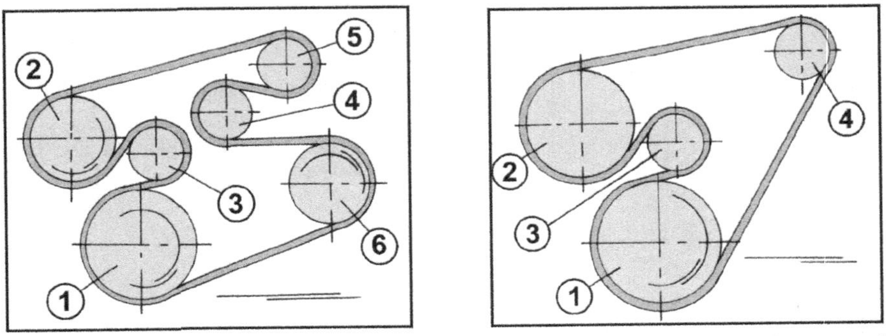

Fig. 1.89 – Layout of the drive belt for an engine without A/C system on the L.H. side and with A/C system on the R.H. side. The numbers are referred to in the description.

Before fitting an old belt, check it for wear, porosity and other damage. All units must be securely attached to the engine before the belt is fitted. Refit the belt as follows, noting the belt layouts shown in Figs. 1.89 and 1.90, depending on which arrangement is used.

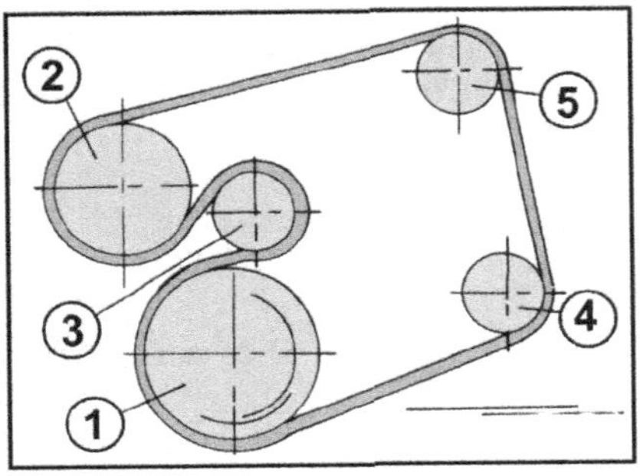

Fig. 1.90 – Layout of the drive belt when two alternators are fitted. The numbers are referred to in the text.

- The belt drive in the L.H. view of Fig. 1.89 is routed over the vibration damper pulley (1), the pulley (2) for the viscous clutch (for the cooling fan), the tensioning pulley (3) and the alternator pulley (4).
- The belt drive arrangement in the R.H. view of Fig. 1.89 is routed around the three pulleys (1) to (3) and additionally around the idler pulley (4), the alternator pulley (5) and the compressor pulley (6).

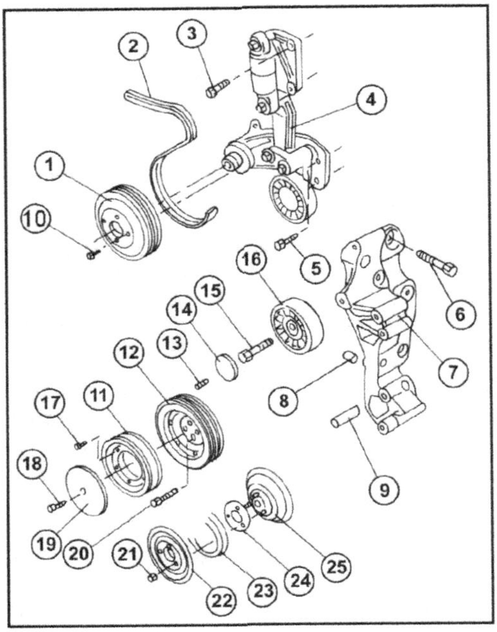

Fig. 1.91 – Details for the removal and installation of the drive belt installation of a four-cylinder engine. The numbers are referred to in the text.

- Fig. 1.90 shows the belt drive when two alternators are fitted. The belt is routed around the vibration damper pulley (1), the pulley (2) for the viscous clutch and the tensioning roller (3). The two alternators (4) and (5) are driven as shown.
- Rotate the bolt as shown in Fig. 1.88 until the belt can be placed over the pulleys, as shown in Figs. 1.89 or 1.90. Slowly release the spanner, checking at the same time that the belt is located properly in all pulleys. No other operations are necessary.

Removal and Installation of Tensioning Device and remaining Parts

Fig. 1.91 shows the parts belonging to the belt drive. Note the following during removal and installation of parts.

- The pulley (1) is bolted to the hub of the tensioning device (4). The bolts (10) are coated with "Loctite" and tightened to 1.5 kgm (10 ft.lb.).

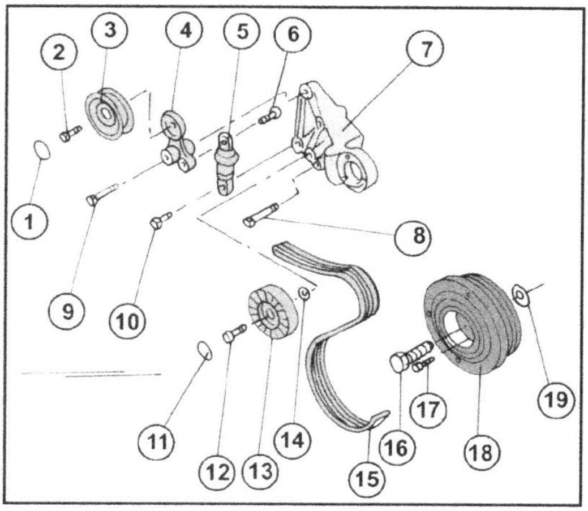

Fig. 1.92 – Details for the removal and installation of the drive belt parts of a five-cylinder engine. The numbers are referred to in the text.

- The ripped V-belt is routed, as shown in Figs. 1.89 and 1.90.
- The tensioning device can be removed after the belt has been taken off. Then, remove the two bolts shown with (1) in Fig. 1.88, unscrew the bolts (5) and remove the tensioning device. The tensioning element (shape of a shock absorber) is marked with "Top" or "Oben" and must be fitted accordingly. Tighten the bolts (3) in Fig. 1.91 to 3.0 kgm (22 ft.lb.). Tighten bolts (5) to 2.0 kgm (14.5 ft.lb.) and from the final position a further 1/8 of a turn (45°).
- The mounting bracket for the alternator (7) is secured with bolts (6), the guide bush (8) and a tensioning sleeve (9). Bolts (6) are tightened to 3.0 kgm (22 ft.lb.).
- The pulley (11) drives the vacuum pump and is secured with bolts (17), tightened to 3.0 kgm (22 ft.lb.). A dust cover (19) is fitted inside the pulley, held in position by bolts (18). The latter bolts are tightened to 1.0 kgm (7 ft.lb.).
- Bolts (20) are used to secure the pulley of the vibration damper and the belt pulley (12). Tighten the bolts to 3.0 kgm (22 ft.lb.) and from the end position by a further quarter of a turn (90°). A tensioning sleeve (13) is fitted into the rear of the pulley. The crankshaft must be prevented from rotating when the vibration damper is removed. The dust cover (19) is secured with the bolts (18). These are tightened to 1.0 kgm (7 ft.lb.).
- The idler pulley (16), if fitted, is secured with bolt (15), covered by a dust cover (14). The bolt is tightened to 2.0 kgm (14. ft.lb.) + 90°.
- The nuts (21) for the two-part pulley for the vacuum pump (22) are tightened to 2.5 kgm (18 ft.lb.). The tension of the belt (23) is adjusted by placing spacer shims (24) between the pulley halves (22) and (25).

Five-Cylinder Engines

Fig. 1.92 shows the belt drive. The necessary parts must be removed to gain access to the belt drive. Fig. 1.93 shows the layout of the belt drives of different versions. Note the following.

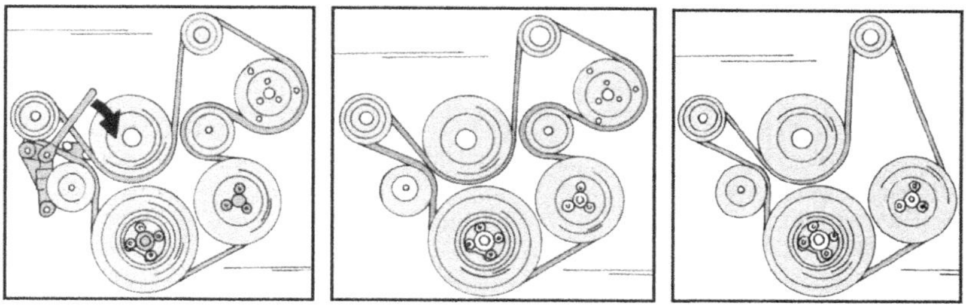

Fig. 1.93 – Layout of the drive belt and details for the removal and installation.

- Lift the tensioning roller in accordance with the L.H. view of Fig. 1.93 and lift the belt from the alternator pulley. The belt is now slack and can be removed. If the belt is to be refitted, mark the running direction into the outside of the belt, using a felt pen.
- Refit the belt (note the running direction if the original belt is fitted). First, place the belt over the pulley of the vibration damper and then over the remaining pulleys. Make sure that the belt engages properly in all pulleys. Lift the tensioning roller as shown in Fig. 1.93 (left) until the belt can be placed around all pulleys. The tensioning roller is now slowly released, but check that the belt engages during this step. The layout of the belt on an engine without A/C system is shown in the centre view of Fig. 1.93. The R.H. view shows the routing when an A/C system is fitted.
- After the belt has been fitted as described above, there are no further inspections necessary. It is, however, a good practice to check the belt frequently for excessive wear.

The tensioning device can be replaced. Refer to Fig. 1.92 for details.

- To remove the tensioning roller (3), take out the dust cover (1) and unscrew the two screws (2). Tighten the screws to 1.0 kgm (7 ft.lb.).
- The bracket (4) for the tensioning element (5) is secured with screws (9). Tighten the bolts to 4.0 kgm (29 ft.lb.).
- The tensioning element (5) is attached with bolts (6) and (10). Tightened to 2.0 kgm (14.5 ft.lb.).
- The bracket (7) carries the viscous fan clutch, the tensioning roller and the tensioning element for the tensioning roller. Tighten the bolts (8) to 4.0 kgm (29 ft.lb.).
- To remove the idler pulley (13), prise out the dust cap (11) and unscrew the bolt (12). Tighten the bolt to 2.0 kgm (14.5 ft.lb.). A spacer washer (14) is situated behind the idler pulley.
- The ribbed belt (15) has already been discussed.
- To remove the vibration damper with the belt pulley (18), unscrew bolt (16) and the small screws (17). The vibration damper must be held against rotation. A special washer (19) is fitted behind the vibration damper, which must always be replaced. Tighten the bolt (16) to 16.0 kgm (115 ft.lb.) and from this position by a further 1/2 of a turn. Screws (17) are secured with 1.0 kgm (7 ft.lb.).

1.4.11. Compression Test

As special adapters are required to check the compression of the cylinders, we recommend that you have the check carried out at a dealer.

1.5. Engine – Tightening Torques

Four-Cylinder Engine

Cylinder Head Bolts:	
1st stage:	3.0 kgm (22 ft.lb.)
2nd stage:	6.0 kgm (43 ft.lb.)
3rd stage:	10.0 kgm (72 ft.lb.)
4th stage:	Angle-tighten by 90°
5th stage:	Angle tighten by 90°
Parts on the cylinder head:	Refer to Fig. 1.10, page 16
Parts of valve/camshaft assembly:	Refer to 1.22, page 21
Cylinder head cover:	1.5 kgm (10 ft.lb.)
Main bearing caps:	2.5 kgm, then a further 1/3 of a turn (120°)
Big end bearing caps:	
- 1st stage:	2.0 kgm (14.5 ft.lb.)
- 2nd stage:	Tighten a further 90°
Banjo bolts for oil splash jets:	1.0 kgm (7 ft.lb.)
Clutch bolts:	2.5 kgm (18 ft.lb.)
Flywheel bolts:	12.5 kgm (90 ft.lb.)
Spacer plate to gearwheel drive housing:	5.0 kgm (36 ft.lb.)
Parts of gearwheel drive:	See Fig. 1.78
Crankshaft pulley to crankshaft:	10.0 kgm (72 ft.lb.)
TDC pointer to cylinder block:	3.0 kgm (22 ft.lb.)
Baffle plate to bottom of cylinder block:	3.0 kgm (22 ft.lb.)
Oil sump to cylinder block:	3.0 kgm (22 ft.lb.)
Vacuum pump pulley:	3.0 kgm (22 ft.lb.)
Dust cover to pulley:	1.0 kgm (7 ft.lb.)
Oil drain plug (sump):	5.0 kgm (36 ft.lb.)
Tensioner roller for ribbed belt:	2.0 kgm (14.5 ft.lb.) + 90°
Tensioner roller bracket to cylinder block:	4.0 kgm (29 ft.lb.) + 45° (1/8 of a turn)
Tensioning element, top and bottom:	3.5 kgm (25 ft.lb.)
Bracket for viscous fan:	2.0 kgm (14.5 ft.lb.) + 45° (1/8 of a turn)
Bracket for tensioning device to block:	3.0 kgm (22 ft.lb.)
Alternator bracket to engine:	3.0 kgm (22 ft.lb.)
Water pump to engine:	3.0 kgm (22 ft.lb.)
Radiator and connections:	Refer to Section 1.7
Clamp bracket for fuel injectors:	4.0 kgm (29 ft.lb.)
Injection pipe union nuts – to pump:	2.7 kgm (19 ft.lb.)
Injection pipe union nuts – to injector:	3.0 kgm (22 ft.lb.)
Oil pressure switch:	1.0 kgm (7 ft.lb.)
Oil pump to engine:	3.0 kgm (22 ft.lb.)
Rear cover to cylinder block:	1.5 kgm (10 ft.lb.)
Parts of exhaust system:	Refer to exploded views

Engine mountings: .... Refer to "Engine Section"
Engine to transmission: .... 8.0 kgm (58 ft.lb.) – M12 bolts
Engine mounting to engine (from below): .... 8.5 kgm (61 ft.lb.)
Propeller shaft to gearbox flange: .... 7.5 kgm (50.5 ft.lb.)
Steering pump to cylinder block: .... 3.0 kgm (22 ft.lb.)

Five-Cylinder Engine

Cylinder Head Bolts:
1st stage: .... 4.0 kgm (29 ft.lb.)
2nd stage: .... 6.0 kgm (43 ft.lb.)
3rd stage: .... Tighten by ¼ of a turn (90°)
4th stage: .... Tighten by ¼ of a turn (90°)
Main bearing caps: .... 6.5 kgm (47.0 ft.lb.)
Big end bearing caps: .... 3.0 kgm (22 ft.1b.), then by a further 90°
Flywheel to crankshaft: .... 6.0 kgm (43.5 ft.lb.), then a further 90°
Clutch plate bolts: .... 2.5 kgm (18 ft. lb.)
Camshaft sprocket:
Bolt marked 8.8: .... 8.5 kgm (61.0 ft.lb.)
Bolt marked 10.9: .... 10.0 kgm (72 ft.lb.)
Camshaft bearing caps: .... 2.0 kgm (14.5 ft.lb.)
Crankshaft timing gear bolt: .... 16.0 kgm (115 ft.lb.) + 90°
Vibration damper to crankshaft timing gear: .... 2.0 kgm (14.5 ft.lb.)
Upper timing belt cover to cylinder block:
- Screw at upper end (see Fig. 1.38): .... 1.0 kgm (7 ft.lb.)
- Screw in centre and at lower end: .... 2.0 kgm (14.5 ft.lb.)
Lower timing belt cover: .... 1.0 kgm (7 ft.lb.)
Cover for injection pump drive belt: .... 1.0 kgm (7 ft.lb.)
Cylinder head cover: .... .1.0 kgm (7 ft.lb.)

Inner cover for Injection Pump Drive Belt:
- M6 bolts: .... 1.0 kgm (7 ft.lb.)
- M8 bolts: .... 2.0 kgm (14.5 ft.lb.)
Inner toothed belt cover: .... 2.0 kgm (14.5 ft.lb.)
Sender wheel to crankshaft (must be replaced): .... 2.5 kgm (18 ft.lb.)
Idler roller for ribbed belt: .... 2.0 kgm (14.5 ft.lb.)
Vacuum pump to cylinder block: .... 2.5 kgm (18 ft.lb.)
Banjo bolt to filter bracket: .... 7.0 kgm (50.5 ft.lb.)
Oil sump to block: .... 1.0 kgm (7 ft.lb.), M6, 2.0 kgm (14.5 ft.lb.), M8
Rear oil seal flange to block: .... 1.0 kgm (7 ft.lb.)
Camshaft bearing caps: .... 2.0 kgm (14.5 ft.lb.)
Cover for injection pump drive belt: .... 1.0 kgm (7 ft.lb.)
Camshaft timing gearwheel:
- Marked with "8.8" in bolt head: .... 8.5 kgm (61 ft.lb.)
- Marked with "10.9" in bolt head: .... 10.0 kgm (72 ft.lb.)
Bolt for timing belt guide roller: .... 1.0 kgm (7 ft.lb.)
Nut for oil cooler to oil filter: .... 2.5 kgm (18 ft.lb.)
Injection pump drive gear to camshaft: .... 16.0 kgm (115 ft.lb.)
Injection pump drive gear to injection pump: .... 9.0 kgm (65 ft.lb.)

Injection pump to cylinder block:
- Bracket to cylinder block: .... 6.5 kgm (47 ft.lb.)
- Injection pump to mounting bracket: .... 2.5 kgm (18 ft.lb.)
Idler roller for injection pump drive belt: .... 2.0 kgm (14.5 ft.lb.)
Tensioning roller for injection pump drive belt: .... 1.5 kgm (10 ft.lb.)
Injection pipe union nuts: .... 2.5 kgm (18 ft.lb.)
Injectors clamp brackets: .... 2.5 kgm (18 ft.lb.)
Glow plugs: .... 3.0 kgm (22 ft.lb.)
Oil pump bolts: .... 1.0 kgm (7 ft.lb.), M6 2.0 kgm (14.5 ft.lb.), M8
Oil drain plug: .... 5.0 kgm (36 ft.lb.)
Oil suction pipe to crankcase: .... 1.0 kgm (7 ft.lb.)
Oil splash jets to crankcase: .... 1.0 kgm (7 ft.lb.)
Oil pump cover: .... 1.0 kgm (7 ft.lb.)
Oil dipstick guide tube to engine: .... 2.0 kgm (14.5 ft.lb.)
Thermostat housing: .... 1.0 kgm (7 ft.lb.)

Water elbow: .... 1.0 kgm (7 ft.lb.)
Relief valve plug (oil pump): .... 4.0 kgm (29 ft.lb.)
Oil pressure switch: .... 1.5 kgm (10 ft.lb.)
Exhaust system: .... Refer to exploded view
Engine mounting bolts: .... 8.5 kgm (61 ft.lb.)
Propeller shaft to gearbox flange: .... 7.0 kgm (50.5 ft.lb.)

1.6. Lubrication System

1.6.1. OIL PUMP

A pressure lubrication system is used on all engines, but the differences must be noted.

Four-Cylinder Engine

The oil pump is driven from the front of the crankshaft and is located in a separate housing. The pump cannot be repaired or overhauled. If the pump fails, a complete pump housing must be fitted. The engine has an oil cooler, which, however, is fitted next to the oil filter to the cylinder block. A cover must be removed to gain access to the oil cooler in the inside of the housing. An oil pressure switch is fitted on the opposite side of the oil filter to control the oil pressure to the pre-set value. Fig. 1.94 shows the attachment of the pump on a removed engine.

Five-Cylinder Engine

A different oil pump arrangement is used on the five-cylinder engine. The pump is located at the lower front face of the engine and is driven directly by the crankshaft. The pump is removed as a complete unit (also with the engine fitted) and can be dismantled after unscrewing the pump cover. Two ring gears are operating inside the pump and produce the oil pressure. Fig. 1.95 shows an exploded view of the oil pump and its associated parts.

The removal and installation of the pump of a four-cylinder engine and a five-cylinder engine is described under separate headings.

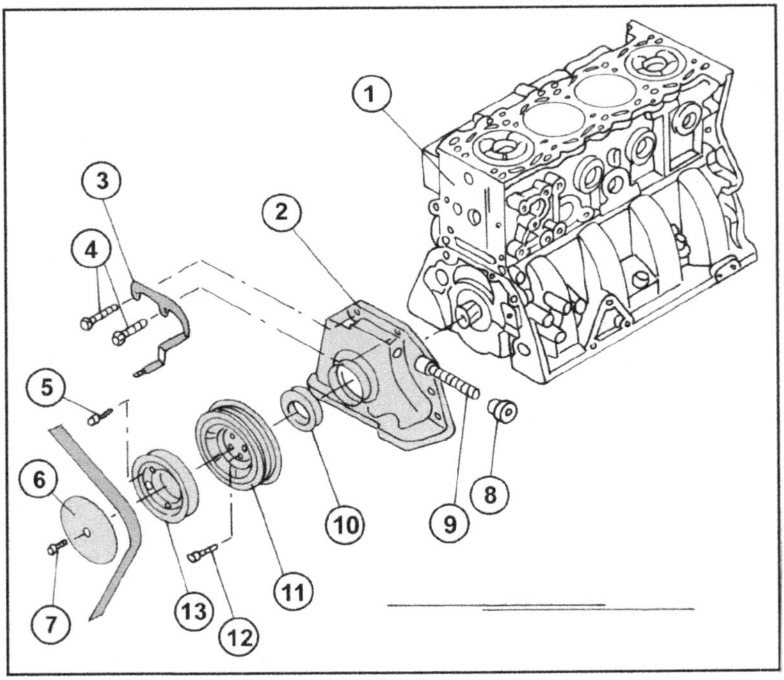

Fig. 1.94 – Oil pump attachment of a four-cylinder engine. The parts shown must be removed.

1 Cylinder block
2 Oil pump assembly
3 TDC pointer
4 Bolts, 3.0 kgm
5 Bolt, 3.0 kgm
6 Dust cover
7 Bolt, 1.0 kgm
8 Plug
9 Coil spring
10 Oil seal
11 Vibration damper
12 Bolt, 10 kgm
13 Vacuum pump pulley

Five-Cylinder Engine

We would like to point out that the crankshaft must be prevented from rotating when the bolt in the centre of the vibration damper is slackened and tightened to the very high torque.

- Remove the drive belt as described in the engine section.
- Remove the upper timing belt cover from the front of the engine.
- Slacken the bolt in the centre of the vibration damper. VW workshops use the restraining lever shown in Fig. 1.39. If not available, engage a low gear, apply the handbrake and slacken the bolt with a socket. Fully remove the bolt and, using two tyre levers, applied at opposite points, lever off the damper.
- Remove the oil dipstick. The timing drive is now removed as described in the relevant section. The engine must not be rotated thereafter.

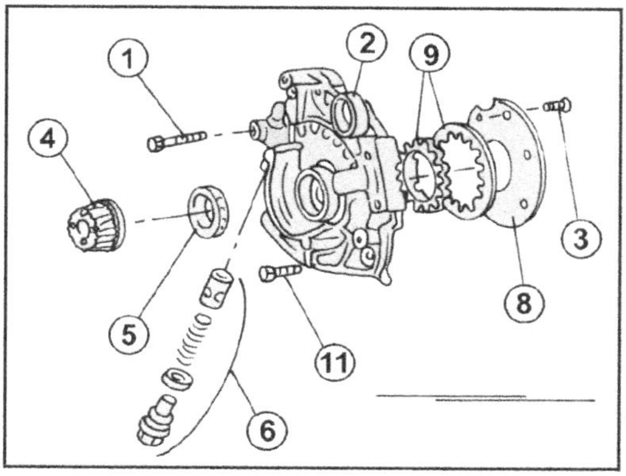

Fig. 1.95 – The oil pump of a five-cylinder engine.
1 Bolt, 1.0 kgm
2 Oil pump
3 Bolt, 1.0 kgm, must be coated with sealing compound
4 Crankshaft timing gear
5 Oil pump oil seal must be replaced
6 Oil pressure relief valve, 4.0 kgm
7 Bolt, 1.0 kgm, must be coated with sealing compound
8 Pump cover
9 Oil pump gearwheels

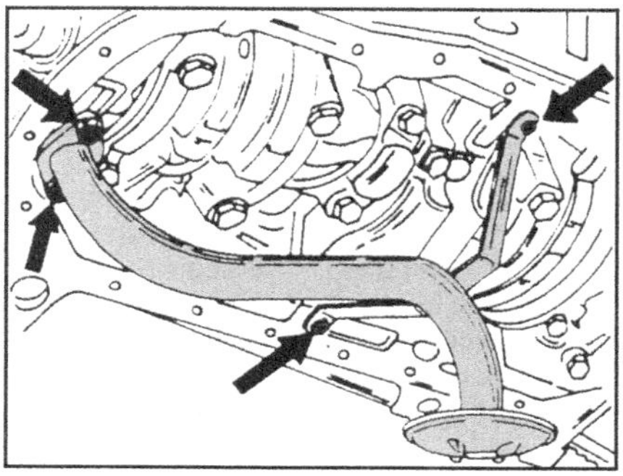

Fig. 1.96 – The arrows show the securing screws for the oil suction pipe at the bottom of the crankcase.

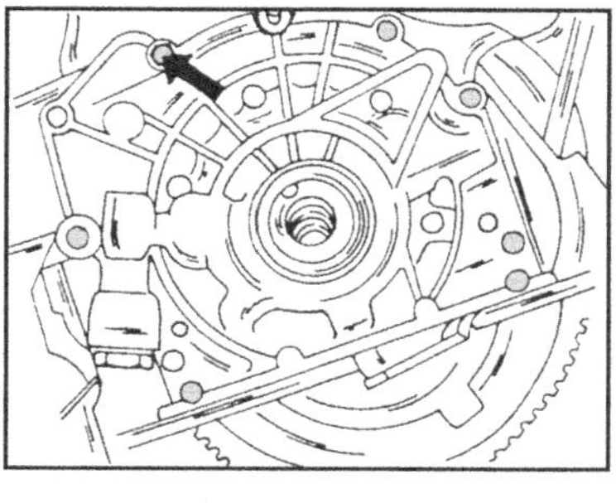

- Remove the oil sump. From the bottom of the crankcase, knock back the lockplates securing the oil suction pipe bolts, unscrew the bolts and remove the suction pipe together with the gasket (Fig. 1.96).
- Unscrew the oil pump from the front of the cylinder block (Fig. 1.97). The crankshaft must not be rotated.

Fig. 1.97 – The arrows show where some of the oil pump bolts are located.

The installation of the oil pump is a reversal of the removal procedure. The following points must be noted. Fig. 1.98 shows the parts which must be re-fitted.

- The oil pump gasket and the crankshaft oil seal in the pump body must always be replaced.
- Tighten the oil pump securing bolts to 2.0 kgm (14.5 ft.1b.) evenly around the body.

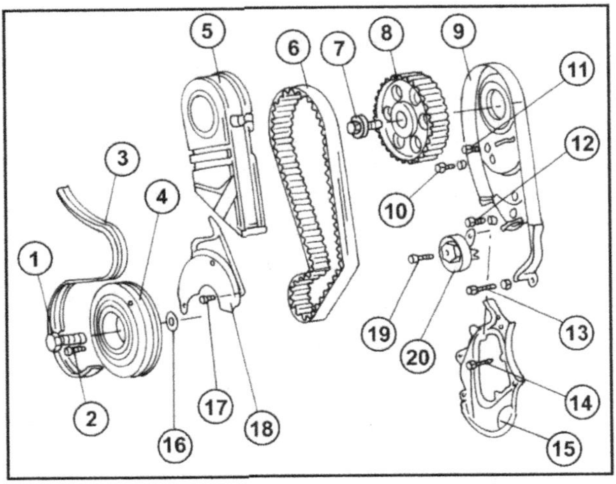

Fig. 1.98 – The parts shown must be removed and refitted when the oil pump of a five-cylinder engine is removed.
1 Bolt, 16 kgm + 180°
2 Bolt, 2.0 kgm
3 Ribbed drive belt
4 Vibration damper
5 Upper belt cover
6 Timing belt
7 Camshaft gear bolt (note different torque)
8 Camshaft timing gear
9 Inner belt cover
10 Bolt, 2.0 kgm
11 Bolt, 1.0 kgm
12 Bolt, 2.0 kgm
13 Bolt, 2.0 kgm
14 Bolt, 1.0 kgm
15 Lower belt cover
16 Washer, replace
17 Bolt, 1.0 kgm
18 Lower cover
19 Bolt, 2.0 kgm
20 Timing belt tensioner

Note: Bolts 11 and 14 must be coated
With sealing compound before installation.

- Replace the oil suction pipe gasket, fit the pipe to the crankcase and tighten the bolts shown in Fig. 1.96 to 1.0 kgm (7.2 ft.lb.). New lockplates must be used and knocked over the bolt heads to secure the bolts in position.
- Refit the timing belt as described for the five-cylinder engine. All timing marks should be in line as the crankshaft has not been rotated. Refit the oil sump and fill the engine with engine oil. Make sure that the selected oil is suitable for diesel engines and is of good quality.
- Start the engine and check for oil leaks around the oil sump and the oil pump.

Four-Cylinder Engine

- Remove the parts shown in Fig. 1.94 in front of the oil pump until the pump is free. Remove the oil sump as described later on.
- Remove the oil pump housing securing bolts. The TDC indicator will come away. Remove the pump from the front of the crankcase.
- Thoroughly clean the sealing faces of crankcase and if necessary from the pump housing face (if the pump is refitted) and coat the pump face with sealing compound. This you will have to obtain from a VW dealer, as a special compound is required. The sealing compound is known as "Loctite Down Corning 780". Your workshop may also be able to tell you which other sealing compound is used. The joint between oil pump and cylinder block face must also be coated. Pump bolt torque is 3.0 kgm.
- The remaining installation is a reversal of the removal procedure. Refer to Fig. 1.94.

1.6.2. Servicing the Oil Pump

Four-Cylinder Engine

The pump cannot be repaired.

Five-Cylinder Engine

The oil pump of the latest engines must be replaced as a complete unit. Before dismantling the pump, check that spare parts, for example, gear sets, are available for your particular engine. The pump gearwheels can be removed after unscrewing the rear cover (12) in Fig. 1.90. If the gearwheels show wear, replace them as a set.

The gearwheels have marks that must be on the outside after the gearwheels have been fitted into the pump housing. The cover can be replaced separately (see note above) if it shows grooves or other damage. If the pump has been removed, unscrew the relief valve plug, withdraw the piston and return spring. The complete pump must be replaced if the piston shows excessive wear. Tighten the plug to 4.0 kgm (30 ft.1b.) on assembly. The cover screws are tightened to 1.0 kgm (7.2 ft.lb.).

If the pump has been dismantled, pour some oil into the gearwheels before the cover is refitted.

1.6.3. Oil Sump

The oil sump can be removed with the engine installed. Remove the undercover from the front of the engine. Transmission and gearbox must be removed in the case of a five-cylinder engine to remove the sump. A conventional gasket is fitted between oil sump and cylinder block face. A different arrangement is used on the four-cylinder engine, as can be seen in Fig. 1.99. The baffle plate fitted to the underside of the crankcase can also be removed. After draining the oil, unscrew the oil sump and take off the gasket. A small socket wrench is required to undo the M6 bolts on a five-cylinder engine.

When installing the oil sump, note the points below.

- Always use a new gasket. No grease or sealers should be applied to gaskets or facings.
- Take care that the gasket and sump are properly positioned before tightening any bolts.
- Tighten the bolts to 3.0 kgm (22 ft.1b.) in the case of a four-cylinder engine, working diagonally and taking several steps to reach the final torque reading. On a five-cylinder engine, fit the M8 bolts with hex. head on the transmission side to 2.0 kgm (14.5 ft.lb.) and the remaining bolts to 1.0 kgm (7.2 ft.lb.).

Fill the engine with the correct quantity of the recommended oil. Run the engine and re-check the oil level. The difference between the "Min" and the "Max." mark on the dipstick is 1 litre (approx. 2 Imp. pints).

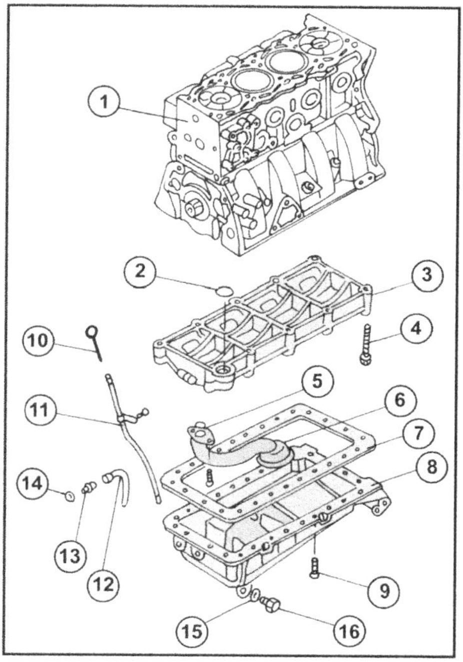

Fig. 1.99 – Oil sump and associated parts fitted to a four-cylinder engine.

1 Cylinder block
2 Oil seal, replace if damaged
3 Oil baffle plate
4 Bolt, 3.0 kgm
5 Oil seal, replace if damaged
6 Oil suction tube
7 Oil sump gasket, replace
8 Oil sump
9 Bolt. 3.0 kgm
10 Oil dipstick
11 Oil dipstick guide tube
12 Oil pipe
13 Threaded connector
14 Oil seal, replace if damaged
15 Oil sealing washer
16 Oil drain plug, 5.0 kgm

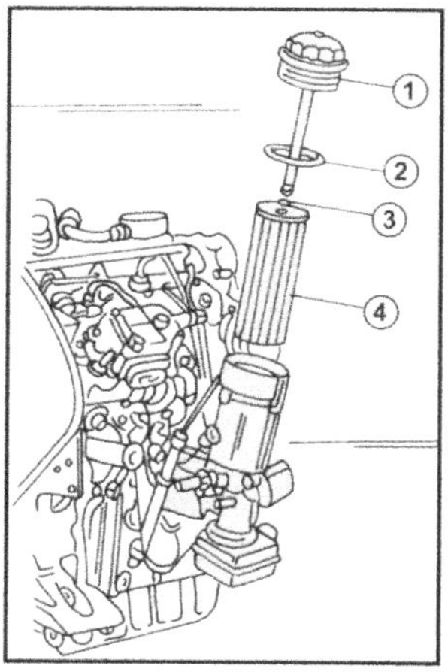

Fig. 1.100 – The oil filter is fitted to the side of the engine as shown.

1 Filter cap
2 Sealing washer (large)
3 Sealing washer (small)
4 Oil filter element

1.6.3. Oil Filter

The oil filter should be replaced in accordance with the service intervals given in the Operator's Handbook and every time the engine has been overhauled.

Fig. 1.101 shows the oil filter as fitted to a five-cylinder engine. The filter is fitted to the upper end of the oil filter bracket, not at the bottom of the oil cooler as was the case of earlier 2.5-litre engines of this type. The oil cooler is fitted to the lower end of the filter bracket and can be removed/refitted without disturbing the oil filter.

To replace the oil filter, unscrew the cap at the upper end. The workshop uses a special wrench (3417), which is placed over the filter cap. A chain filter wrench cap can, however, be used. Otherwise, a large pair of pliers may do the trick. Remove the cap and the oil filter element. Normally, it will stick to the cap. Thoroughly clean the inside of the filter cap and the inside of the housing.

Fit the large and the small sealing ring over the large bolt of the filter cap and insert the new filter element into the housing. Fit the cap and tighten it, using the hands only.

Fill the engine with oil, start it up and let it run for a while. After the engine has reached its operating temperature, check the filter for oil leaks.

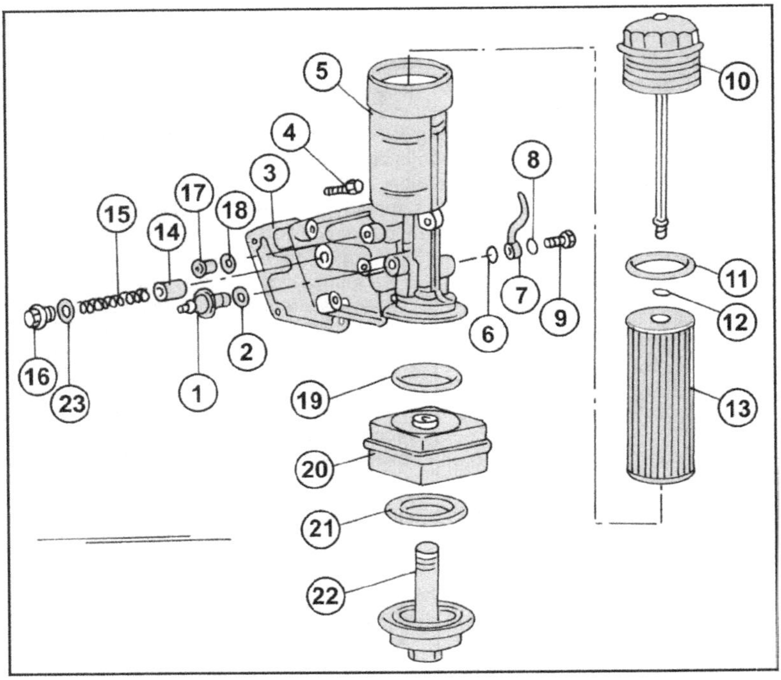

Fig. 1.101 – Oil filter and oil cooler in the case of a 2.5-litre engine (5 cyl.)

1 Oil pressure switch (grey)
2 Sealing ring, replace
3 Gasket, replace
4 Bolt, 1.5 kgm + 90°
5 Oil filter bracket
6 Sealing ring
7 Oil pipe to turbocharger
8 Sealing washer, replace
9 Banjo bolt, 2.0 kgm
10 Filter cap
11 "O" sealing ring
12 "O" sealing ring
13 Oil filter element
14 Relief valve plunger
15 Return spring
16 Plug, do not remove
17 Plug, 1.0 kgm
18 Seal, replace
19 Sealing ring
20 Oil cooler
21 Sealing ring
22 Plug, 2.5 kgm
23 Sealing washer

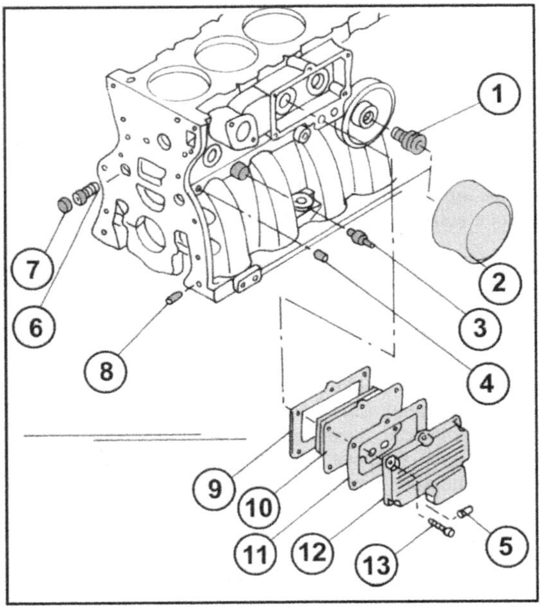

Fig. 1.102 – Oil filter and oil cooler arrangement of the four-cylinder engine.

1 Threaded sleeve
2 Oil filter
3 Oil pressure switch
4 Closing plug
5 Non-return valve
6 Pressure relief valve
7 Cover
8 Tensioning sleeve
9 Gasket, replace
10 Oil cooler
11 Gasket, replace
12 Oil cooler cover
13 Bolt, 3.0 kg

Four-Cylinder Engine

The oil filter of this engine is located next to the oil cooler on the side of the cylinder block, as shown in Fig. 1.102. To remove the oil filter, use a chain-type filter wrench.

The fitting instructions are printed on a label attached to the filter body and must be followed. Always use a new rubber seal, slightly lubricated with engine oil or grease. Use the hand only to fit the new filter element. Finally, check the oil level in the oil sump and check the filter connection for leaks.

1.6.4. Checking the Oil Pressure

The oil pressure of the engine lubrication system can only be checked by means of a suitable pressure gauge, which should have provision for the connection of the oil pressure switch. We, therefore, recommend to have the oil pressure checked in a workshop.

1.6.5. Oil Level and Oil Change

Check the oil level at regular intervals, especially when venturing on longer journeys. Make sure that the vehicle is standing on a level surface when the oil is checked. The difference between the "Min" and "Max" mark on the dipstick is approx. 1 litre. It is therefore possible to judge without difficulties the amount of oil that may be missing. Never overfill the engine.

The oil change is a straightforward operation. To speed up the draining of the oil, allow the engine to run for a few minutes but make sure that it cannot overheat. Place a container of adequate size underneath the oil drain plug on the oil sump to collect the oil. The five-cylinder engine has more than 7.5 litres. The four-cylinder engine has approx. 8.5 litres.

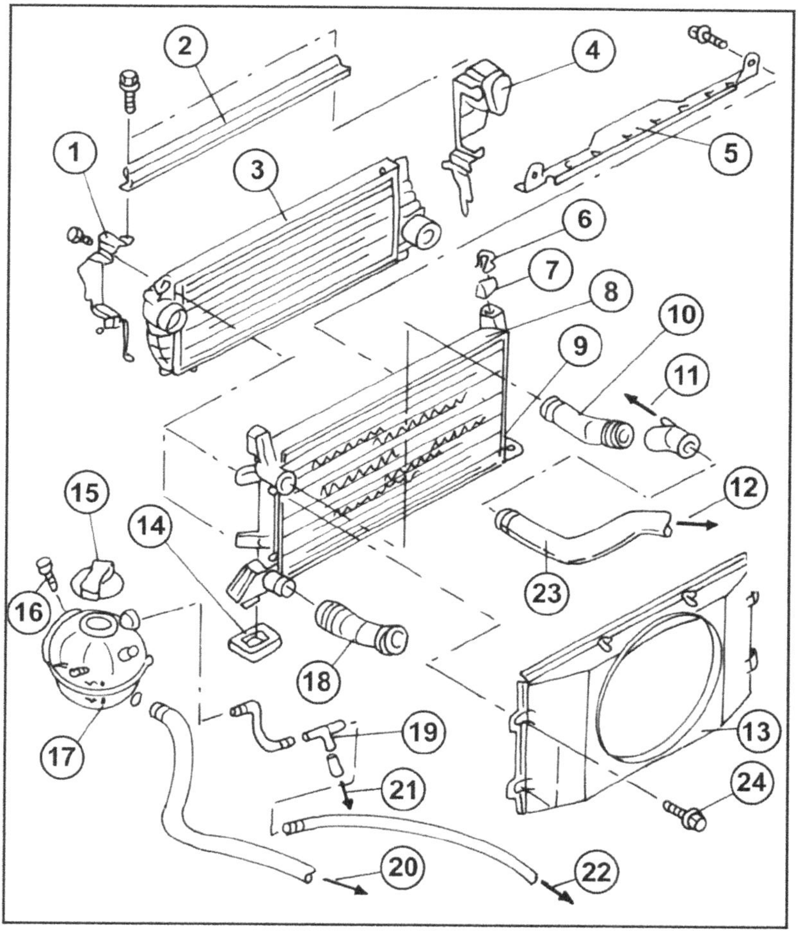

Fig. 1.103 – Radiator and associated parts in the case of a 5-cylinder engine.

1 Air guide, L.H. front
2 Upper air guide
3 Charge air cooler
4 Air guide, R.H. front
5 Lower air guide
6 Radiator retaining clip
7 Bearing bush
8 Radiator
9 Drain plug
10 Upper radiator hose
11 To elbow on engine
12 To elbow on engine
13 Fan surround
14 Bearing bush
15 Expansion tank cap
16 Screw, 1.0 kgm
17 Expansion tank
18 Lower radiator hose
19 "T" connector
20 To coolant pipe
21 To upper hose
22 To cylinder head
23 Connecting hose
24 Screws

Note the different filling capacities for the four- and five-cylinder engines (see page 7).

1.7. Cooling System

1.7.0. Technical Data

System type:	Pressurised, with impeller-type water pump, thermostat and separate expansion tank. Cooling fan operated by viscous clutch fan.
Water pump drive - four-cylinder engine:	Through gear drive assembly at rear of engine.
- five-cylinder engine:	Driven by timing belt
System capacity:	See Section 0.3 (page 7)
Thermostat:	
Opens at:	87 °C
Fully open at:	102 °C
Anti-freeze capacity:	See Section 1.7.4

1.7.1. Radiator – Removal and Installation

The radiator together with the associated parts of a 2.5-litre five-cylinder engine is shown in Fig. 1.103. Differences in some of the parts, mainly in the hose connections, in the case of a four-cylinder engine can be seen in Fig. 1.104. The arrangement of the individual parts is otherwise similar on all engines. Study the layouts to get familiar with the system of your particular vehicle.

To remove the radiator proceed as follows, note that the operations are more or less identical on all models. Fig. 1.103 in conjunction with 1.104 should be referred to the following.

- Disconnect the battery.
- Drain the cooling system as described in the next section. Remove the expansion tank cap to speed up the draining operation.
- Slacken the clamps of the upper and lower radiator hoses. Withdraw the hoses from the connections and push them to one side without kinking them. Also remove the hose between radiator and expansion hoses can either be secured by spring clamps or screw clamps. Screw clamps can be used during installation.

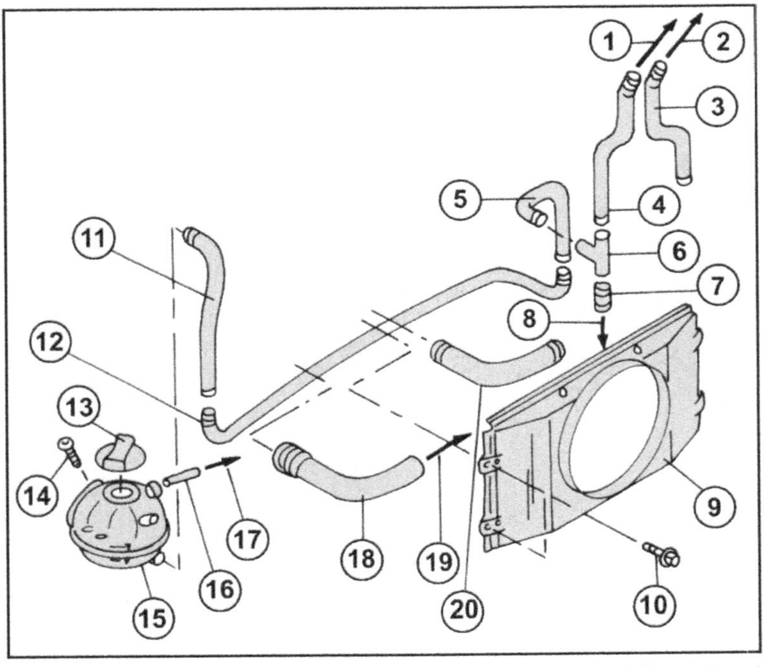

Fig. 1.104 – Hose connections in the case of four-cylinder engine. The remaining parts are similar, as shown in Fig. 1.103.

1 To heater (outlet)
2 To heater (inlet)
3 Coolant hose, heater, inlet
4 Coolant hose, heater, outlet
5 Coolant hose
6 "T" piece, hoses
7 Coolant hose
8 Connector on water pump
9 Radiator shroud
10 Screws, shroud
11 Coolant hose
12 Coolant hose
13 Expansion tank cap
14 Screw, 1.0 kgm
15 Expansion tank
16 Connection hose
17 Connection to elbow
18 Lower radiator hose
19 To thermostat on engine
20 Upper radiator hose

- Disconnect the hoses from the charge air cooler.
- Refer to Fig. 1.105 and remove the screw securing the air intake hose (arrow) and withdraw the hose (1). Remove the hose (2) from the radiator shroud (cowling) and the retaining clips (3) for the radiator on the L.H. and R.H. sides. The clips are shown by item "6" in Fig. 1.103. The radiator is attached in the same manner on all engines, irrespective of the fitted engine.

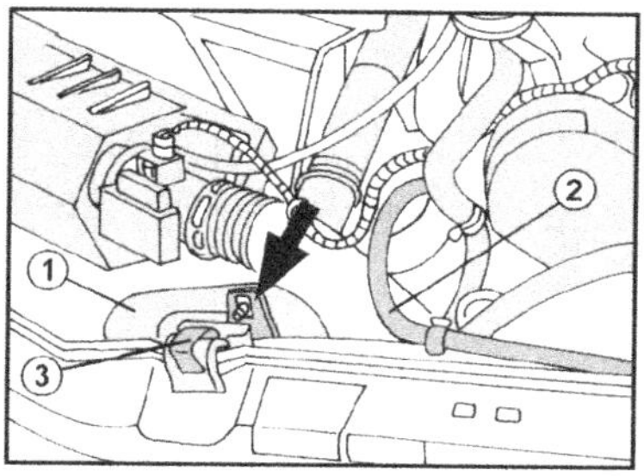

Fig. 1.105 – Details for the removal or the radiator (refer to the text).

- Unscrew the expansion tank from the lock carrier (refer to Fig. 1.1) and place the assembly to one side.
- Remove the lock carrier with the attached parts as described in the relevant section (body). Also remove the steps on the L.H. and R.H. sides. The radiator can now be removed together with the shroud towards the front.

The installation is a reversal of the removal procedure. Make sure that all coolant hoses are properly connected. Finally, fill the cooling system. The headlamp adjustment must be checked after the completed installation.

1.7.1.0. DRAINING AND FILLING THE COOLING SYSTEM

Set the heater control lever in the dashboard to the "Hot" position and remove the expansion tank filler cap. Careful when the engine is hot.

Drain the coolant by either removing the lower water hose or unscrewing the drain plug. Fig. 1.106 shows the location of the drain tap.

NOTE: If the anti-freeze mixture is to be re-used, place a container underneath the drain points to collect as much as possible.

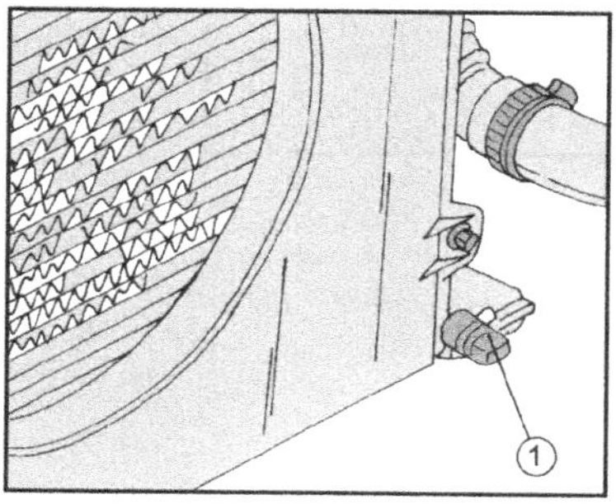

Fig. 1.106 – The drain plug (1) is located at the bottom of the radiator (all models).

To ensure that the cooling system is filled without any air pockets, follow the instructions given below.

- Check that the heater control lever is still in the "Hot" position and reconnect the lower hose to the radiator. Close the drain plug if it has been unscrewed.
- Fill the expansion tank to the upper edge with anti-freeze solution (see Fig. 1.107 and Section 1.7.4) and fit the filler cap. This may take up to 5 minutes.

Start the engine and run it at idle speed until the operating temperature is reached, i.e., the cooling fan will switch on. Check the coolant level in the expansion tank and if necessary fill in additional coolant. The coolant level should be between the "Min" and "Max" marks on the expansion tank (Fig. 1.107), *only when the engine is cold.*

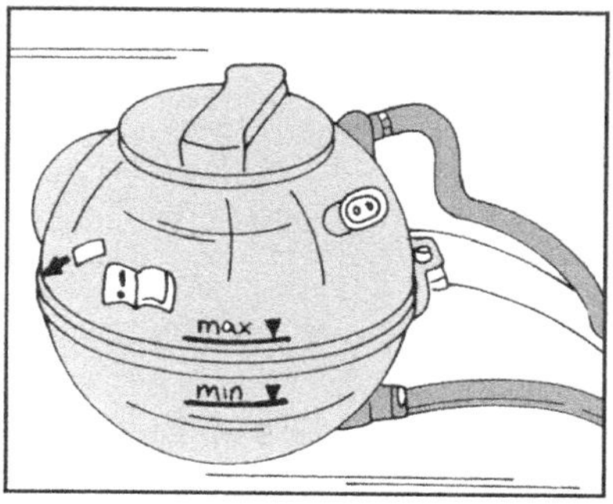

Fig. 1.107 – View of the coolant expansion tank.

1.7.2. Coolant Pump

1.7.2.0. REMOVAL AND INSTALLATION

The water pump is fitted to the cylinder block but cannot be seen from the outside.

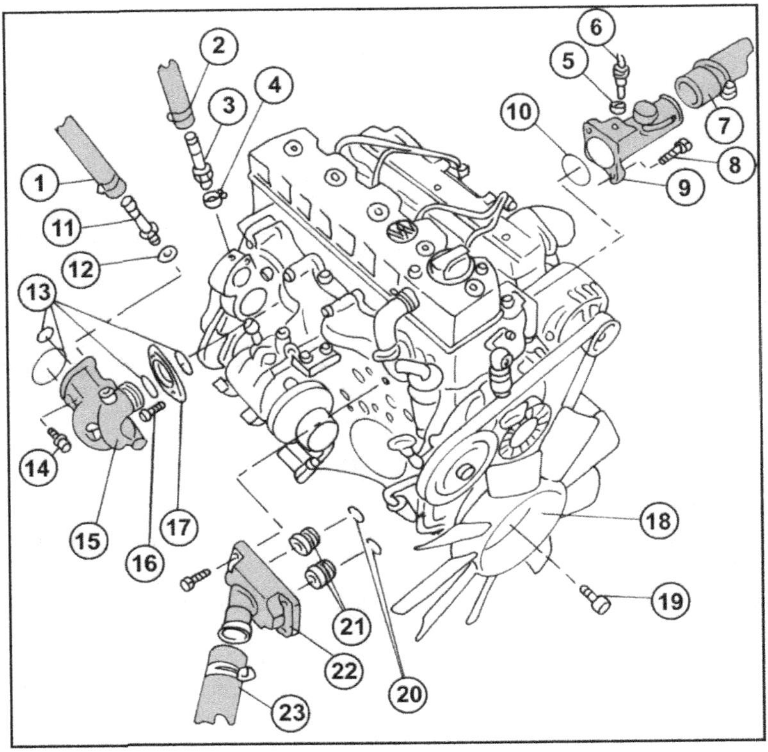

Fig. 1.108 – The parts of the cooling system on the cylinder block (4-cyl.).
1 Coolant hose to heater, outlet
2 Coolant hose to heater, inlet
3 Hose connector
4 Seal, replace
5 Seal, replace
6 Coolant temperature sender unit
7 Upper radiator hose
8 Screw, 3 kgm
9 Connector elbow
10 "O" seals, replace
11 Connector
12 "O" seal, replace
13 "O" seal,
14 Screw, 3 kgm
15 Coolant pump
16 Water pump bolts, tighten bolt 14 first
17 Flange
18 Viscous fan
19 Screw, 4.0 kgm
20 "O" seals, always
21 Thermostat
21 "O" seals, always replace
22 Thermostat housing cover
23 Coolant hose, connected to (22)

The pump is driven either by the toothed timing belt (5-cyl.) or from the gear drive on the flywheel end (4-cyl.). The water pump cannot be repaired or overhauled and must be replaced, also in the case of leaks. Note the following when the water pump must be removed.

Four-Cylinder Engine

Fig. 1.108 shows the component parts of the cooling system associated with the water pump on the cylinder block. All operations can be carried out by referring to the illustration.

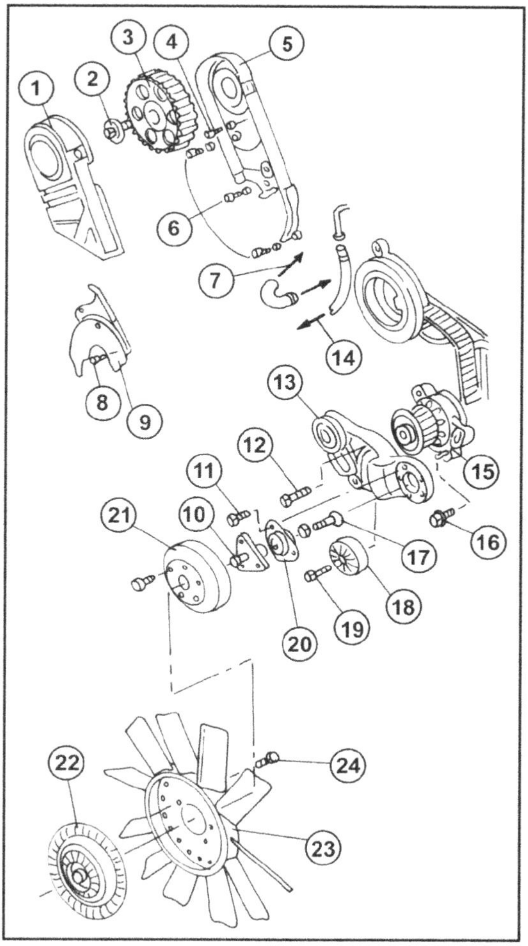

Fig. 1.109 – Details for the removal and installation of the water pump (five-cylinder engine).

1 Upper timing belt cover
2 Camshaft sprocket bolt (note different tightening torques)
3 Camshaft timing gear
4 Screw, 1.0 kgm, coat with sealing compound
5 Rear timing belt guard
6 Screw, 2.0 kgm, coat with sealing compound
7 Hose to cylinder block
8 Screw, 1.0 kgm
9 Lower timing belt cover
10 Bolt, flange shaft
11 Bolt, on coupling bearing
12 Bolt, 4.0 kgm
13 Bracket (see Fig.1.92)
14 To oil cooler
15 Water pump
16 Screw, 2.0 kgm
17 Screw, 2,0 kgm
18 Belt pulley
19 Bolt, 2.0 kgm
20 Fan clutch bearing
21 Pulley, cooling fan
22 Fan coupling
23 Fan blade assembly
24 Bolt, fan blade assembly

Five-Cylinder Engine

Fig. 1.109 shows the parts to be removed. Operations to be carried out to remove the pump are described under separate headings and should be referred to when necessary.

- Drain the cooling system as already described. Collect the anti-freeze if still in good condition.
- Remove the drive belt from the front of the engine as described for the five-cylinder engine on page 49.
- Remove the hydraulic pump (steering pump) without disconnecting the hoses and place it on one side.
- Remove the timing belt covers and the timing belt as described earlier. Remove the mounting bolts and remove the complete pump. Fig. 1.110 shows the fitted pump. The pump will have to be turned slightly before removal.

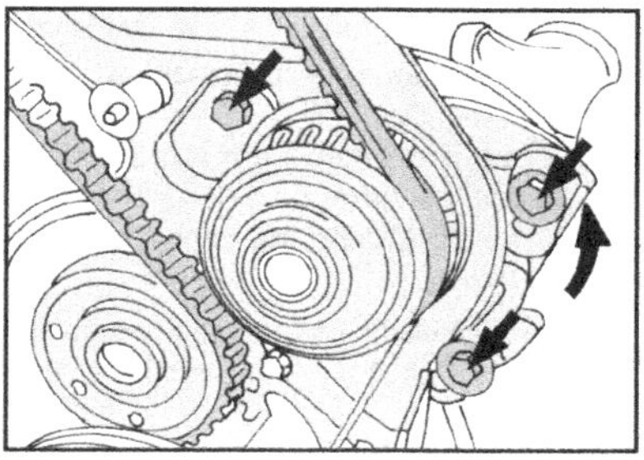

Fig. 1.110 – Water pump attachment (five-cylinder engine).

The installation of the water pump is a reversal of the removal procedure. Always use a new "O" ring between the pump and the engine face. Tighten the pump bolts to 2.0 kgm (14.5 ft.lb.). Refit and tension the drive belt. Refill the cooling system, as described in Section 1.7.1.0.

1.7.3. Thermostat

The location of the two thermostats fitted to the four-cylinder engine is shown in Fig. 1.108. The two units are located underneath a cover at the side of the cylinder block, where the lower cooling hose is connected. The thermostats can be removed after draining the cooling system, disconnecting the hose and unscrewing the cover (22) in Fig. 1.108. Remove the two "!O" rings (20) and immediately check them for wear.

The thermostat of the five-cylinder engine is located underneath a connecting elbow on the side of the cylinder block, where the cooling hose is connected. To remove the thermostat, drain the cooling system and disconnect the hose after slackening the hose clip. Unscrew the cover and lift out the thermostat, noting its location.

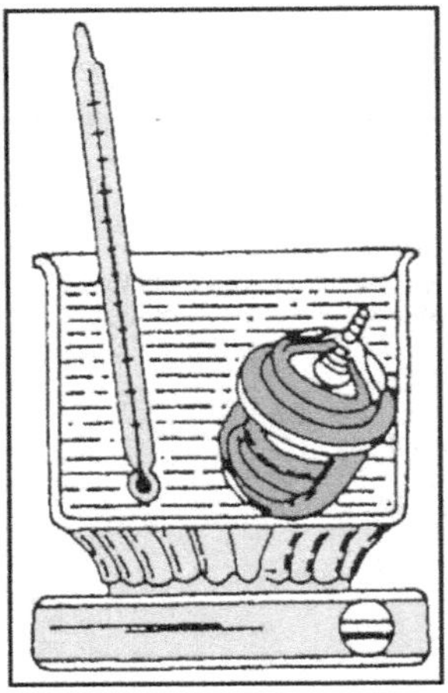

Fig. 1.111 – Checking a thermostat.

A thermostat cannot be repaired or adjusted and must be replaced if faulty. A thermostat can be tested by immersing it in a container of cool water and carefully raising the temperature to check that it opens properly. Suspend the thermostat on a piece of wire so that it does not touch the sides or the bottom of the container. The arrangement is shown in Fig. 1.111. The opening temperature is marked in the case and can be used as a guide during the test. The thermostat should open at around 79 °C in the case of a four-cylinder engine or 87 °C in the case of a five-cylinder engine. The thermostat must be fully open at 94° (4-cyl.) or 102 °C (5-cyl.). The thermostat pin must travel 7 mm during this check (both engines).

Use a new "O" sealing (5-cyl.) or two new rings (4 cyl.) when fitting the thermostat(s). Check the hose clamps before re-using them. Finally, re-fill the cooling system and check the system for leaks.

Note: An overheating engine can be due to a thermostat sticking in the closed position.

1.7.4. Anti-Freeze Solution

The cooling system is filled with anti-freeze and the mixture should be left in the system throughout the year. If a fresh mixture is prepared, use anti-freeze and mix it with water (50% anti-freeze and 50% water will protect the system down to –30 °C). We advise you only to use the anti-freeze marketed by Volkswagen (Standard TL VW 774D), as the solution has been specially formulated for use with your engine. If the system is topped up with plain water, remember that the anti-freeze solution will be diluted. It is always best to mix anti-freeze solution with water, even when topping up. The following should also be noted when dealing with anti-freeze.

- If the radiator, heater radiator or cylinder head gasket have been replaced, fill the system with fresh anti-freeze.
- Adjust the amount of anti-freeze in accordance with the temperatures to be expected. Up to a temperature of –25 °C, use 4.6 litres of anti-freeze and 6.9 litres of water. To protect the engine to –35 °C, use 5.75 litres of anti-freeze and 5.75 litres of water. **Never use more than 60% of anti-freeze.** The total capacity depends on the fitted equipment. The above is a guideline only.
- Anti-freeze solution and coolant containing anti-freeze must be disposed of sensibly in order to protect the environment.

1.7.5. Coolant Run-On Pump

The electrically operating pump is fitted to the five-cylinder engine and is fitted at the lower end at the rear of the engine. To replace **a** faulty pump, pull back the rubber retainer, withdraw the connector plug and disconnect the coolant hoses. Then, unscrew the pump.

1.7.6. Viscous Fan and Drive Coupling (Clutch)

To remove:

- Disconnect the battery (observe the precautions) and remove the radiator as already described. Also remove the drive belt (see the section "Engine").
- Remove the fan blade pulley from the drive clutch. The clutch must be prevented from rotating. This normally requires the use of a special wrench and the operations are not the same on all engines.
- **Four-Cylinder Engine:** Remove in the manner shown in Fig. 1.112 by slackening the centre bolt and remove together with the fan blade.

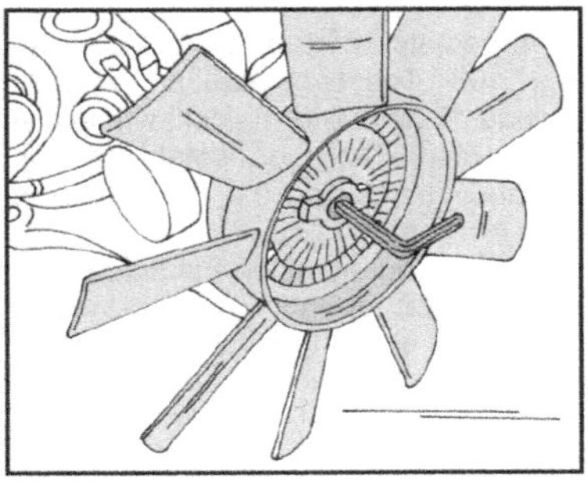

Fig. 1.112 – Removal of the viscous fan blade assembly in the case of a four-cylinder engine. Use an Allen key as shown. Tighten the bolt to 4.0 kgm (29 ft.lb.).

Refit in reverse order. Tighten the clutch bolt to 4.0 kgm (29 ft.lb.).

- **Five-Cylinder Engine:** Prevent the drive clutch (2) in Fig. 1.113 from rotating as shown in the inset and slacken the union nut (1) with an open-ended spanner of 36 mm A/F. The nut has R.H. thread.

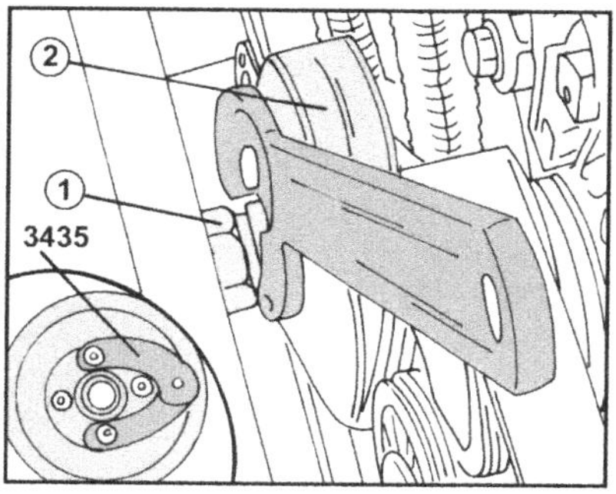

Fig. 1.113 – Removal of the viscous fan in the case of a five-cylinder engine. The union nut (1) is slackened when the pulley (2) is held against rotation.

- The operation is more difficult when the pulley is attached with three bolts (Fig. 1.113 shows a pulley with four-bolt attachment). In the case of a three-bolt attachment, remove two of the bolts and attach the holder with two new, but longer bolts as shown in the illustration. You are warned that the space available is very limited.

The installation is a reversal of the removal procedure. Note the following tightening torques: fan to bearing – 4.5 kgm, fan blade assembly to clutch with new bolts – 0.8 kgm (7 ft.lb.), and viscous clutch to pulley with special bolts – 2.0 kgm (14.5 ft.lb.).

1.8. Diesel Injection System

Absolute cleanliness is essential during any repairs or work on the diesel fuel injection system, irrespective of the nature of the work in question. Thoroughly clean union nuts before unscrewing any of the injection pipes.

The fuel injection pump must be repaired or overhauled by a specialist fuel injection engineer, and an exchange pump or a new pump may be fitted in case of malfunction or damage.

The adjustment of the injection timing and also the removal and installation of the injection pump requires certain special tools and these operations should not be undertaken if these are not available. The following text describes these operations, in case the listed special tools can be obtained or hired.

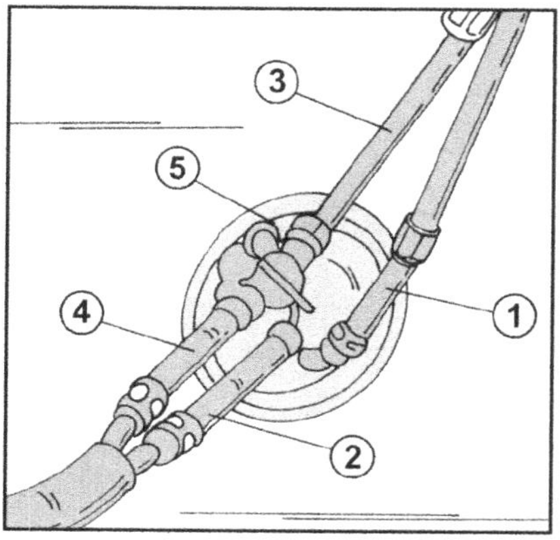

Fig. 1.114 – Connections at the top of the fuel filter.
1 Hose from fuel tank
2 Line to injection pump
3 Return line from fuel pump
4 Return line to tank (marked with arrow
5 Retaining clip

1.8.1. Fuel Filter

The fuel filter should be replaced every 20,000 miles (30,000 km) but must be drained approximately every 10,000 miles (15,000 km) or once a year to empty the water collected through condensation. The filter is located in the engine compartment. Fig. 1.114 shows the connections on the top of the filter. All engines use the same filter.

Fig. 1.115 – The location of the bleeder plug at the bottom of the fuel filter.

The front of the vehicle must be on chassis stands to gain access to the drain plug. To drain the water, open the drain plug (arrow in Fig. 1.115) at the bottom of the filter housing and allow about 100 c.c. to drain off. Then, close the drain plug. If no fuel can be seen coming from the plug opening, slacken one of the filter pipes to allow air into the filter.

A control valve is fitted to the top of the filter. This is held in position by a retaining clip (5) in Fig. 1.114. When the filter must be replaced, remove the clip and withdraw the control valve together with the hoses towards the top. The filter is held in position by means of a clamp. The valve can be released from the filter to bleed the fuel system.

To replace the filter, mark the two connecting hoses on both sides and remove them after opening the clamps. Slacken the filter mounting clamp and withdraw the filter upwards.

Install the filter in reverse order to the removal procedure. Refer to Fig. 1.114 when connecting the hoses.

Start the engine and accelerate a few times. The fuel should be visible in the transparent hose. The fuel system can be bled by removing the retaining clip for the control valve and withdrawing the valve out of its location. The function of the valve is given under item (4) in Fig. 1.116.

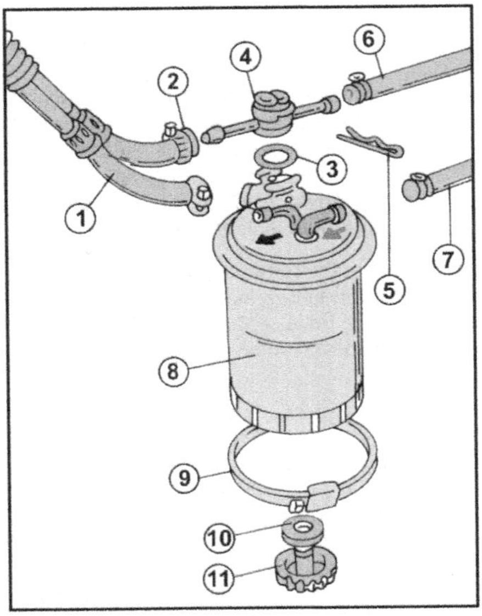

Fig. 1.116 – View of the fuel filter.
1 Transparent supply line to injection pump
2 Return pipe, blue, from injection pump
3 "O" sealing ring, replace
4 Control valve, below 15 °C flow to filter open, above 31 °C flow to filter closed
5 Retaining clip
6 Return pipe, blue to fuel tank
7 Supply pipe, black, from fuel tank
8 Fuel filter
9 Securing strap
10 Sealing washer
11 Drain plug

1.8.2. Fuel Injection Pump

The special tools given in the following text must be available to remove and install the injection pump. Required are the puller for the injection pump gear and the arrester pin for the gearwheel. The injection timing must be checked and adjusted after installation of the pump. This may be an operation for a workshop.

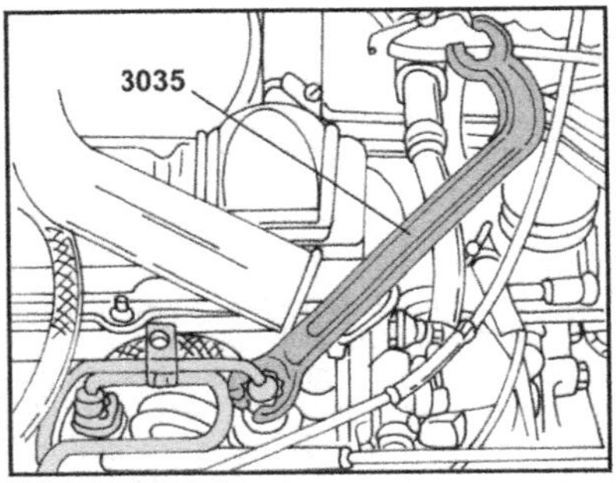

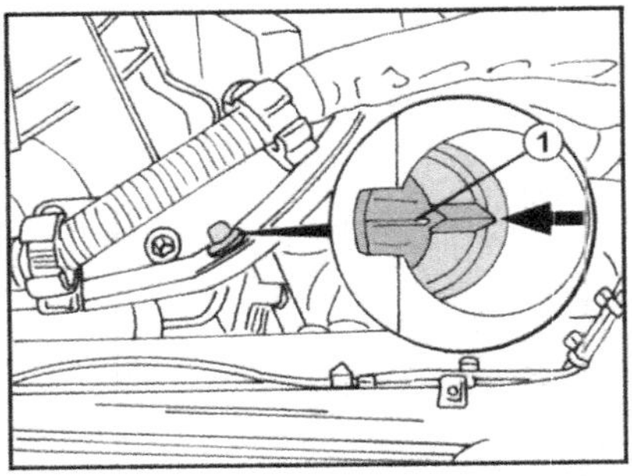

Fig. 1.117 – Using an opened ring spanner to remove injection pipe union nuts on the L.H. side. Rotate the crankshaft until the TDC mark (1) in the flywheel is in line with the notch in the intermediate flange at the position shown by the arrow in the R.H. view.

To preserve the excellent performance of your engine, have the timing checked after installation of the pump, if in any doubt about the running of the engine.

1.8.2.1. Removal and Installation – Four-Cylinder Engine

- Disconnect the injection pipes from the injection pump. To slacken the union nuts, use a ring spanner with a slot cut into one side, as shown in Fig. 1.117 on the L.H. side. Similarly, unscrew the injection pipes from the injectors. Note: the pipes must always be removed as one assembly without bending them. Protect the open ends to prevent entry of foreign matter.
- Remove the cylinder head cover as described in the engine section.
- Remove the dust cover from the centre of the vibration damper and place a socket onto the bolt in the inside of the pulley to rotate the crankshaft. Remove the closing plug in the flywheel housing above the front axle.
- Turn the engine so that the piston of the No. 1 cylinder (flywheel end) is at the top dead centre. The position is obtained when the mark in the flywheel is in line with the notch in the intermediate flange as can be seen in Fig. 1.117 on the R.H. side. To make sure that the correct position is obtained, check the mark on the crankshaft pulley. The mark shown by the arrow in Fig. 1.118 (L.H. view) must be in line with the edge of the pointer.

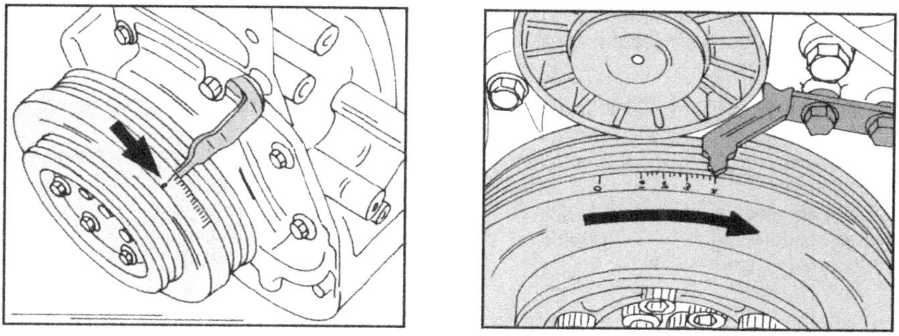

Fig. 1.118 – Checking the TDC position on the vibration damper (L.H. view) and the position of the damper pulley after the crankshaft has been rotated as specified in the text.

- Rotate the crankshaft by approx. two revolutions in the direction of rotation until the pulley mark (vibration damper) is 25° before the top dead centre (TDC). Compare the position by referring to the R.H. view of Fig. 1.118.
- Withdraw the cable connector plug from the fuel shut-off valve on the injection pump and disconnect the accelerator cable from the pump.
- The injection pump is now removed by referring to Fig. 1.119. First, remove the two brackets at the positions shown by the arrows. Then, unscrew the bolts (1) and the nut (2). The injection pump can now be lifted out. If a new or exchange pump is to be fitted, remove and refit the drive gear as described below.

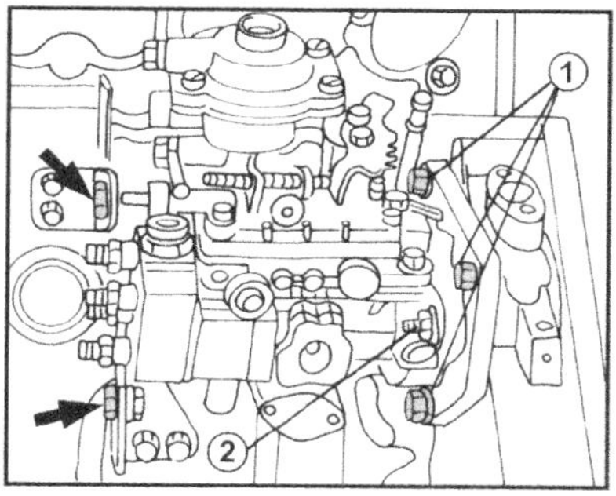

Fig. 1.119 – Attachment of the injection pump in the case of the four-cylinder engine. The bolts shown by the arrows are fitted to the mounting bracket. Bolts (1) and nut (2) retain the pump on the other side.

The injection pump cannot be repaired or overhauled. A new or exchange pump must be fitted in case of malfunction. Otherwise, obtain the advice of a specialist workshop to find out more about the pump.

The installation of the pump is carried out in the manner described below, but we would like to point out that the engagement of the pump gear with the drive gear on the drive gear assembly is not easy, due to the design of the gear teeth. Read the instructions paragraph by paragraph to ensure correct installation.

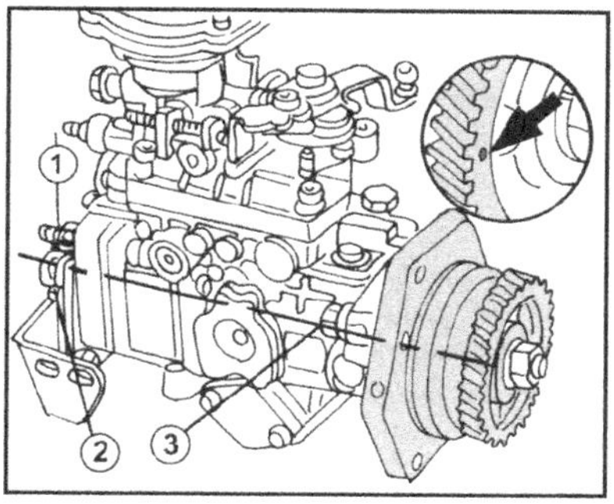

Fig. 1.120 – Aligning the drive gear of the injection pump (four-cylinder engine). Align the mark (arrow in inset) so that it is positioned in the centre of an imaginary line drawn through the threaded connector (1) and (2) for the injection pipes.

- Check that the crankshaft is still in the position shown in Fig. 1.118 on the R.H. side. Otherwise, rotate the crankshaft as necessary.
- Slacken the three securing nuts (3) in Fig. 1.120 and slowly rotate the injection pump until the studs are in the centre of the elongated holes. In this position, tighten the nuts finger-tight to the intermediate flange.
- Align the marking on the injection pump drive gear (shown by the arrow in the inset) until it is between the two threaded connections for the injection pipes of cylinders No. 1 (1) and No. 4 (2), as shown by the interrupted line.
- Turn the injection pump with the drive gear into the direction of cylinder No. 1 until it reaches its stop.

The installation of the pump is now critical.

- Insert the injection pump with a new "O" sealing ring into the timing housing. The pump must be guided to line up with the bores for the securing bolts.
- To obtain the engagement of the skew gear, turn the pump by a maximum of 1/3 of the bores towards the engine. To complicate this operation further, you will have to look from above onto the pump. In this position, tighten the securing bolts (1) in Fig. 1.119 to the timing housing.
- The point of injection commencement must now be adjusted as described below.

Note: If you carry out the adjustment yourself and the value given below cannot be obtained, there is a certain possibility that the gear tooth engagement is out by one tooth. In this case, remove the pump and re-install it as described.

- Tighten the nut (2) in Fig. 1.120 to 2.0 kgm (14.5 ft.lb.), followed by the bolts (2) to the timing housing and finally the bolts shown by the arrows. All are tightened to the same torque.
- Fill the injection pump through the connector for the fuel return line with diesel fuel. Never allow a dry pump to operate.
- Re-connect all injection pipes and fuel lines. All other operations are carried out in reverse order to the removal procedures. Further information is given under the heading "Adjusting the Point of Injection Commencement", given below.

Removal and Installation of the Injection Pump Drive Gear

If the injection pump is replaced, you will have to transfer the drive gear from the old pump to the new pump. A puller is required for this operation. The Woodruff key must also be re-used.

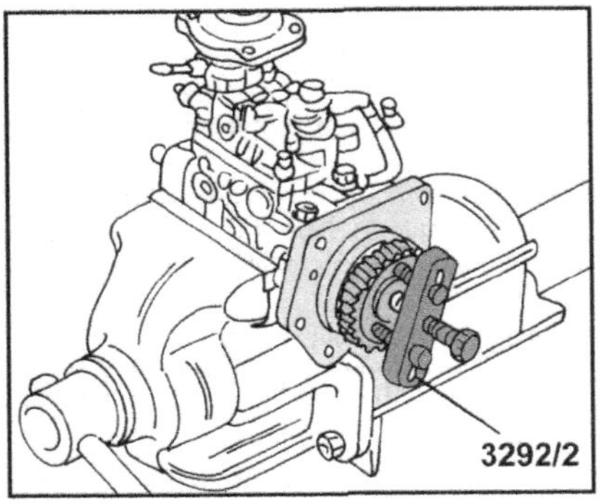

Fig. 1.121 – Removal of the injection pump drive gear (4-cyl.). The illustration shows the VW puller, but a similar puller can be used.

- Counterhold the drive gear in a suitable manner against rotation and slacken the gearwheel nut a few turns.
- Attach a suitable puller to the drive gear. A suitable tool is shown in Fig. 1.121, which is attached with two 8-mm bolts to the drive gear. Tighten the centre bolt until the puller is under tension and then use a hammer and give a short blow to the end of the puller bolt. The drive gear should come away from the pump shaft. Remove the puller and the shaft nut and take off the gearwheel. Use a side cutter to remove the Woodruff key.

Note the following during installation of the drive gear.

- The inside of the gearwheel and the outside of the shaft cone must be free of oil and/or grease.
- The Woodruff key must be properly installed.
- Drive the gearwheel over the pump shaft without dislodging the Woodruff key, fit the nut and tighten it to 9.0 kgm (65 ft.lb.). The gearwheel must be prevented from rotating.
- Refit the injection pump as described above.

Adjusting the Point of Injection Commencement

The following description assumes that the crankshaft is in the position specified above, i.e., flywheel and crankshaft vibration damper are aligned at TDC, as shown in the relevant illustrations. As already mentioned, you will need special tools for the adjustment. An adapter, No. 3313 and a dial gauge are used to adjust the pump. Provided that you can obtain the tools and have some experience, proceed as follows.

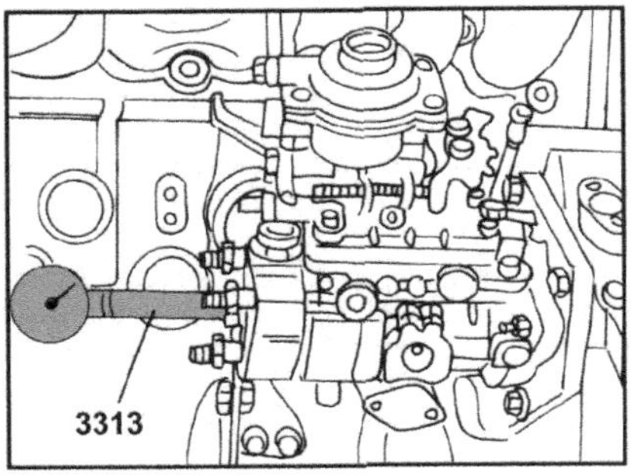

Fig. 1.122 – Fit the dial gauge and the adapter into the rear of the injection pump after removal of closing plug.

- Remove the closing plug from the injection pump head and fit the adapter and the dial gauge into the pump, as shown in Fig. 1.122. Set the dial gauge to a pre-load of 1.0 mm.
- Slowly rotate the crankshaft anti-clockwise (against the normal direction of rotation) until the dial gauge needle no longer moves. In this position, set the dial gauge to "Zero" (0) but retain the pre-load of 1 mm.
- Now, rotate the crankshaft slowly in a clockwise direction (direction of rotation) until the mark in the flywheel and the notch in the intermediate flange are in line (see Fig. 1.117 on the R.H. side). The dial gauge should show 1.70–1.80 mm. If an adjustment is necessary, set the value to 1.70 mm, with a tolerance of ±0.4 mm. If the value has not been obtained, adjust the pump as follows.
- To adjust the pump setting, slacken the nuts (2) in Fig. 1.119 (3 nuts) and also the bolts from the rear console bracket (shown by the arrows). The pump must be rotated until the above dimension is obtained. The setting will have to be re-checked as described above, i.e., rotate the pump in small steps, re-tighten the bolts/nuts and re-check.

- After correct adjustment, tighten the nuts (2) and the bolts on the other side to 2.0 kgm (14.5 ft.lb.).
- Remove the dial gauge and the adapter, refit the closing plug and tighten it 2.0 kgm (14.5 ft.lb.).
- Refit the dust cover to the vibration damper (1.0 kgm/7.2 ft.lb.).
- Refit the injection pipes. Pipes can only be replaced as a complete bundle. Tighten the union nuts on the injection pump side to 2.7 kgm and on the injector side to 3.0 kgm.
- Finally, refit the cylinder head cover.

1.8.2.2. Removal and Installation – Five-Cylinder Engine

The adjustment of the point of injection commencement can only be carried out with the special equipment from Volkswagen. If you are, however, have no option but to remove the pump, mark its fitted position and refit the pump in the original position. The exact adjustment must, however, be entrusted to a workshop dealing with diesel injection equipment, but you should be able to start the engine without problems and drive to the nearest workshop. Proceed as follows.

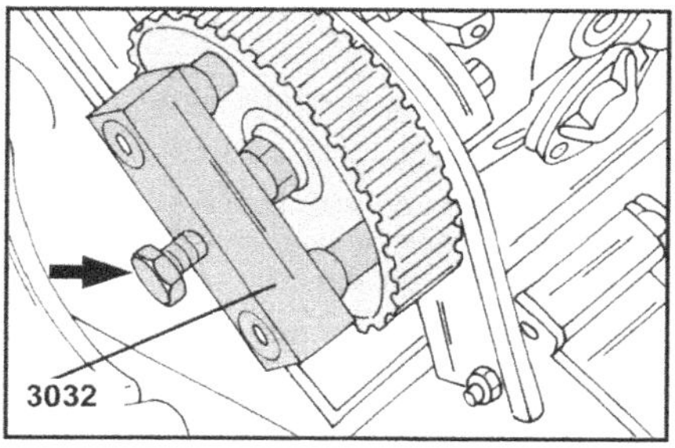

Fig. 1.123 – Removal of the injection pump drive gear with the special puller in the case of a five-cylinder engine. Tighten the centre bolt (arrow) and then tap against it with a hammer to free it of the shaft.

- Remove the injection pump-toothed belt as described in the engine section. Also remove the idle pulley for the injection pump drive belt.
- Remove the injection pump drive wheel. To do this, slacken the pump shaft nut and unscrew the nut until the end face is flush with the end of the shaft. To extract the drive wheel, a special puller VW 3032 is required, which is shown in Fig. 1.123 in applied condition. Insert the two claws of the puller into the holes of the drive wheel at right angles to the centre spindle and counterhold the drive wheel with one hand. Tighten the centre spindle, and when under tension, tap the end of the centre spindle with a short blow of a hammer. The impact will loosen the drive wheel from its cone seat. Remove the puller, fully unscrew the nut and withdraw the drive gear.
- Unscrew all injection pipe union nuts at the pump end and use suitable caps to close off the pipes, whilst the pump is removed from the engine.

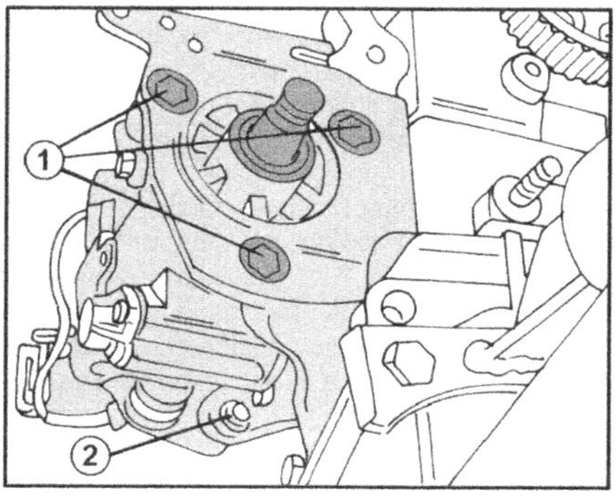

- Disconnect the connector plugs from the injection pump. The plug can either have 10 or 7 pins, depending on the version. Also withdraw the plugs from the fuel shut-off valve and the injection amount adjuster from the pump. Free the cable and plug from the retainer.
- Mark the fitted position of the pump in the most convenient fashion to have a reference mark on the pump and the mounting bracket for ease of installation. Even if a new pump is fitted, you will have a guide of the approximate position during installation. Remove the bolts shown in Fig. 1.124. Three bolts (1) and a nut (2) must be removed. The pump can now be lifted out.

Fig. 1.124 – The securing bolts (1) of the injection pump of a five-cylinder engine. A nut (2) is located at the bottom.

The installation of the injection pump is a reversal of the removal procedure, making sure that all marks made during removal are aligned. Check that the Woodruff key is in position in the pump shaft and carefully tap the drive wheel over the key and shaft. Fit the shaft nut and tighten to 9.0 kgm whilst restraining the pump wheel. Have the injection timing checked as soon as possible.

1.8.3. Adjustments on the Injection System

1.8.3.1. Throttle Cable Adjustment

Vehicles with a five-cylinder engine have a sensor for the throttle pedal position, which is located on the throttle pedal, and the adjustment is only possible in a workshop.

The throttle operating mechanism in the case of a vehicle with four-cylinder engine can be adjusted as follows.

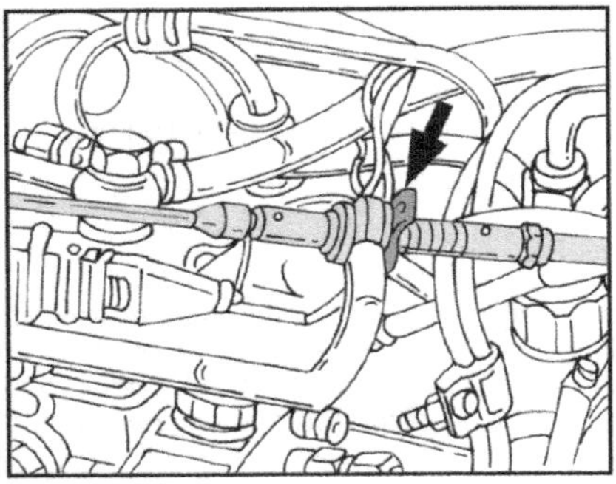

Fig. 1.125 – The arrow shows where the retaining clip for the throttle cable adjustment is located.

- Have the accelerator pedal pressed to the floor by a second person and check that the lever on the injection pump is resting against the stop without being under tension.
- If necessary, remove the retaining clip of the cable at the mounting bracket (Fig. 1.125) and re-insert it into a different groove of the outer sleeve.
- Re-check the adjustment and if necessary carry out further adjustments.

1.8.4. Injectors

NOTE: The injection pressure of the injectors should only be tested in a specialist shop. Never attempt to carry out this operation yourself.

The attachment of the injectors on these engines is different when compared with conventional diesel engines. Each injector is attached by means of a nut and a clamp bracket. A heat shield is inserted between injector and cylinder head. Fig. 1.126 shows in injector with the relevant parts.

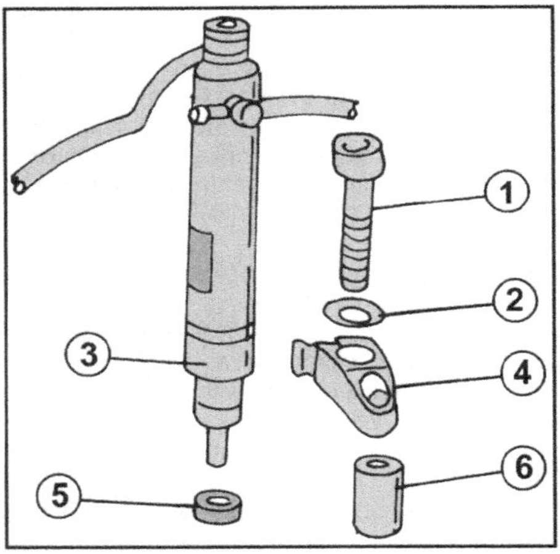

Fig. 1.126 – Injector of a five-cylinder engine.
1 Screw, 2.5 kgm
2 Concave washer
3 Injector, No. 5 with needle lift sensor
4 Clamp bracket
5 Heat shield, always replace
6 Bearing spacer (not 4-cyl.)

To find a faulty injector, unscrew the union nuts one alter the other at the injector connection, with the engine running. Special wrenches are available to unscrew injector pipe union nuts. These have the shape of a ring spanner with a slot to place over the injector pipe. The spanner can then be placed over the union nut. This prevents damage to the hexagon, which can occur when an open-ended spanner is used. Fig. 1.117 shows how such a spanner is used. Run the engine at increased speed. If the engine noise does not change after a certain injection pipe has been disconnected, the faulty injector has been found.

Injectors must always be replaced in sets, but it should be noted that the injector No. 5 of a five-cylinder engine has a so-called needle lift sensor, which must only be fitted to this cylinder. Injectors for a four-cylinder engine have a rubber bush at the lower end, which must be fitted in a certain position, i.e., note the installation during removal. The bearing (6) in Fig. 1.126 is only fitted to injectors of a five-cylinder engine.

Injection pipes must always be replaced in sets.

To remove an injector, disconnect the injection pipes and remove the complete pipe bundle. This will facilitate the installation.

Replace the heat shield between the injector and the cylinder head when fitting the injectors. Tighten the clamp bracket with the bolt to 2.5 kgm (five-cylinder) or 4.0 kgm (four-cylinder). In the latter case, make sure that the rubber bush is fitted correctly to the injector and the position of the clamp bracket on the cylinder head. Union nuts for the injection pipes are tightened to 2.5 kgm in the case of a five-cylinder engine. Note the difference on a four-cylinder engine, 2.7 kgm on the side of the injection pump and 3.0 kgm on the side of the injectors.

Note: Faulty injectors are indicated by the following symptoms:

- Misfiring
- Knocking in one or more cylinders
- Overheating of the engine
- Loss of performance
- Black exhaust smoke
- High fuel consumption
- Very blue exhaust smoke when the engine is started in cold condition.

1.8.5. Glow Plugs

The glow plugs are used to pre-heat the combustion mixture during starting. Glow plugs cannot be repaired and must be replaced in case of malfunctioning.

Automatically operating glow plugs are fitted, together with a glow plug relay that controls the automatic operation of the glow plugs. The glow plugs start to operate immediately after the starter key is switched "on" and the pre-heating period is determined by the temperature of the engine at the moment of starting.

The automatic glow plugs operate as follows.

When the ignition/starter key is switched "on", the voltage is fed to the glow plug relay and the automatic glowing process commences. A warning light lights up to show you that the glow plugs operate.

The duration of the glow plug operation and therefore the time the warning light remains "on" depends on the coolant temperature of the engine, which is detected by the temperature sender unit. As soon as the warning light goes "off", the plugs will glow another 15 seconds and it is during this period that the engine must be started.

If one of the glow plugs is faulty, it can normally spotted with the engine cold. As the fuel in the remaining cylinders will ignite, you may notice that the cylinder with the faulty plug cuts in later.

1.8.5.0. Glow Plug – Removal and Installation

The removal of the glow plugs requires the use of a jointed universal spanner (No. 3220) in the case of a five-cylinder engine or a straight socket (No. T5001) in the case of a four-cylinder engine. It will, therefore, be rather difficult to remove a glow plug without the correct spanners.

1.8.6. Air Cleaner

The air cleaner is fitted with a dry paper element. The removal and installation are simple. Remove and install the air cleaner element as follows.

- Open the securing clip, push the locking handle downwards and disengage from its location.
- Lift off the air cleaner housing and take out the element. Clean the housing thoroughly.
- Fit the new air cleaner element and refit the air cleaner housing. The plastic tongue on one side of the filter element must engage into the opening of the retainer.
- Secure the filter housing and lock it in position with the securing clip.

1.8.7. Turbocharger

All five-cylinder engines, with the exception of the engine with code letters "AGX", are fitted with a turbocharger. A modified turbocharger system is fitted to the engine with code letters "ANJ". All four-cylinder engines are fitted with a turbocharger.

If you notice that the expected engine performance can no longer be obtained, it may be due to malfunction of the turbocharger. A faulty turbocharger must be replaced. Before replacing the turbocharger, check the inlet and exhaust sides of the turbocharger system for entry of unwanted air. In this case, check the system as follows.

- Check the air hose between the turbocharger and the inlet manifold and the hose between the inlet manifold and the injection pump. Hoses can be detached or not sealed properly.
- Air can enter the engine between the inlet manifold and the cylinder head.
- Air enters the engine between the inlet manifold flanges and the cylinder head.
- The clamp of the waste gate valve is not tightened properly or the gasket requires replacement.
- A leak has developed between the exhaust manifold and the cylinder head or between the manifold and the turbocharger.
- The charging pressure can only be checked in a workshop with a special pressure gauge.

1.8.7.1. Removal and Installation of the Turbocharger

The removal and installation is a fairly complicated operation and is not the same on all engines.

Removal and Installation – Five-Cylinder Engine

Before removal, you will have to know the engine code letters of your particular engine in order to refer to the differences in the following text.

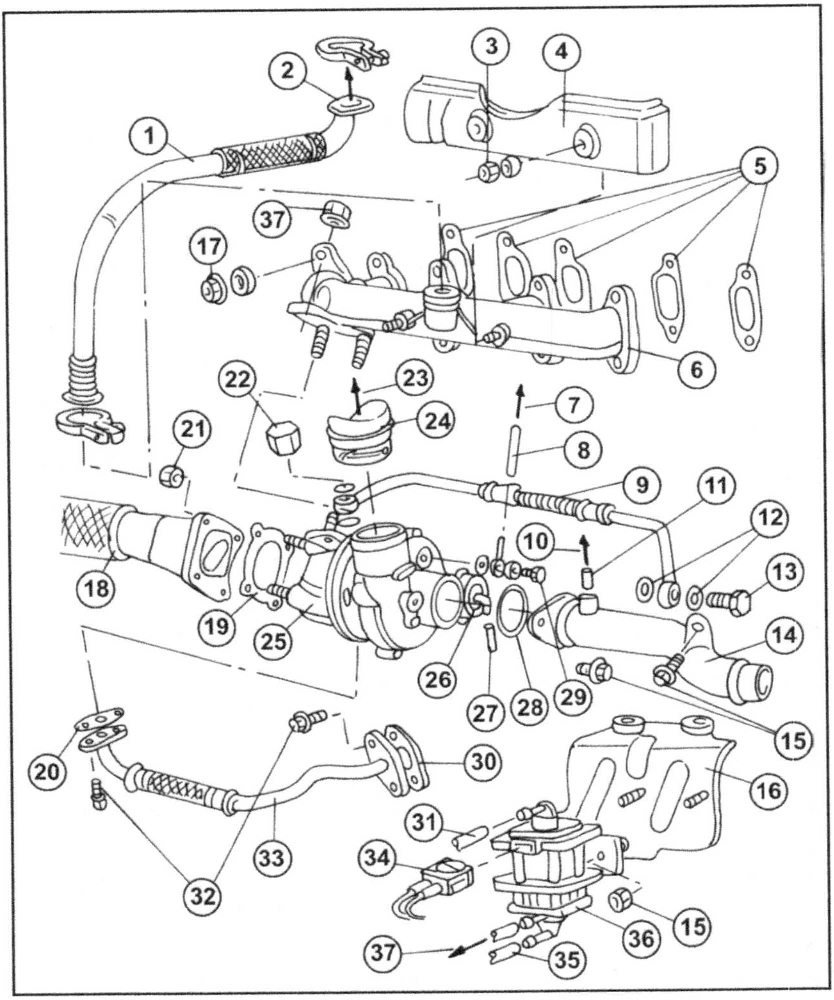

Fig. 1.127 – Turbocharger system – AHD, APA, BBE and BBF engines.

The removal can be carried out by referring to Fig. 1.127. The illustration shows the installation in the case of engine codes AHD, APA, BBE and BBF. The following information will help you to understand the various items.

- The connecting pipe (1) is connected with the flange (2) to the EGR valve.
- The nut (3) is tightened to 2.5 kgm (18 ft.lb.).
- The heat shield (4) protects the inlet manifold from excessive heat.
- The gaskets (5) for the exhaust manifold (6) must always be replaced if the manifold has been removed. Observe the correct installation.
- The vacuum hose (8) is of blue colour and leads with the end (7) to the vacuum hose (31), also of blue colour.
- The oil feed pipe (9) must be filled with engine oil before the turbocharger is fitted.
- The vacuum hose (11) is of yellow colour and is connected with the end (10) with the vacuum hose (35), also of yellow colour.
- The two sealing rings (12) must always be replaced and are fitted to the ring connection of the oil feed pipe (9). The banjo bolt (13) is tightened to 3.0 kgm (22 ft.lb.).
- The connecting pipe (14) is fitted between the turbocharger and the connecting pipe for the air mass meter and is attached to the turbocharger and the oil filter bracket. The two bolts and the nut (15) are tightened to 1.0 kgm (7.2 ft.lb.).

- The retaining bracket (16) carries the electrically operated exhaust gas return valve (EGR valve) – electrical identification N18, and the solenoid valve for the charge air limit (N75). The nut (17) is tightened to 2.6 kgm (18 ft.lb.).
- The front exhaust pipe (18) is connected to the catalytic converter. Gaskets (19) and (20) must always be replaced. The nuts (21) and the dome nut (22) are tightened to 2.5 kgm (18 ft.lb.).
- The air charge hose (24) is connected with the end (23) to the pipe of the charge air cooler.
- The turbocharger (25) is removed as described below.
- The charge air control valve (26) is a part of the turbocharger and cannot be separated.
- A control pipe (27) is of red colour.
- The sealing ring (28) must only be replaced if it is damaged.
- The banjo bolt (29) is tightened to 1.6 kgm (11.5 ft.lb.).
- The gasket (30) for the oil return pipe (33) must always be replaced. The pipe is attached to the cylinder block. The bolts (32) are tightened to 3.0 kgm (22 ft.lb.).
- The hose (31) is of blue colour.
- The connector plug (34) is fitted to the solenoid valve for the charge air control. The valve (36) has the electrical identification N75.

The removal and installation is carried out as follows. Refer to Fig. 1.127 and the above details during the installation of the turbocharger.

- Switch off the ignition and disconnect the battery.
- Refer to Fig. 1.128 in the case of an engine with code letters **APA, BBE** and **BBF**. Disconnect the charge air hose (1) from the turbocharger, withdraw the blue vacuum hose (2) from the turbocharger, the yellow vacuum hose (3) from the connecting pipe and the red control pipe (4) from the charge air control valve.

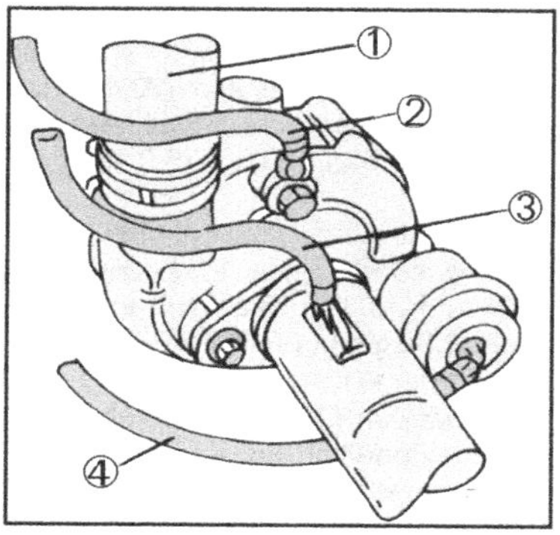

Fig. 1.128 – Details for the removal and installation of the turbocharger (engines APA, BBE and BBF). The numbers are referred to in the text.

- Disconnect a connection pipe from the turbocharger and the two pipes shown in Fig.1.129. These are the oil feed pipe (1) and the oil return pipe (2).
- In the case of an **ANJ** engine, remove the charge air hose and the connecting pipe from the turbocharger and unscrew the two oil pipes from the turbocharger. Also disconnect a vacuum hose from the bottom of the turbocharger.

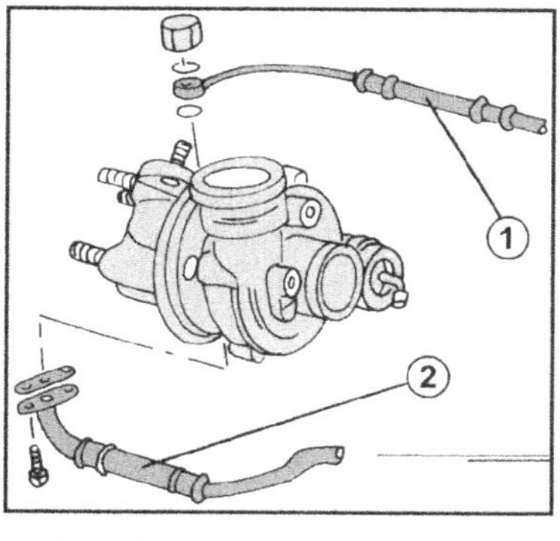

Fig. 1.129 – The location of the oil feed pipe (1) and the oil return pipe (2) on the turbocharger of a five-cylinder engine. On the four-cylinder engine, the pipes are fitted to the L.H. and R.H. sides.

- Remove the access cover inside the cab to gain access to the upper bolt for the exhaust manifold. In the case of **AHD, APA, BEE and BBF** engines, remove the exhaust manifold and the nuts between the exhaust manifold and the turbocharger. In the case of an **ANJ** engine, proceed similarly, but additionally remove a strut between the turbocharger and the starter motor.
- In the case of all engines, remove the starter motor and take out the turbocharger.

The installation of the turbocharger is carried out in reverse order to the removal procedure, noting the following points.

- Coat the contact faces of nuts and bolts between exhaust manifold and turbocharger with heat-resistant grease. The workshop uses grease with Part No. G 000 500. Try to obtain some. Bolts and nuts (always new) are tightened to 6.0 kgm (43.5 ft.lb.) in the case of the **AHD, APA, BEE and BBF** engines or 2.5 kgm (18 ft.lb.) in the case of the **ANJ** engine.
- Fit the oil return pipe with a new gasket and tighten the bolts to 3.0 kgm (22 ft.lb.).
- Fill the turbocharger through the connection opening for the oil feed pipe with engine oil, fit the pipe with a new gasket and tighten the bolts to 2.5 kgm (18 ft.lb.).
- Re-connect all vacuum hoses and the control pipe in accordance with the colours and Fig. 1.128 and tighten the hose clamps. Fig. 1.129 should also be referred to.
- All other information can be taken from the description of the exploded view on page 70.
- After the installation, start the engine and allow to idle for approx. 1 minute. Never accelerate the engine immediately to allow the oil to circulate through the turbocharger.

Removal and Installation – Four-Cylinder Engine

The parts associated with the turbocharger system are shown in Fig. 1.130. Before the removal of the turbocharger, note the following notes for the item numbers in the illustration.

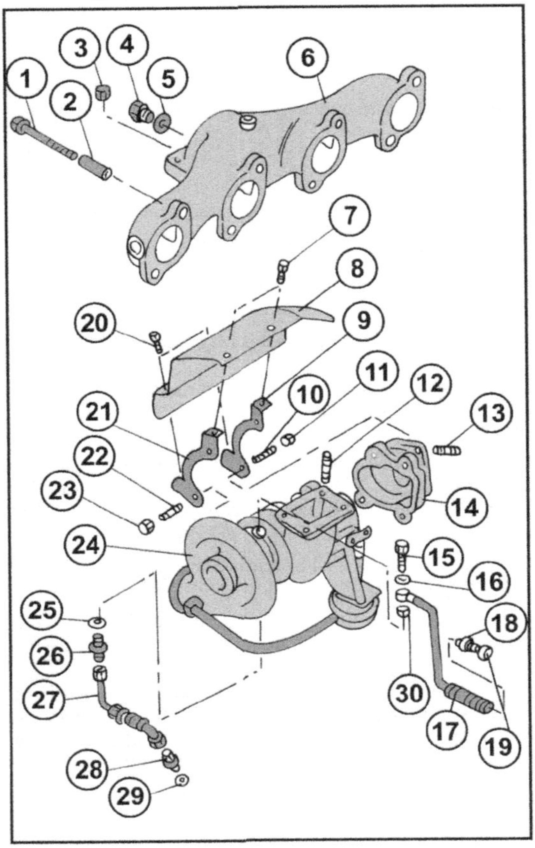

- The screw (1), together with the spacer bush (2), tightened to 6.0 kgm (43.5 ft.lb.).
- The nut (3) must always be replaced if removed. Tighten to 6.0 kgm (43.5 ft.lb.). Coat thread and contact face with heat-resistant grease.
- The nut (4) is fitted with a sealing washer (5). Tighten to 3.0 kgm (22 ft.lb.). Secure the exhaust manifold (6).
- Screws (7) and (2) are tightened to 0.5 kgm. Secure the heat shield (8),
- The bracket (9) is secured with a stud (10) and the nut (11). Tighten to 1.5 kgm (11 ft.lb.).
- The two studs (12) and (13) secure the connecting elbow (14).

Fig. 1.130 – Turbocharger installation.

- The banjo bolt (15) is fitted with a sealing washer (16) on one side and (30) on the other side. The bolt secures the oil feed pipe (17). Tighten the banjo bolt to 3.0 kgm (22 ft.lb.).
- Item (18) is fitted with a new seal (19).
- Bracket (21) corresponds to bracket (9). Studs (22) and nut (23) are used to secure the bracket [1.5 kgm (11 ft.lb.)].
- The turbocharger is identified by (24).
- The sealing washer (25) is fitted underneath the threaded connector (26) and are used to connect the oil return pipe (27). At the other end of the pipe is a further connector (28) with a sealing washer (29). Sealing washers must always be replaced if the pipe is removed.

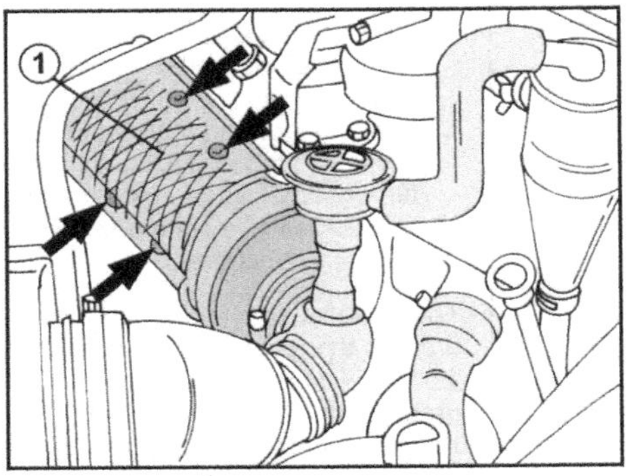

Fig. 1.131 – The arrows show where the heat shield is attached (4-cyl.).

The turbocharger can be removed as follows.

- Switch off the ignition and disconnect the battery earth cable.
- Remove the heat shield (1) shown in Fig. 1.131. Remove the screws shown by the arrows.

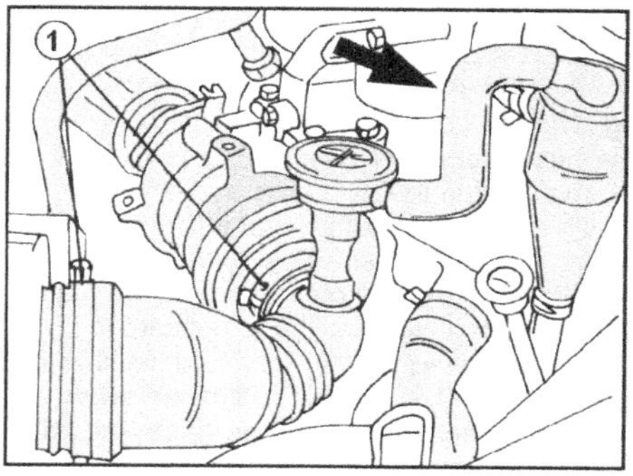

Fig. 1.132 – Disconnecting the hose from the turbocharger during removal (4-cyl.).

- Remove the hose clamps (1) in Fig. 1.132, disconnect the hose (arrow) from the oil separator and remove all other hoses.
- Disconnect the oil feed pipe and the oil return pipe from the turbocharger. These are connected similarly as shown in Fig. 1.129, but on the L.H. and R.H. sides of the turbocharger.
- Remove the front exhaust pipe and then the turbocharger from the exhaust manifold (nuts). The turbocharger can now be lifted out.

The installation of the turbocharger is carried out in reverse order. Note the following points.

- Coat the contact faces of nuts and bolts between exhaust manifold and turbocharger with heat-resistant grease. The workshop uses grease with Part No. G 000 500. Try to obtain some. Bolts and nuts (always new) are tightened to 6.0 kgm (43.5 ft.lb.).
- Fit the oil return pipe with a new gasket and tighten the bolts to 1.0 kgm (7 ft.lb.).
- Fill the turbocharger through the connection opening for the oil feed pipe with engine oil, fit the pipe with a new gasket and tighten the bolts to 2.5 kgm (18 ft.lb.).
- Fit the exhaust pipe with a new gasket and tighten the bolts (2.5 kgm/18 ft.lb.) evenly across.
- All other information can be taken from the description of the exploded view on Page 71. Note all tightening torques. The description of the various items also specifies when gaskets, sealing washers, nut and bolts must be replaced when necessary.
- After the installation, start the engine and allow to idle for approx. 1 minute. Never accelerate the engine immediately to allow the oil to circulate through the turbocharger.

1.8.8. Charge Air Cooler (Intercooler)

Figs. 1.133 (5-cyl.) and 1.134 (4-cyl.) show the component parts belong to the charge air cooler installation. All work on the cooler can be carried out by referring to the illustrations and the following information.

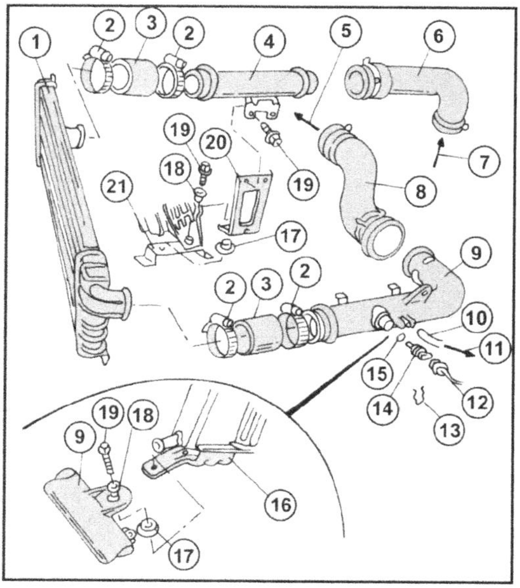

Fig. 1.133 – Charge air cooler installation for a five-cylinder engine. The various items are referred to below.

To Fig. 1.133:

- The air charge cooler (1) is attached to the water radiator, as already shown in the section "Cooling System".
- The clamps (2) secure the pressure hose (3) on both sides. The hose is connected to the charge air pipe (4).
- The end (5) of the charge air pressure tube (8) is connected to the EGR valve.
- The charge air hose (9) is secured with parts (2) and (3) to the cooler.
- The black pressure pipe (10) is fitted with the end (11) with the electronic control unit for the diesel injection. A cable connector (12) is fitted to the temperature sender unit for the inlet manifold. The plug can either have 2 or 4 pins, depending on the number of pins on the control unit. If replacement is necessary, make sure to obtain the correct part. The sender unit (14) is pushed in position and secured with a retaining clip (13). Make sure that the clip engages properly. The sealing washer (15) must always be replaced.
- The bracket (16) is used to secure the reservoir for the steering fluid and is attached by means of the rubber grommet (17), a spacer (18) and the bolt (19). Tighten the bolt to 1.0 kgm (7.2 ft.lb.).
- The bracket (20) secures the charge air tube to the longitudinal member (21) as shown.

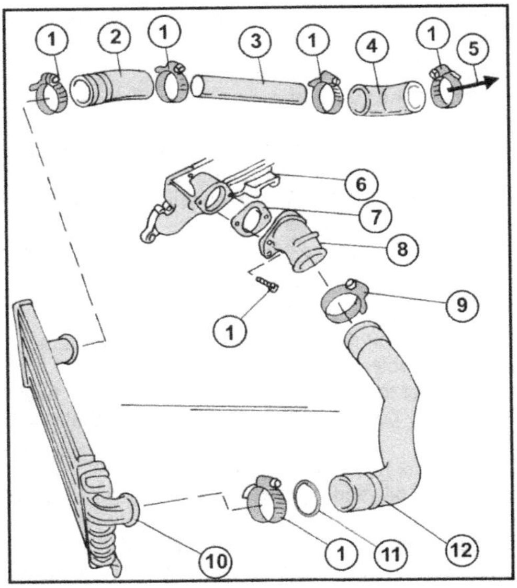

Fig. 1.133 – Charge air cooler installation for a four-cylinder engine. The various items are referred to below.

To Fig. 1.134:

- The hose clamps (1) secure the pressure hose (2) on one side to the charge air cooler (10) and on the other side to the charge air tube (3). A further pressure hose (4) is fitted on the other side of the charge air pressure tube, the end (5) of which is connected to the turbocharger.
- A connecting elbow (8) is fitted to the inlet manifold (6). The gasket (7) must always be replaced if the connection has been separated. The connecting elbow is tightened with the bolts (9) to 3.0 kgm (22 ft.lb.).
- The hose (12) is fitted to the charge air cooler (10) by means of the hose clamp (1). A sealing ring (11) is used to connect the parts.

1.8.9. Exhaust Gas Re-Circulation System (EGR System)

The system is fitted to five-cylinder engines and should not require any attention. Not all engines have the same system. The system is controlled by the fault diagnosis system and only your dealer can check the system.

2. CLUTCH

2.0. Technical Data

Type: .... Single dry plate diaphragm clutch
Operation .... Hydraulic system
Clutch diameter .... 240 mm

2.1. Removal and Installation

To remove the clutch unit, it will be necessary to separate the transmission from the engine, with the assembly fitted to the vehicle. The flywheel must be prevented from rotation when the clutch is unscrewed, but due to the different arrangement of the flywheel, there is a difference between the four- and five-cylinder engines.

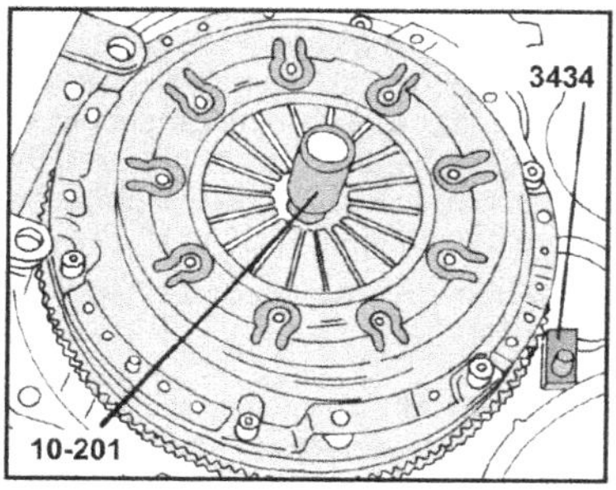

Fig. 2.1 – Locking the flywheel with a "flywheel brake" (five-cylinder engine), which is inserted into the teeth of the flywheel ring gear. The centring drift in inserted into the driven plate.

- Remove the transmission (Section 3.1).
- Mark the clutch in its fitted position on the flywheel if there is a possibility that the clutch unit is re-used. To remove the clutch, unscrew the six bolts securing the pressure plate to the flywheel and lift off the flywheel and then the driven plate, now tree. To prevent the flywheel from rotating, a "flywheel brake", as shown in Fig. 2.1 can be used or you can insert a tyre lever into the teeth of the flywheel ring gear. This is not possible on a four-cylinder engine. In this case, the vibration damper must be blocked. The workshop uses a lever, as shown in Fig. 2.2. This is a type of ring spanner, which is placed over the centre bolt inside the vibration damper (remove the dust cover first). A long ring spanner or a socket of suitable size can therefore be used. Before removing the driven plate, note the position of the longer part of the driven plate hub, as the driven plate must be refitted in the same way. Note that the clutch pressure plate is located on dowel pins, which can prevent immediate removal. If this is the case, insert the tip of a strong screwdriver underneath the clutch plate and carefully lever it off. The component parts of the clutch are shown in Fig. 2.3.

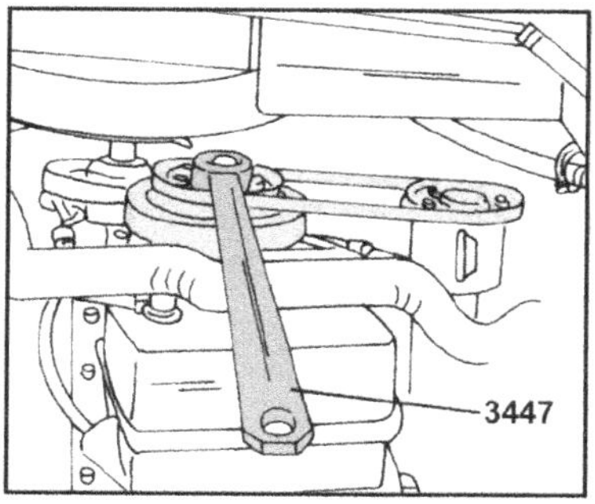

Fig. 2.2 – The crankshaft of a four-cylinder engine can be prevented from rotating in the manner shown. The spanner is applied to the bolt inside the vibration damper.

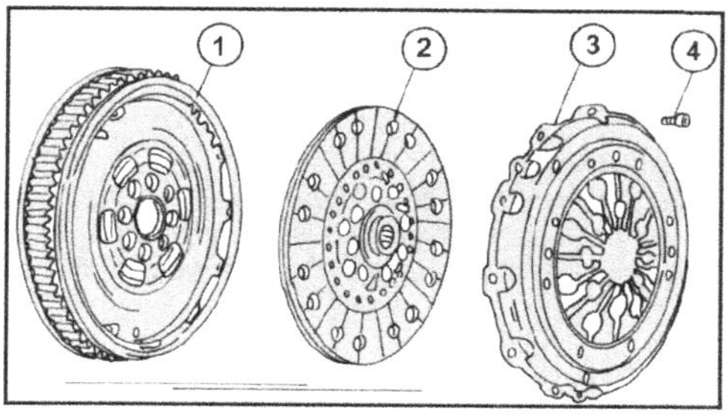

Fig. 2.3 – The component parts of the clutch.
1 Flywheel
2 Driven plate
3 Pressure plate
4 Bolt, 2.5 kgm

- Clean the inside of the flywheel. If the driven plate linings have worn down to the rivet heads, there is a possibility that the flywheel face or the clutch plate face is scored. In this case, replace the flywheel (or the clutch pressure plate).

Install in the reverse sequence to removal, noting the following points.

- If the old clutch unit is fitted, align the marks made before removal. A new clutch can be fitted in any position.
- A centring mandrel is required to centre the clutch-driven plate inside the flywheel. Tool hire companies normally have sets of mandrels for this purpose. An old transmission (clutch) shaft, which you may be able to obtain from a VW workshop, can also be used. Experienced D.I.Y. mechanics will also be able to align the clutch plate without the help of a mandrel. Fig. 2.1 shows the mandrel in position. The same mandrel is used on all models.
- Fit and tighten the six clutch to flywheel bolts to a torque reading of 2.5 kgm (18 ft.lb.). The flywheel must be locked against rotation when the clutch bolts are tightened. Refer to Figs. 2.1 or 2.2 for details.
- Finally, refit the transmission to the engine.

2.2. Servicing

The cover assembly – pressure plate and diaphragm spring – must not be dismantled. Replace, if necessary with a complete assembly from your dealer or distributor.

Inspect the driven plate and the linings, replacing the complete plate if the linings are worn down close to the rivets. A driven plate with the linings contaminated with grease or oil cannot be cleaned successfully and should also be replaced. All rivets should be tight and the torsion springs should be sound and unbroken. Check the condition of the driven plate splines. Clamp the driven plate between the centres of a lathe and apply a dial gauge to the outside of the plate as shown in Fig. 2.4, at a diameter of approx. 220 mm. The max. run/out of the driven plate should be no more than 0.8 mm (0.03 in.).

Fig. 2.4 – To check the driven plate for run-out, clamp it between the centres of a lathe and check with a dial gauge.

Check the rivet fastening of the clutch pressure plate and replace the plate, if loose rivets can be detected.

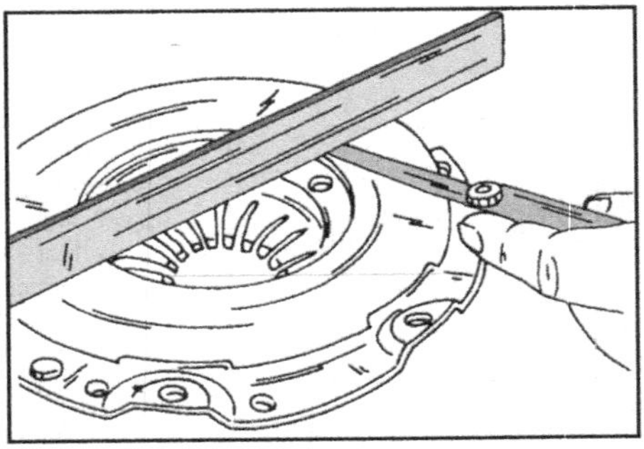

Fig. 2.5 – Checking the clutch pressure plate for distortion. The gap should not be more than given below.

Place a straight edge (steel ruler) over the friction face of the pressure plate and insert feeler gauges between the ruler and the surface. If the gap at the innermost spot of the friction face is no more than 0.08 mm (0.003 in.), the plate can be re-used. Fig. 2.5 shows this check.

2.2. Clutch Release Mechanism

Engagement and disengagement of the clutch is by means of the slave cylinder push rod, acting on the clutch release fork and sliding the ball bearing-type release bearing along a guide tube on the clutch shaft of the transmission. The release system is free of play, as the wear of the clutch linings is compensated automatically.

2.2.0. REMOVAL AND INSTALLATION

The transmission must be removed to replace the release bearing. Remove the bearing from the bearing sleeve on the front transmission housing cover, as shown in Fig. 2.6. To remove the release fork, refer to Fig. 2.7 and move it in the direction of arrow (a) and then pull it from the ball pin in the clutch housing in direction of arrow (b).

Thoroughly grease the guide sleeve on the front transmission cover, the ball pin and all of the parts of the release mechanism in contact with the release bearing with long-term grease. Push the release lever in the reverse direction of arrow (b) over the ball pin until the spring clip of the release lever engages with the ball pin. Check for secure fitting. Then, move the lever in the reverse direction of arrow (a) until the slave cylinder push rod is engaged with the ball-shaped cut-out in the release lever.

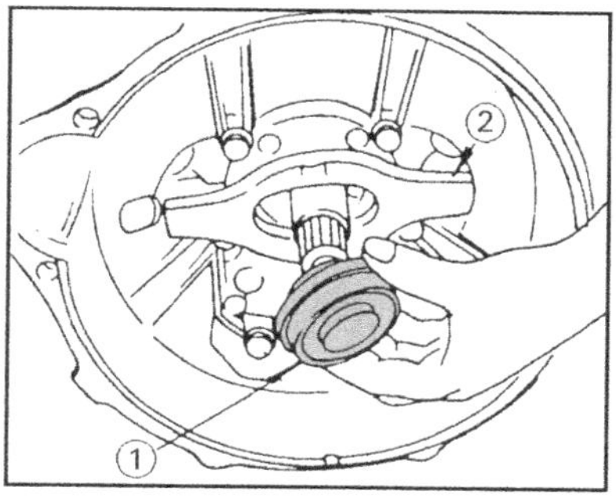

Fig. 2.6 – Removal and installation of the clutch release bearing.
1 Release bearing
2 Release fork

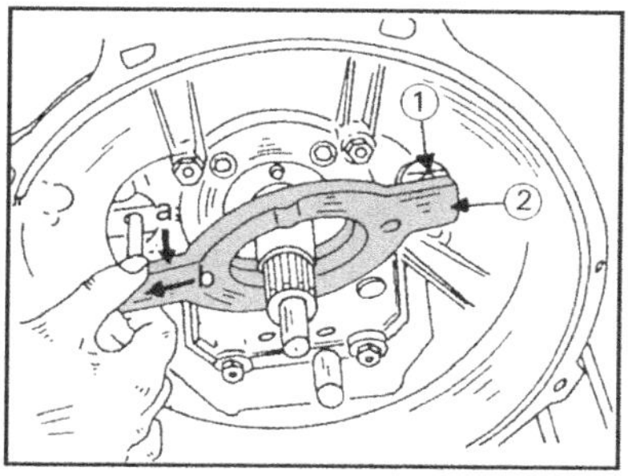

Fig. 2.7. – Removal and installation of the clutch release lever (see the text).
1 Ball pin 2 Release lever

Grease the release bearing on the inside and on both sides at the rear, where it rests against the release lever, and slip the bearing over the guide sleeve. Rotate the bearing until it snaps in position into the release lever. Check that the bearing is properly fitted and refit the transmission.

2.3. Clutch Master Cylinder

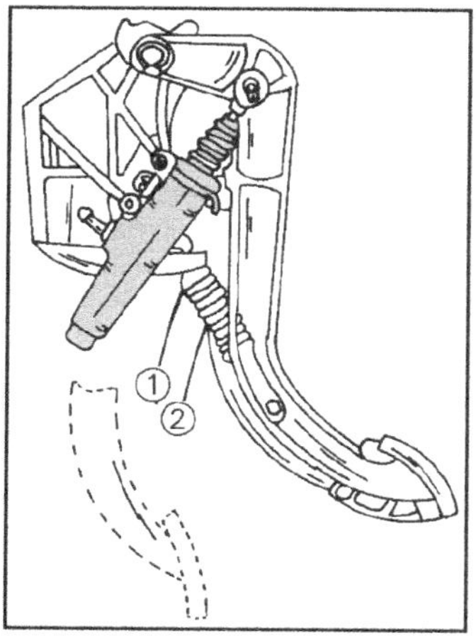

2.3.0. REMOVAL AND INSTALLATION

Fig. 2.8 shows where the clutch master cylinder is fitted near the clutch pedal. Remove the cylinder as follows.

- Remove the cover under the instrument panel, above the pedals.
- Remove the floor mat.
- Use a syringe and draw enough fluid out of the fluid reservoir to bring the fluid level below the "Min." mark. Otherwise, use a bleeder hose, push it over the slave cylinder bleeder screw, open the screw and operate the clutch pedal until the system is empty (collect the fluid in a glass jar).
- Disconnect the fluid hose from the combination fluid reservoir (for brake and clutch) and unscrew the metal pipe from the clutch master cylinder. Disconnect the electrical plug from the brake light switch.

Fig. 2.8 – Clutch master cylinder in fitted position.
1 Off-centre point spring
2 Spring retainer
A Pedal travel

- Unscrew the master cylinder from the pedal carrier. Pull back the clutch pedal and remove the spring retainer, the spring and the washer from the push rod. Using a pair of circlip pliers, remove the circlip securing the push rod (at the top of the clutch pedal), as shown in Fig. 2.9.
- Remove the connecting hose with the push connector and lift out the cylinder.

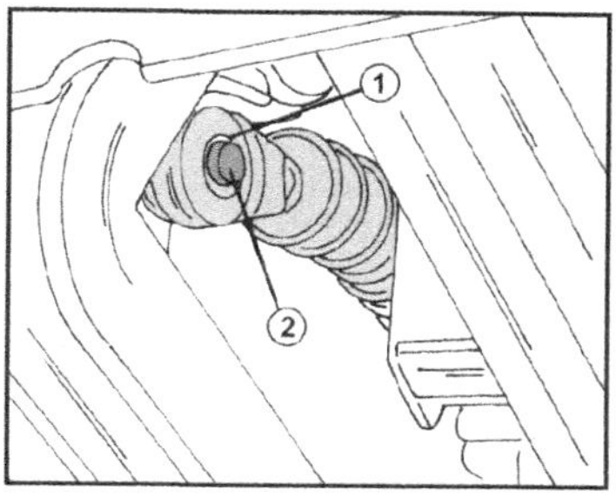

Fig. 2.9 – Removal of the clutch master cylinder. Push out the fulcrum pin (2) after removal of the circlip (1).

The installation of the master cylinder is a reversal of the removal procedure. Make sure that the push in connector for the connecting hose is properly in position. Fit the push rod with the flange facing towards the clutch pedal and make sure that the circlip engages into its groove (refer to Fig. 2.9). Finally, bleed the clutch system (Section 2.5).

Start the engine. Depress the clutch pedal and engage the reverse gear. Grating noises will immediately indicate a faulty release of the clutch operation.

Fully depress the clutch pedal and check that it reaches its stop. Otherwise, the clutch cannot fully disengage.

2.4. Clutch Slave Cylinder – Removal and Installation

Cylinders for four- and five-cylinder models are not the same. Cylinders are either made of cast iron (black colour) or aluminium (silver colour). This must be noted when a new cylinder must be fitted. Proceed as follows during replacement.

- Unscrew the fluid pipe from the slave cylinder, using an open-ended spanner. Close the end of the pipe suitably to prevent fluid leakage (rubber cap for bleeder screw). Remove the two cylinder securing screws and take off the cylinder. Note that a cast iron cylinder has a sealing washer between cylinder and transmission.
- Fit the slave cylinder, engaging the push rod into the ball-shaped cut-out of the clutch release lever, and insert the two screws. Tighten the screws. Finally, bleed the clutch system as described in the next section.

2.5. Bleeding the Clutch System

A pressure bleeder is used by VW workshops. The following description involves the brake system and is therefore to be treated with caution. A transparent hose of approx. 1 metre (3 ft.) in length is required. Proceed as follows.

- Fill the brake/clutch fluid reservoir.
- Remove the dust cap of the bleeder screw on the clutch slave cylinder, push the bleeder hose over the bleeder screw and open the bleeder screw. Fig. 2.10 shows how the pressure bleeder is connected to the slave cylinder, but the hose "A" can be inserted into a glass jar as described.

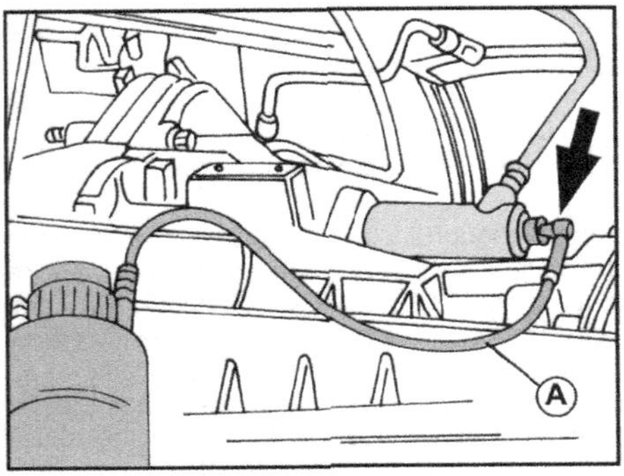

Fig. 2.10 – Bleeding the clutch system. The bleeder hose "A" is connected to the bleeder screw (arrow).

- Ask a second person to operate the clutch pedal until the hose is completely filled with brake fluid and no more air bubbles can be seen. Make sure that the reservoir retains enough brake fluid to prevent air from being drawn in, i.e., the reservoir must be continuously refilled during the bleeding process.
- Close the bleed screw on the slave cylinder and remove the hose. Refit the dust cap (easily forgotten).
- Check the fluid level in the reservoir and, if necessary, top it up to the "Max." mark. Start the engine, depress the clutch pedal and engage reverse. No grating noises should be heard.

2.6. Clutch Pedal Switch (5 cyl.)

If the switch fitted to the clutch pedal on the above models is suspect, we recommend to have it checked in a workshop. We also recommend to have the switch replaced in a workshop, should this be necessary.

3. Manual Transmission

The vehicles covered in this manual are fitted with a five-speed transmission, but two different types, depending on the engine, are used. Transmission type 000.1 can be fitted to models with four- or five-cylinder engine, and transmission type 000.2 is used together with the five-cylinder AGX (55 kW). There are, however, differences between the various transmissions. Each transmission is identified by three letters that are stamped into a type identification plate, attached to the transmission. If, for example, an exchange transmission is fitted, check the code letters in accordance with Figs. 3.1 or 3.2. The type identification plate also contains the day and month of production. The transmission type is also quoted in the vehicle identification plate.

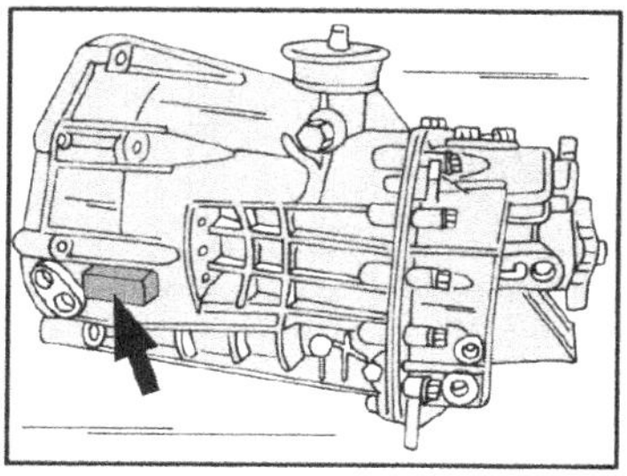

Fig. 3.1 – Location of the type identification plate on a type 0.001 transmission (at lower end).

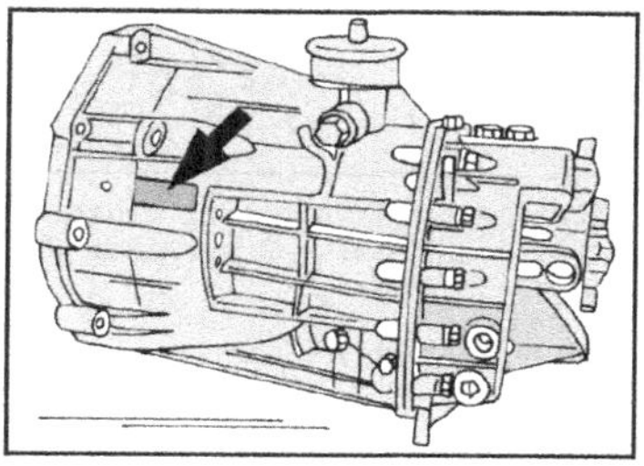

Fig. 3.2 – Location of the type identification plate on a type 000.2 transmission (at upper end).

The overhaul of the transmission is not described in this manual. The description in the overhaul section is limited to some minor repair operations, not involving the gear train or the gear shafts. If the transmission appears to be damaged or faulty, try to obtain an exchange unit. Transmission overhaul is now limited to specialised workshops that are equipped with the necessary special tools.

3.0. Technical Data

Fitted Transmission:
- Five-speed transmission:........ Type 0.001 or 0.002

Oil capacity:
- Type 0.001........ 2.2 litres
- Type 0.002........ 1.8 litres

Type Identification:
- Type 0.001 – 2.8-litre AGK, ATA:........ DDX, DLA, DLB, DLC, DLD
- Type 0.001 – 2.5-litre APA, AHD, ANJ........ DKW, AKX, DKZ
- Type 0.002 – 2.5-litre AGX........ DEL

Lubrication Oil........ Gearbox oil G 052 011 A 1 or SAE 75W W90 (synthetic oil)

Gear Ratios:	**Type 0.001**	**Type 0.002**
- 1st gear:	3.643:1	3.615:1
- 2nd gear:	1.875:1	1.870:1
- 3rd gear:	1.097:1	1.097:1
- 4th gear:	0.721:1	0.738:1
- 5th gear:	0.565:1	0.578:1
- Reverse gear:	3.429:1	3.385:1

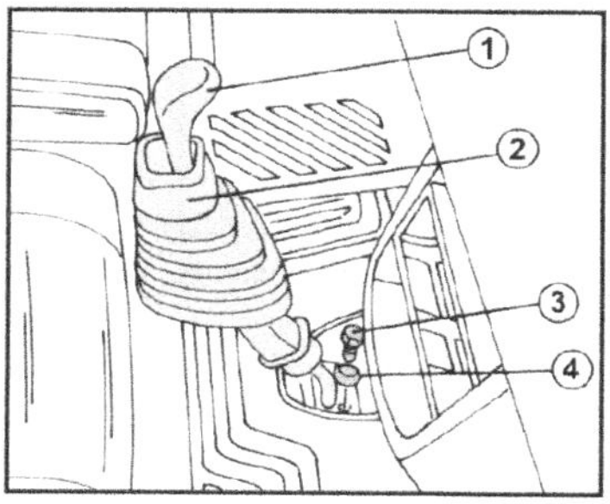

3.1. Removal and Installation

The following text describes the general removal and installation operations. The transmission is heavy and the necessary precautions must be taken when it is lifted out.

Fig. 3.3 – Attachment of the gearchange lever inside the cab.
1 Gearchange lever
2 Gearchange lever gaiter
3 Bolt, 6.0 kgm
4 Plain washer

- Disconnect the battery earth cable from the battery.

- In the inside of the cab, pull up the rubber gaiter around the gearchange lever and slacken the clamp screw shown in Fig. 3.3, detach the gaiter and remove the lever together with the gaiter. Remove the upper transmission to engine bolts through the opening for the gearchange lever.

Fig. 3.4 – Remove the safety bracket (1) for the propeller shaft after unscrewing the bolt on one side. The propeller shaft flange is unbolted from the transmission during removal (R.H. view). Secure the shaft in the horizontal position as described on the next page.

- Place the vehicle on secure chassis stands.
- Drain the transmission oil as described later on.
- Remove the "U"-shaped bracket, shown in Fig. 3.4, underneath the propeller shaft. Remove the bolt on one side and slacken the bolt on the other side. Swivel the bracket around, away from the propeller shaft.
- At the rear of the transmission disconnect the propeller shaft (Fig. 3.4, R.H. view). Mark the flanges of the shaft and the transmission before removal to refit the shaft in the original position. Hold the bolt heads with a ring spanner, unscrew the nuts on the other side with a socket. Rotate the shaft to reach all bolts. To retain the shaft in horizontal position refer to Fig. 3.5. Swivel the safety bracket (2) to one side and place the shaft (1) into the bracket.

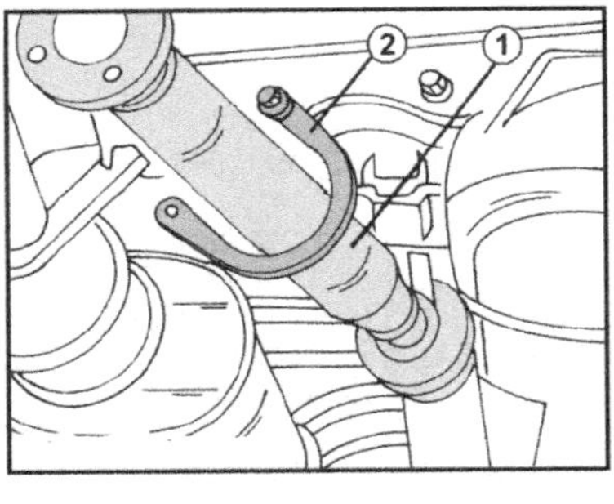

Fig. 3.5 – Keep the propeller shaft (1) in the horizontal position by placing it into the safety bracket (2) as shown.

- Make up a piece of wood, 90 mm long, 30 mm high and 20 mm wide, and insert it between the engine oil sump and the crossmember. This will retain the engine in position after the transmission has been removed.
- Remove the rear engine crossmember. The transmission must be supported suitably. Separate the mounting in the centre of the crossmember shown in Fig. 3.6 and then remove the two bolts on the outside of the crossmember on the L.H. and R.H. sides, as shown in the R.H. view of Fig. 3.6.

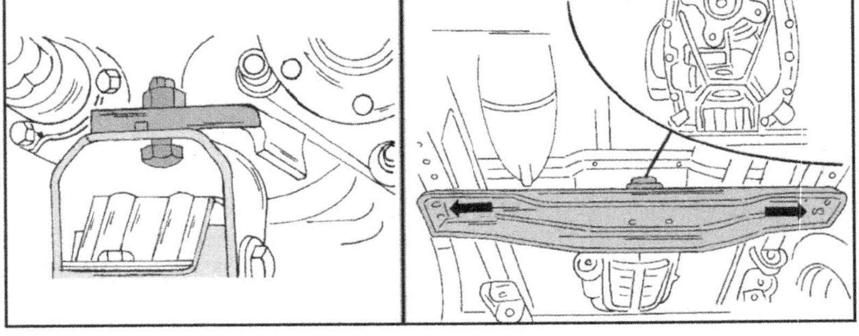

Fig. 3.6 – The attachment of the transmission to the centre of the crossmember. The crossmember member is secured on the outside to the longitudinal members of the chassis as shown in the R.H. view.

- Unscrew the clutch slave cylinder from the transmission housing without disconnecting the fluid line. If a cast iron cylinder (black colour) is fitted, remove the sealing washer between cylinder mounting face and transmission. There is no washer when the cylinder is made of aluminium (silver colour). Use a piece of wire to attach the cylinder to the chassis. Do not allow the cylinder to hang down under its own weight.

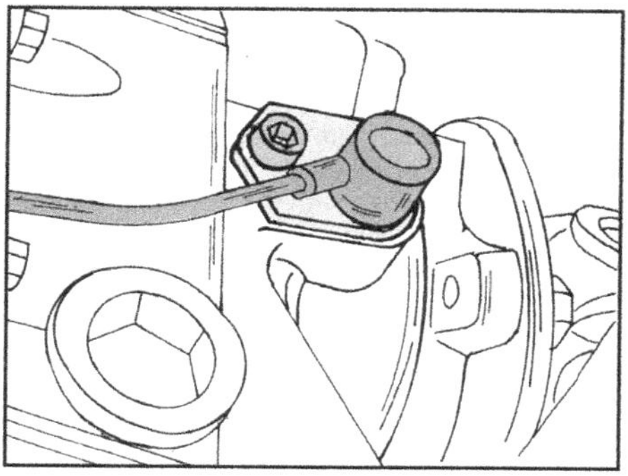

Fig. 3.7 – The connector for the speedometer sensor on the transmission.

- Follow the various electrical cables to the transmission and disconnect the plugs, amongst them the plug for the reversing light switch and the sensor for the speedometer. The latter has the shape shown in Fig. 3.7. Remove the screw shown on the L.H. side to remove the complete assembly.
- Make sure that the transmission is adequately supported. If necessary, use a wooden block between jack head and transmission. Jack up the transmission until it is under slight tension. Remove the bolts between engine and transmission around the connecting flange. Also remove the earth cable.
- Remove the starter motor from the transmission.
- Slowly withdraw the transmission from the engine and lower it on the jack. Take care not to rest the weight on the clutch shaft, as this may bend the shaft or damage the clutch-driven plate. You will need a helper to disengage and lower the transmission, as it must remain on the jack until it is free and can be withdrawn from underneath the vehicle – **Attention: Danger of accidents.**

When the transmission is removed for any reason, always check the condition of the clutch mechanism. Remove the clutch and check as described in the previous chapter. Also check the clutch release bearing.

The installation is a reversal of the removal procedure, but the following points must be noted.

- Lightly grease the centring lug and the splines of the clutch shaft with long-term grease. Place the transmission onto the jack and lift it into the approximate fitting position.
- Engage a gear (not reverse), align the transmission with the engine and push the transmission against the engine. Rotate the drive flange at the end of the transmission to and fro until the clutch shaft has engaged with the driven plate. Fully push the transmission against the engine. Note that the two dowel sleeves must engage in the locating holes in the transmission.
- Insert the bolts into the transmission, tighten them evenly until the gap between engine and transmission is nearly closed and then tighten the bolts to 5.5 kgm (40 ft.lb.) or to 8.0 kgm (58 ft.lb.) when an engine with code letters AGX, AHD, APA or ANJ is fitted. The same torque is also valid for the earth cable of all transmissions.
- Refit the starter motor.
- Refit the rear crossmember from below and fit the transmission mounting in the centre and the bolts on the outside. Tighten the bolts to 4.5 kgm (32.5 ft.lb.). The bolt and nut of the centre mounting in Fig. 3.6 are tightened to 9.0 kgm (65 ft.lb.). Remove the piece of wood inserted underneath the oil sump.
- Refit the speedometer sensor (Fig. 3.7) and tighten the screw to 0.8 kgm (6 ft.lb.). Also connect the plug to the reversing light switch.
- Fit the clutch slave cylinder with the gasket (if fitted) against the transmission, fit the two bolts and tighten them evenly to 2.5 kgm (18 ft.lb.).
- Connect the propeller shaft flange to the drive flange of the transmission, insert the bolts, fit the nuts and tighten the nuts to 7.0 kgm (50.5 ft.lb.). The bolt heads must be held against rotation, using a ring spanner.
- Fit the propeller shaft safety bracket. Tighten the bolts to 9.0 kgm (65 ft.lb.).
- Refit the starter motor (4.5 kgm).
- Coat the cone seat of the gearchange lever with heat-resistant grease and connect it with the lever in the transmission. Fit the bolt and tighten to 6.3 kgm (45 ft.lb.). Refit the gearchange lever gaiter.
- All other operations are carried out in reverse order. If the oil has been drained, fill the transmission with the correct quantity of the recommended oil (see Section 3.0.).
- If a radio with anti-theft code is fitted, enter the code. The clock must also be reset.

3.2. Transmission Repairs

Many repairs can be carried out without dismantling the transmission completely. The following sections described some of the jobs which do not need special tools. Complete dismantling of the transmission requires special tools. A damaged transmission should be replaced by an exchange unit.

3.2.0. CLUTCH SHAFT OIL SEAL

The same instructions apply when the guide sleeve for the clutch release bearing must be removed (replaced).

The transmission must be removed from the vehicle to replace the oil seal inside the guide sleeve. After separating the clutch release lever from the ball pin, unscrew the guide sleeve from the inside of the transmission. The guide sleeve is at the same time the front cover. The sleeve is secured, as shown in Fig. 3.8.

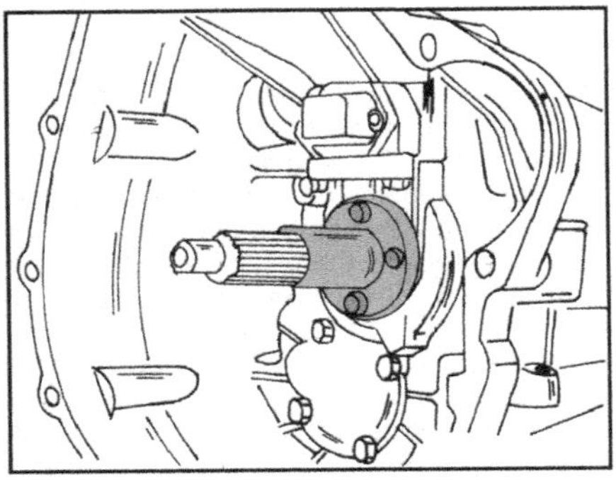

Fig. 3.8 – The oil seal for the clutch shaft (main drive shaft) is located underneath the clutch release bearing guide sleeve.

Remove the oil seal, using a screwdriver and fit a new seal. Fill the space between the sealing lips with M.P. grease. A suitable drift must be available to drive in the oil seal until the outer face is flush with the bore in the cover. Refit the guide sleeve and evenly tighten the bolts to 1.0 kgm (7.2 ft.lb.). Coat the splines of the drive shaft with long-term grease and refit the clutch release fork.

3.2.1. REAR TRANSMISSION OIL SEAL

Again, oil leaks may be the reason for the removal of the rear transmission cover. A puller is required to remove the transmission drive flange from the shaft. A suitable drift is required to fit the new oil seal. This oil seal can be replaced with the transmission fitted to the vehicle.

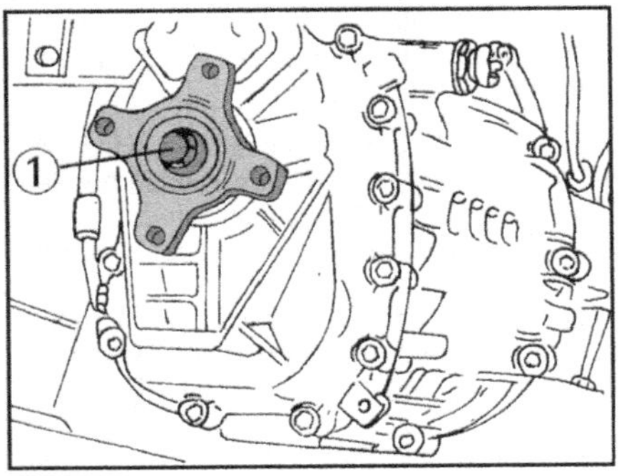

Fig. 3.9 – The flange must be held against rotation when the nut (1) in the centre is slackened or tightened.

- Place the front end of the vehicle on secure chassis stands.
- Disconnect the propeller shaft at the rear end of the transmission and unbolt the propeller shaft safety bracket as described during the removal of the transmission.
- Insert a suitable rod into one of the drive flange holes, turn the flange until the rod rests against the transmission case and the nut (1) in Fig. 3.9. Unscrew the bolt a few turns without removing it completely.
- Use a two-arm puller to withdraw the flange (3) in Fig. 3.10 from the end of the transmission. The puller claws are placed underneath two of the flange ends, with the puller centre bolt resting against the bolt (1), which is still in position. Remove the flange, fully unscrew the bolt (1) and the washer (2) and remove the flange from the shaft.

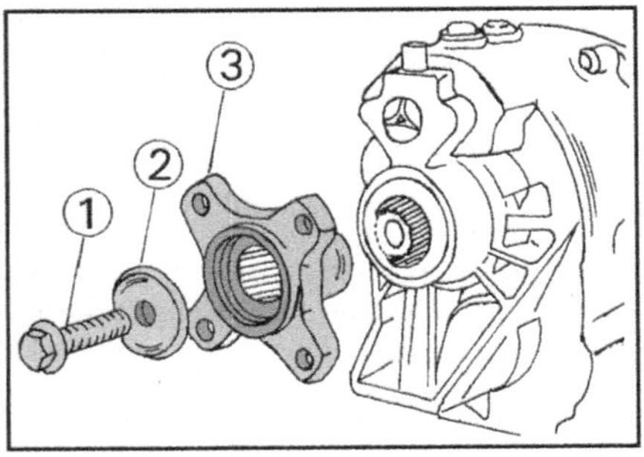

Fig. 3.10 – Attachment of the drive flange.
1 Bolt
2 Plain washer
3 Drive flange

- Remove the oil seal from the inside of the transmission. To prevent damage, screw two self-tapping screws on opposite sides into the oil seal and use a pair of pliers and a screwdriver to remove the oil seal.
- Fit a new sealing ring into the opening of the transmission. The space between the sealing lips should be filled with M.P. grease. Use a suitable drift and drive the oil seal into the transmission until the outer face is flush with the transmission case. Excess grease must be removed.
- Knock the flange over the serrations of the transmission shaft until fully home, place the washer in position and fit the bolt. Hold the flange against rotation and tighten the bolt to 11.0 kgm (80 ft.lb.).
- Re-connect the propeller shaft and tighten the bolts to 7.0 kgm (50.5 ft.lb.).
- Fit the "U"-shaped safety bracket underneath the propeller shaft and tighten to 8.5–9.0 kgm (61–65 ft.lb.).

3.2.2. GEARCHANGE LEVER

The attachment of the gearchange lever is shown in Fig. 3.3. Remove the lever as follows.

- Release the lever gaiter (2) in Fig. 3.3 and pull it upwards towards the lever knob (1).
- Remove the screw (3) securing the gearchange lever to the gearlever stump in the transmission and remove the washer (4).
- Withdraw the gearchange lever (4) upwards. Do not a hammer to help during removal, as internal parts of the gearchange mechanism could be damaged.

Grease the cone seat of the gearchange lever with heat-resistant grease before installation. Joint the gearchange lever and the lever in the transmission, fit the bolt with the washer and tighten the bolt to 6.3 kgm.

3.3. Gearbox Oil Level/Oil Change

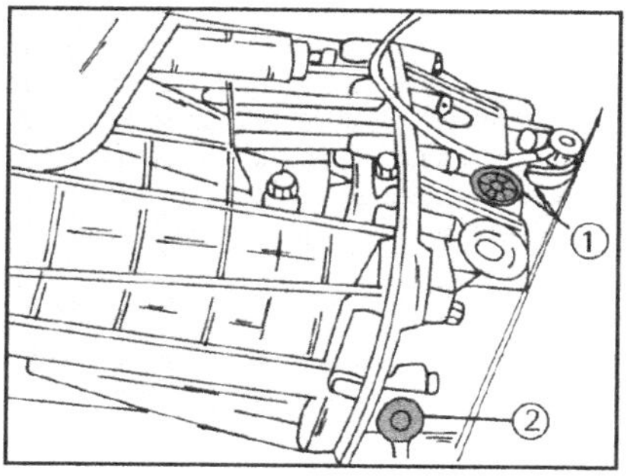

The gearbox is filled with quantity and the oil given in Section 3.0. Fluid or oil can only be lost if a leak has developed. To check the oil level, remove the oil filler plug (1) in Fig. 3.11 from the side of the transmission (Allen key required) and check that the fluid/oil is up to the lower edge of the plug bore. If the oil cannot be seen, insert the forefinger to check whether you can reach it. If necessary, top up with the specified oil. A grease gun (thoroughly cleaned) can be used to fill in the required oil. Refit and tighten the filler plug (5.0 kgm/36 ft.lb.).

Fig. 3.11 – The location of the oil filler plug (1), also used to check the oil level, and the oil drain plug (2).

To change the transmission oil, place a container underneath the transmission and remove the drain plug (2) in Fig. 3.11. Allow the oil to drain completely. If the vehicle is driven a short distance, the draining of the oil will be speeded up. The filler plug should be removed.

Fill the transmission with the quantity given for the type in question until the correct level is obtained. Fit the two plugs and tighten them to the torque given above.

3.4. Gearbox – Tightening Torques

Front gearbox cover (guide sleeve) .... 1.0 kgm (7 ft.lb.)
Drive flange bolt .... 11.0 kgm (80 ft.lb.)
Gearchange lever to gear lever stump .... 6.5 kgm (45 ft.lb.)
Oil level/filler plug .... 5.0 kgm (36 ft.lb.)
Oil drain plug .... 5.0 kgm (36 ft.lb.)
Propeller shaft flange bolts .... 7.0 kgm (50.5 ft.lb.)
"U"-shaped safety bracket .... 9.0 kgm (65 ft.lb.)
Sensor for speedometer to transmission .... 0.8 kgm (6 ft.lb.)

Transmission to engine:
- All engines, except below: .... 5.5 kgm (40 ft.lb.)
- AGX, AHD, APA, and ANJ engines: .... 8.0 kgm (58 ft.lb.)

Transmission crossmember to sidemembers .... 4.5 kgm (32.5 ft.lb.)
Transmission mounting to crossmember (centre) .... 9.0 kgm (65 ft.lb.)
Clutch slave cylinder to transmission: .... 2.5 kgm (18 ft.lb.)
Ball joint pin for clutch release lever: .... 2.5 kgm (18 ft.lb.)
Starter motor to transmission: .... 6.0 kgm (43.5 ft.lb.)

4. Front Axle

The following section deals with the actual front axle, including the wheel bearings. All parts belonging to the front suspension, i.e., front springs and stabiliser bar, are described in a separate section.

Compared to the previous models of similar construction before 1997, you will find many differences in the arrangement of the front suspension. The front suspension consists of damper struts with integral shock absorbers, a lower suspension arm and a stabiliser bar (not all models). The front wheels are sprung by means of a leaf spring, fitted in traverse direction to the vehicle. The stabiliser bar is connected by means of connecting rods on each side to a suspension carrier plate. The connecting rods have rubber bushes and are pushed over the ends of the stabiliser bar on one end and over spigots on the carrier plates on the other end. No nuts are used to secure the connecting rods. The centre of the bar is bolted to the front of the chassis.

The two lower suspension arms are attached at the inside to the chassis frame and at the outside to the wheel bearing housing (steering knuckle) by means of a bolted-on ball joint. The connection between the damper strut and the steering knuckle is held together by means of four bolts. The upper end is bolted to the body panel. Fig. 4.1 shows the front suspension on one side.

As in the case of the rear axle (see separate section), different front axles are used in the various models. The vehicle type is important when new parts are to be fitted. If in doubt, we advise you to contact your parts supplier, stating model, model year, etc.

Note – Wheel bolts/nuts: When wheel bolts are mentioned, we also refer to the wheel nuts used on the LT46 model. In this case, the wheel bolts are fitted to the detachable flange of the wheel hub.

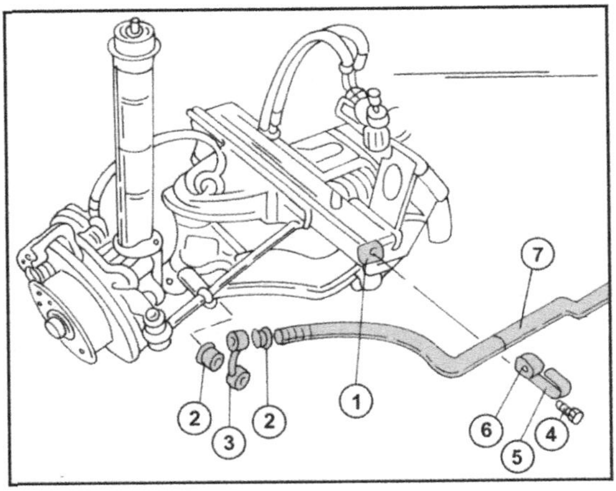

Fig. 4.1 – View of the front suspension on one side (without front spring). Bushes (2) and (6) must be coated with oil during installation.

1 Nut, 2.5–2.8 kg
2 Rubber bearing for connecting rod
3 Connecting rod
4 Bolt
5 Mounting clamp
6 Rubber bearing
7 Stabiliser bar

4.1. Front Axle

4.1.0. Removal and Installation

The following text describes the removal of the complete front axle. In general, this will not be necessary, as many parts of the front axle can be removed without taking out the complete assembly. The necessary descriptions follow later in this section. Fig. 4.2 shows the attachment of the damper struts and other parts of the suspension and can be referred to. Fig. 4.3 shows the remaining parts of the suspension without the front spring. Remember that the front axle is a fairly heavy unit, i.e., the jack must have the necessary capacity. Proceed as follows with the removal. Note that the number references apply to Fig. 4.2.

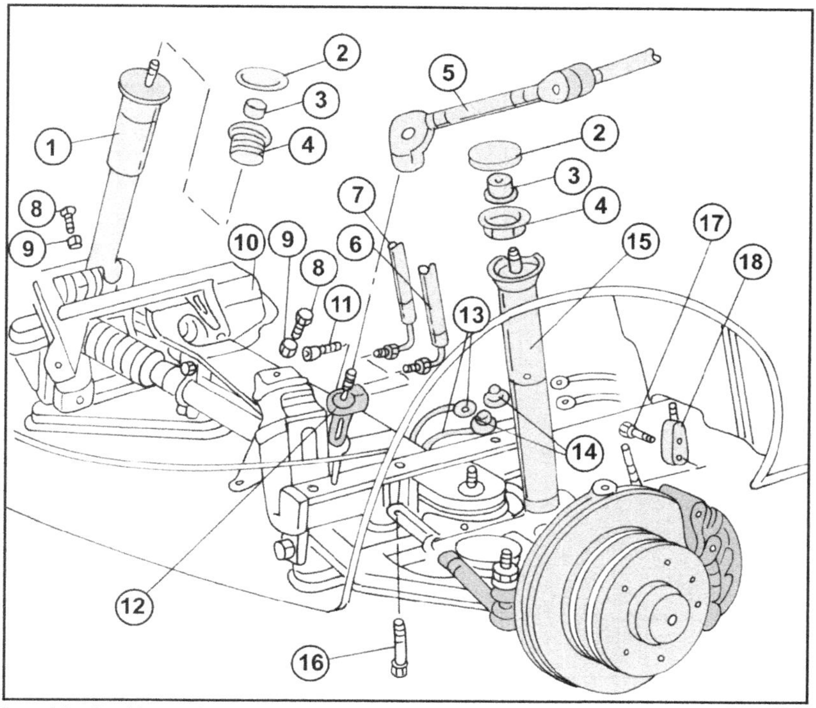

Fig. 4.2 – View of the fitted front axle.

1 R.H. damper strut
2 Closing cap
3 Damper strut nut, 10 kgm
4 Rubber bearing, damper
5 Steering intermediate shaft with universal joint
6 Fluid pressure pipe (steering)
7 Fluid return line (steering)
8 Bolt, engine mounting, 8–9 kgm
9 Plain washer
10 Front axle
11 Socket head screw, 3–3.5 kgm, steering universal joint to steering pinion
12 Steering unit wear indicator
13 Brake pipe
14 "T" connector
15 L.H. damper strut
16 Bolt, 9–10 kgm,
17 Screw, cable for brake pad wear indicator
18 Cable for brake

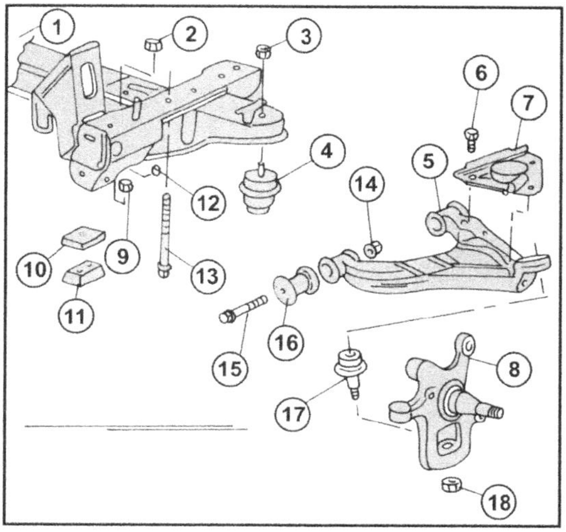

Fig. 4.2 – The component parts of the front suspension without spring. All self-locking nuts must be replaced.

1 Front axle
2 Self-locking nut, 12.5 kgm
3 Self-locking nut. 4.5 kgm
4 Rebound rubber
5 Suspension arm
6 Bolt, 6.5 kgm
7 Spring stop plate
8 Steering knuckle
9 Self-locking nut
10 Bearing plate, L.H. and R.H. different
11 Upper spring bearing. Note fitted position
12 Self-locking nut
13 Bolt, 12.5 kgm
14 Self-locking nut
15 Bolt, 14 kgm, tighten
16 Rubber bush
17 Suspension ball joint
18 Self-locking nut, 20 kgm

- Slacken the wheel bolts and jack up the front end of the vehicle. Make sure that the chassis stands cannot obstruct any of the parts to be removed. Fig. 4.3 shows where chassis stands can be placed under the vehicle. Turn the front wheels in the straight-ahead position, and remove the ignition key in order to lock the steering lock.

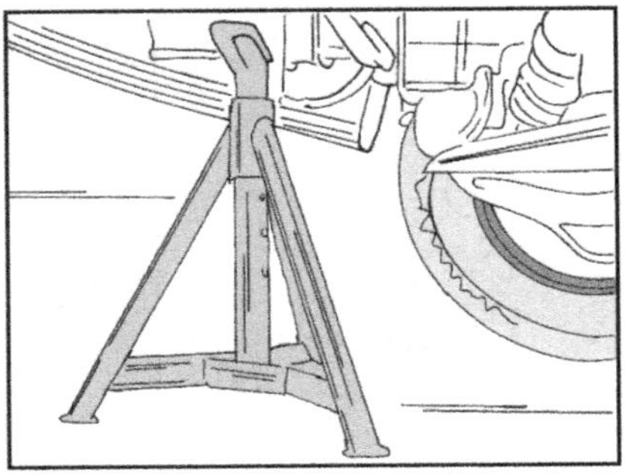

Fig. 4.3 – Place chassis stands on the L.H. and R.H. sides of the vehicle after jacking up the front end.

Note: The steering wheel must not be rotated after disconnecting the steering column universal joint when an airbag is fitted to prevent damage of the spring underneath the steering wheel.

- Disconnect the battery earth cable.
- Unscrew the union nuts for the brake pipes (13) at the two "T" pieces (14) and withdraw the pipes. Fig. 4.4 shows where the pipes are connected. Mark the connecting point of each pipe, if you are not sure where they come from. Close the pipe ends suitably to prevent ingress of dirt or other foreign matter. Remove the two "T" pieces.

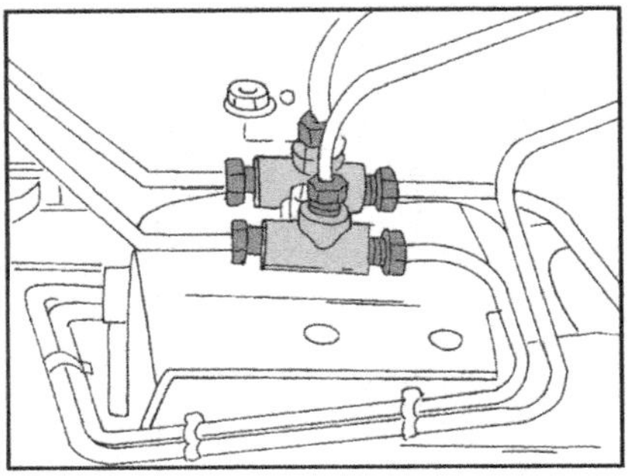

Fig. 4.4 – The brake pipes (13) in Fig. 4.2 are connected to the two "T" pieces (14) at the positions shown.

- Remove the clamp securing the cable for the brake pad wear indicator (18). To do this, remove the screw (17) and withdraw the connector plug from the cable connection.
- If ABS is fitted, disconnect the cable from the wheel speed sensor on the steering knuckle. The sensor can also be removed from the steering knuckle and remain on the cable. In this case, remove the single screw and withdraw the sensor.
- Remove the clamp bolt (11) for the steering shaft universal joint from the steering pinion with an Allen key of suitable size and withdraw the universal joint (5) from the steering pinion (12). Fig. 4.5 shows the connection between rack and pinion steering and universal joint.

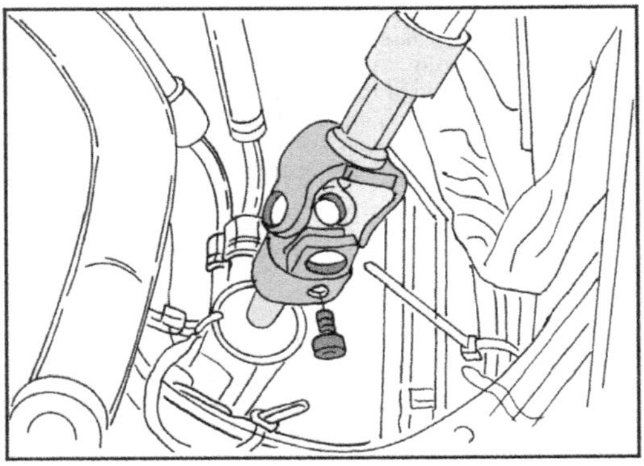

Fig. 4.5 – The connection between the steering pinion and the universal joint.

- Clamp the hoses (8) and (7) for the steering fluid with suitable clamps (to minimise the loss of fluid) and unscrew the union nuts from their connections. Collect any fluid that may be running out in a suitable manner. Protect the disconnected hoses against entry of foreign matter.
- Remove the nuts (3) for the damper struts inside the passenger compartment after removal of the closing covers (2). On the L.H. side, you will find the cover next to the foot pedal.
- Remove the air intake hose between the air cleaner and the intake manifold after slackening the hose clips or remove the air cleaner, depending on the version.
- With a suitable hand crane or hoist, lift the power unit until the engine mountings are under slight tension. Connect the ropes or chains to the lifting eyes.
- Place a trolley jack underneath the front axle and lift the assembly until it is under slight tension.
- Remove the engine mounting on the L.H. and R.H. side (bolt 8 and washer 9). Fig. 4.6 shows the attachment of the mounting on the engine side. The other mounting is secured in a similar manner. Check once more the secure suspension of the power unit before the mountings are removed.

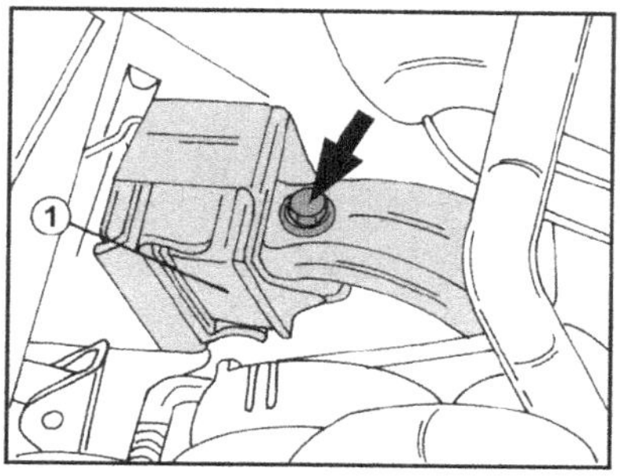

Fig. 4.6 – The engine mounting on the engine side.

- Remove the front axle on the L.H. and R.H. sides from the sidemembers. Fig. 4.7 shows where you can find the bolts. These are bolts (16) in Fig. 4.2.
- Slowly lower the front axle on the mobile jack. A helper is required to steady the axle on one side. The axle must not be rested on the two splash shields for the front disc brakes.

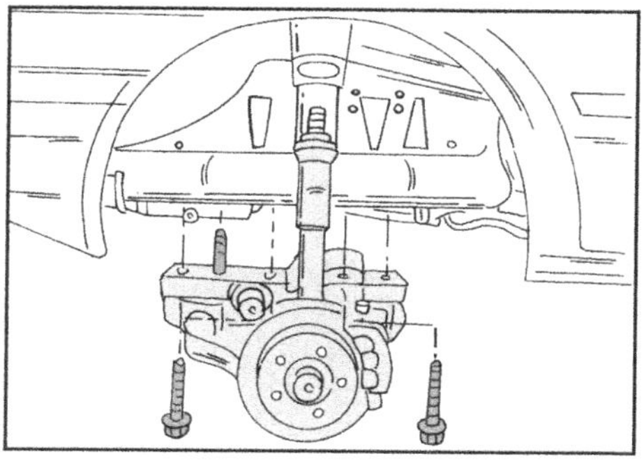

Fig. 4.7 – Removal of the front axle. Two bolts are inserted at the positions shown. The other side is attached in the same manner.

The installation of the front axle is carried out as follows.

- Place the front axle onto the mobile jack and lift it into position. The axle must engage with the locating studs. At the upper end, engage the two damper struts into the holes in the body panels. The rubber mountings must already be in position on the damper struts. Fully lift the jack until the axle is under tension against the mounting points.
- Insert the two bolts shown in Fig. 4.7 on each side (bolts 16 in Fig. 4.2) and tighten them evenly. The final tightening torque is between 9 and 10 kgm (65 and 72 ft.lb.). Make sure not to forget any of the bolts. First, tighten one side and then the other.
- Fit the steering inter-mediate shaft universal joint over the steering pinion (see Fig. 4.5), align the bolt holes, insert the bolt and tighten the bolt to 2.5 kgm (18 ft.lb.). This is bolt (11) in Fig. 4.2.
- Lower the power unit into the engine mountings on the front axle carrier. Insert the bolt with the washer (Fig. 4.6) and tighten the bolt to 8.5 kgm (61 ft.lb.). The operations must be carried out on both sides. Remove the lifting equipment after the engine mountings are refitted.
- Refit the air intake hose or the air cleaner.
- Re-connect the fluid hoses to their connections (union nuts) and tighten them. Do not forget to remove the clamps from the hoses.
- Tighten the upper damper strut mountings (nuts) to 9–10 kgm (65–72 ft.lb.) and fit the closing covers.
- Reconnect the brake pipes to the "T" pieces in Fig. 4.4. Tighten the union nuts to 1.2–1.6 kgm (9–11.5 ft.lb.).
- Re-connect the cable for the brake pad wear indicators and fit the cable clamp. Tighten the bolt to 1.0 kgm (7 ft.lb.). Check that the cables are properly routed.
- Refit the wheels and tighten them finger-tight. The wheel bolts are tightened when the vehicle is back on its wheels. Tighten torque 16–18 kgm (115–130 ft.lb.). Refit the wheel embellishers. Wheel bolts should be re-tightened after the vehicle has been driven for approx. 30 miles.

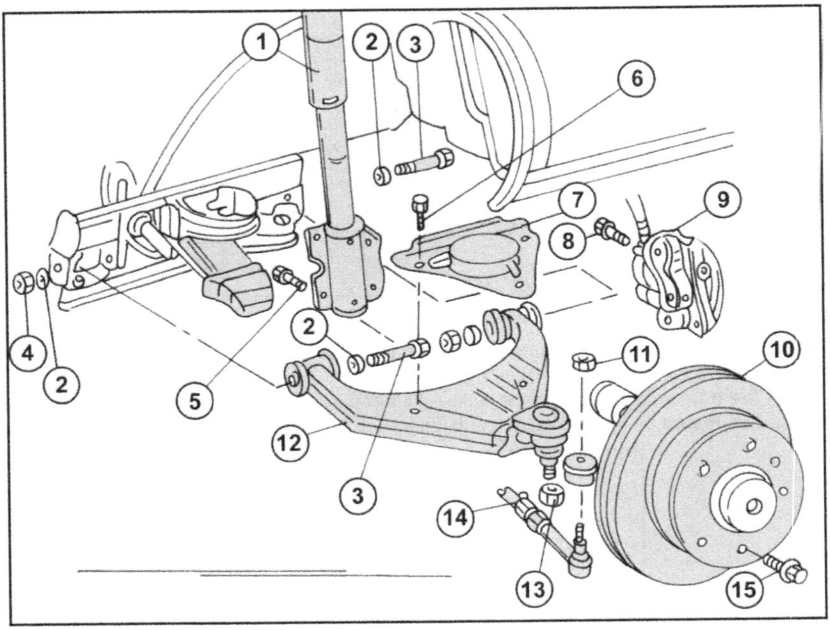

Fig. 4.8 – Details for the removal and installation of a suspension arm.

1 Damper strut
2 Plain washer
3 Bolt, M14
4 Nut, 13 – 15 kgm
5 M14 bolt, 30–32 kgm
6 M10 bolt, 4–5 kgm
7 Upper bracket
8 Caliper bolt, 20.5 kgm
9 Brake caliper
10 Steering knuckle
11 M16 nut, 12–13 kgm
12 Suspension arm
13 M20 nut, 20–30 kgm
14 Track rod
15 Wheel bolt

- Bleed the brake system as described in the section "Brakes".
- Start the engine and run it at idle speed for a few minutes. Move the steering wheel to and fro a few times. Stop the engine and check the fluid level in the steering fluid reservoir. Correct as necessary.

4.2. Lower Suspension Arm

Fig. 4.8 shows the attachment of a lower suspension arm. A suspension arm can be removed without taking out the front axle. If the axle has been removed, follow the instructions below in the same manner. The following description assumes that the steering knuckle has been removed, as described in Section 4.3.

The tightening torques are very high and a solid torque wrench must be available. You will also need a puller to separate the track rod ball joint and the suspension ball joint pin at the lower end of the steering knuckle.

4.2.0. REMOVAL OF A SUSPENSION ARM

- Remove the steering knuckle, as described in Section 4.3.

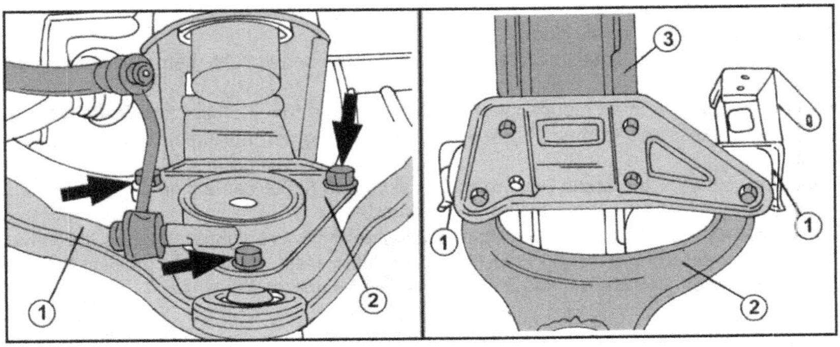

Fig. 4.9 – Details for the removal and installation of a suspension arm. The numbers are referred to in the description.

- Refer to Fig. 4.9. Remove the spring stop plate (2) from the suspension arm (1) after removal of the bolts shown by the arrows (L.H. view). Refer to the R.H. view and remove the bolts (1) for the suspension arm (2) from the front axle carrier (3) and take out the suspension arm. Fig. 4.10 shows the attachment of the suspension arm on one side. The other side is secured in the same manner. Make a note of the fitting direction of the bolts. **Attention: The arm will drop as soon as the second fulcrum bolt is removed.** We recommend that a helper holds the arm until the last bolt has been knocked out. The arm is then free and can be removed.

4.2.1. INSTALLATION OF A SUSPENSION ARM

- Place the suspension arm onto the jack and lift it into the approximate position. Insert the suspension arm into the mounting brackets. Insert the fulcrum bolt into the front mounting from the side shown in Fig. 4.10 and fit the washer and the nut. Do not tighten the nut at this stage. Remove the jack.

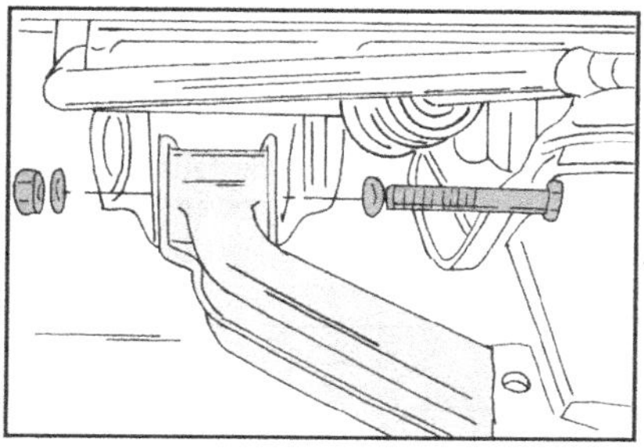

Fig. 4.10 – The attachment of a suspension arm to the front mounting bracket.

- Pull the suspension arm towards the rear and insert it into the mounting bracket. Insert the fulcrum pin from the outside towards the inside and drive it in position through the mounting bracket and the arm. On the other side, fit the washer and the nut. Do not tighten the nut fully at this stage.
- Refit the steering knuckle as described later on. Tighten the suspension arm bolts to 14.0 kgm (101 ft.lb.) when the vehicle is resting with the wheels on the ground to pre-tension the rubber bearings in the operating position.
- Refit the wheels, lower the vehicle to the ground and tighten the wheel bolts to 16–18 kgm (115–130 ft.lb.).

Replacing the Suspension Arm Ball Joint(s)

The suspension ball joints have a press fit in the suspension arms and can be replaced. As, however, special mandrels and a hydraulic press are required to remove and replace the ball joints, we recommend that you take the removed suspension arm to a dealer to have the ball joint replaced. At the end of this chapter, you will find the operations to check suspension ball joints.

Replacing the Suspension Ball Joint Rubber Bushes

The rubber bushes in the suspension arms can be replaced with suitable mandrels and a press. Again, it is possible to take the removed suspension arm to a workshop to have the bushes replaced.

4.3. Steering Knuckle – Removal and Installation

The wheel hub must be removed to remove the steering knuckle on its own. The removal is described in the following section. Once removed, you can refer to Fig. 4.11 to remove the steering knuckle. Ball joint pullers for the track rod ball joint and the suspension ball joint are required.

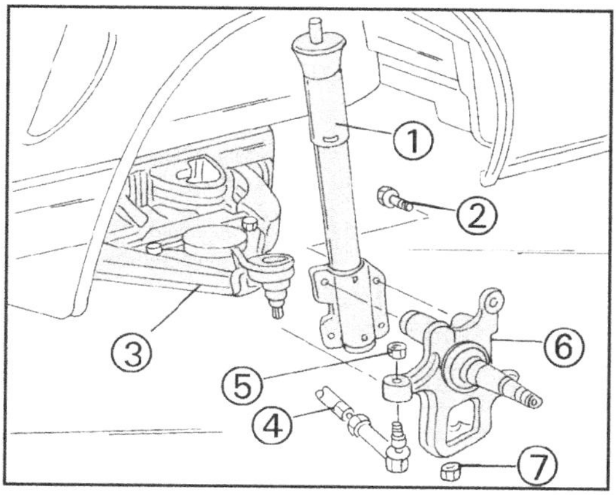

Fig. 4.11 – Details for the removal and installation of a steering knuckle.
1 Damper strut
2 Bolt, 19 kgm
3 Suspension arm
4 Track rod
5 Nut, 13 kgm
6 Steering knuckle
7 Nut, 20 kgm

- Place the front end of the vehicle on chassis stands and remove the front wheel.
- Remove the brake caliper without disconnecting the brake hose. Use a piece of wire to suspend the caliper. Do not allow it hang down on the hose.
- In the case of a vehicle with ABS referring to Fig. 4.12, remove the rubber grommet (1) from the bracket (2) on the damper strut. Remove the wheel speed sensor (1) in the R.H. view of Fig. 4.12 by means of a screwdriver out of the steering knuckle.

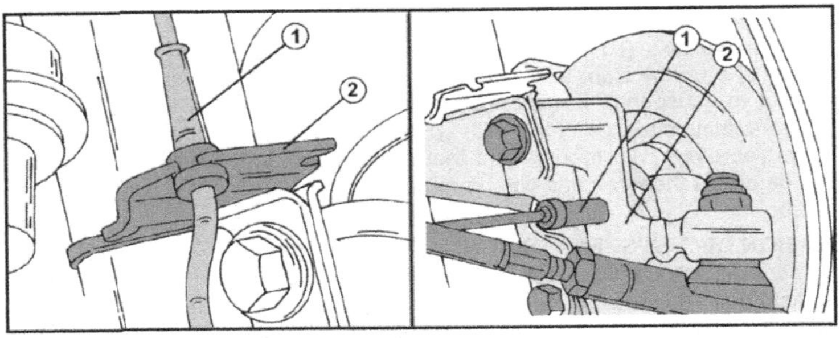

Fig. 4.12 – Removal of parts on models with ABS. The numbers are referred to in the description.

- Remove the track rod ball joint nut and separate the connection with a suitable ball joint puller in the manner shown in Fig. 4.13. The LT46 model has a protective shield for the track rod ball joint. Do not damage the joint rubber dust cap when separating the joint.
- Place a jack underneath the suspension arm and lift it up about 4 in. to release the tension of the damper strut. Now, unscrew the four bolts shown in Fig. 4.13 and separate the damper strut from the steering knuckle.
- Remove the ball joint at the bottom of the steering knuckle.

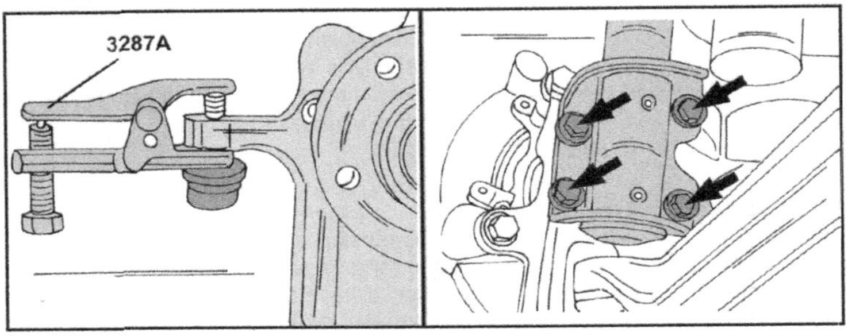

Fig. 4.13 – The L.H. view shows the separation of the track rod ball joint from the steering knuckle. The suspension ball joint is removed in a similar manner. The R.H. view shows the attachment of the damper strut to the steering knuckle. Note that bolts of different diameter (27.5 and 25.5 mm) are used. Both types are tightened to 18.5 kgm.

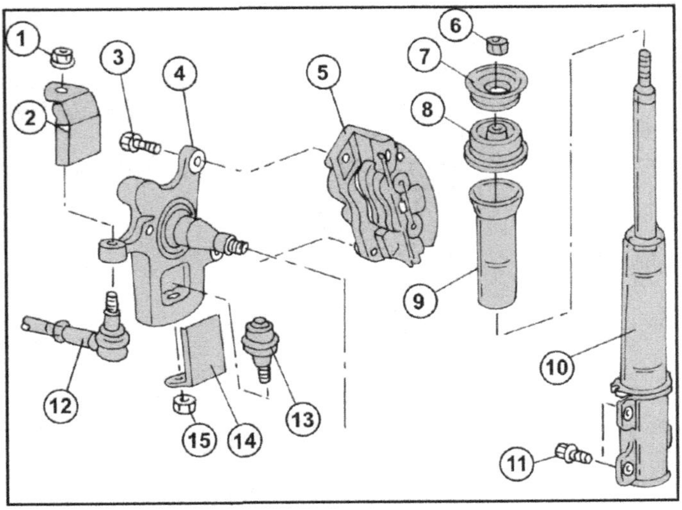

Fig. 4.14 – Details for the removal and installation of the steering knuckle and the damper strut.

1 Nut, 13 kgm, replace
2 Protective shield, LT46
3 Bolt, 19 kgm
4 Steering knuckle
5 Brake caliper
6 Nut, 11 kgm
7 Upper rubber/metal bearing
8 Lower rubber/metal bearing
9 Protective sleeve
10 Damper strut
11 Hexagonal bolt
12 Track rod
13 Ball joint
14 Protective shield
15 Nut, 29 kgm (self-locking)

- Remove the steering knuckle.

The installation is a reversal of the removal procedure. The various tightening torques have already been specified during the installation of the lower suspension arm and this section should be referred to during the installation of the various parts. Fig. 4.14 shows the component parts of the steering knuckle with additional tightening torques. Remember to tighten the wheel bolts to the correct torque when the wheels are resting on the ground.

4.4. Front Wheel Hubs and Wheel Bearings

4.4.0. WHEEL BEARINGS – ADJUSTMENT

- Place the front end of the vehicle on chassis stands and remove the front wheel(s).
- Push the brake pads away from the brake disc. If necessary, remove the brake caliper, as the brake disc must rotate freely for a proper check.

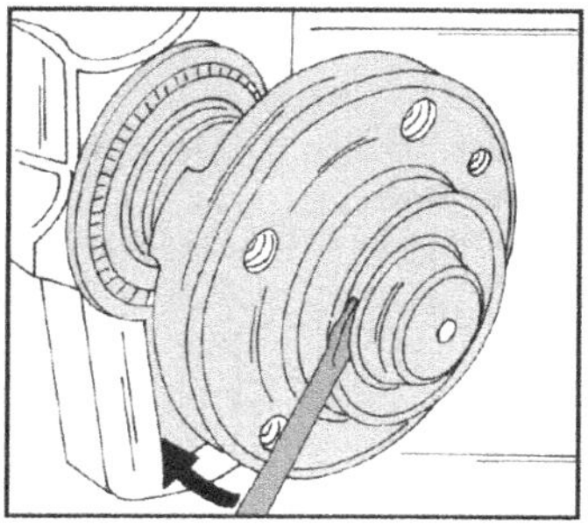

Fig. 4.15 – Removal of the hub grease cap with a screwdriver. Move the screwdriver in the direction shown by the arrow.

- Remove the hub grease cup. Use a screwdriver, applied as shown in Fig. 4.15.

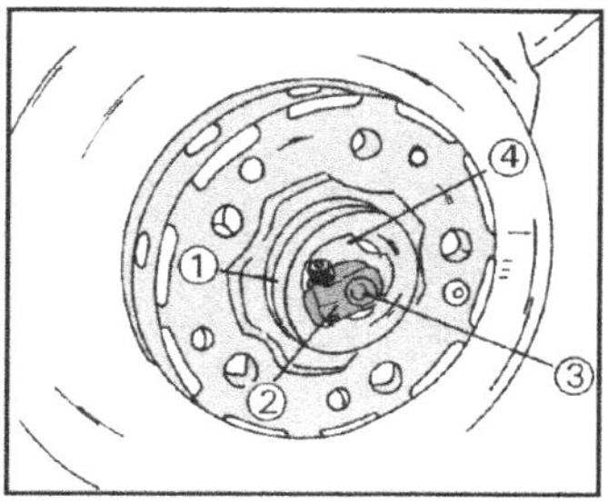

Fig. 4.16 – Details for the front wheel bearing adjustment.
1 Front wheel hub
2 Clamp nut
3 Axle stump
4 Thrust washer

- Refer to Fig. 4.16 and slacken the socket head clamp bolt for the nut (2). Rotate the wheel hub evenly and at the same time tighten the clamp nut until the hub can just be rotated.
- From this position, slacken the nut by 1/3 of a turn. With a rubber or plastic mallet, knock against the end of the stub axle (3) to release any tension in the wheel bearings.
- Attach a dial gauge to the front wheel hub, as shown in Fig. 4.17. Set the dial gauge to a pre-load of 2.0 mm. The stylus must rest on the centre of the stub axle. Grip the wheel hub with both hands and move it in and out. Before each check, rotate the wheel hub a few turns, but not during the actual measurement. The dial gauge reading must be between 0.02 and 0.04 mm. If this is the case, tighten the socket head clamp bolt to 1.0–1.6 kgm (7–11.5 ft.lb.), as shown in Fig. 4.18. Re-check the setting. If the wheel bearing clearance is properly adjusted, it must still be possible to move the washer between the clamp nut and the outer wheel bearing with a small screwdriver.

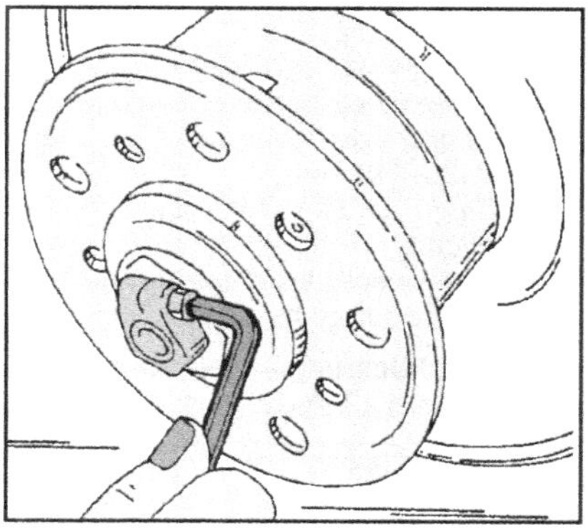

Fig. 4.18 – The wheel bearing adjusting clamp nut is tightened with an Allen key.

- Fill the grease cap to the inner edge with wheel bearing grease and fit over the wheel hub. Refit the wheel and lower the vehicle to the ground.

4.4.1. WHEEL HUB – REMOVAL AND INSTALLATION

In the case of the LT46, the flange fitted in front of the brake disc must be removed before the wheel hub can be removed. Otherwise, the operations are the same for all models.

- Jack up the front end of the vehicle and place chassis stands underneath the front axle. Remove the front wheel on the side in question and remove the hub grease cap as described in the last section or in other suitable manner (refer to Fig. 4.15).
- Have the brake pedal depressed by a helper and slacken the brake disc securing screw (cross-headed screw). The disc can also remain on the hub, if desired.
- Unscrew the brake caliper from the steering knuckle and suspend it with a piece of wire to the front suspension. Do not allow the caliper to hang down on the brake hose. Push the caliper in direction brake disc to push the piston back into its bore. This will facilitate the removal of the caliper.
- Fully unscrew the brake disc securing screw and remove the brake disc.
- Slacken the clamp nut (2) in Fig. 4.16 with an Allen key as shown in Fig. 4.18, unscrew the nut and remove the thrust washer underneath it.
- Remove the wheel hub from the axle stump. A tight wheel hub can be removed with a slide hammer in a similar manner or a puller is used. The outer wheel bearing can drop on the floor – careful!
- The inner wheel bearing and the oil seal will remain on the wheel hub and must be removed with a suitable puller, as described later on.
- Thoroughly clean all parts of the wheel hub and the axle stump and check for wear.

If there is no need to replace the wheel bearings, refit the hub as described below. Otherwise, refer to the next section to replace the wheel bearings.

The wheel hub is refitted as follows.

- Slightly coat the running area for the oil seal on the axle stump with wheel bearing grease, push the assembled wheel hub over the stump and drive in position, using a plastic mallet.
- Insert the outer, well greased bearing into the hub and over the axle stump. Again, tap the bearing in position. Fit the thrust washer and screw on the clamp nut.

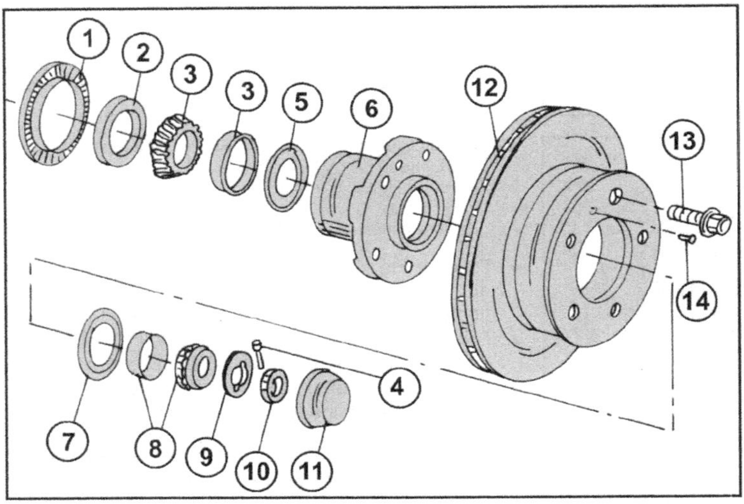

Fig. 4.19 – The component parts of a front wheel hub.

1	Rotor for speed sensor	6	Wheel hub	11	Hub grease cap
2	Oil seal	7	Outer grease washer	12	Brake disc
3	Inner wheel bearing	8	Outer wheel bearing	13	Wheel bolt, 18 kgm
4	Clamp bolt, 1.3 kgm	9	Thrust washer	14	Screw, 2.0 kgm
5	Inner grease washer	10	Clamp nut		

- Adjust the wheel bearings as already described.
- Fit the brake disc over the wheel hub and refit the wheel brake caliper. Tighten the bolts to 17.5–20.5 kgm (126–148 ft.lb.).
- Fill the grease cap with a little bearing grease and knock the cap in position, using a rubber or plastic mallet only.
- Refit the wheels, lower the vehicle to the ground and tighten the wheel nuts to 18 kgm (130 ft.lb.). Re-tighten the bolts after approx. 30 miles.

4.4.2. Wheel Bearings – Replacement

Refer to Figs. 4.19. A flange is fitted in front of the brake disc in the case of an LT46 model.

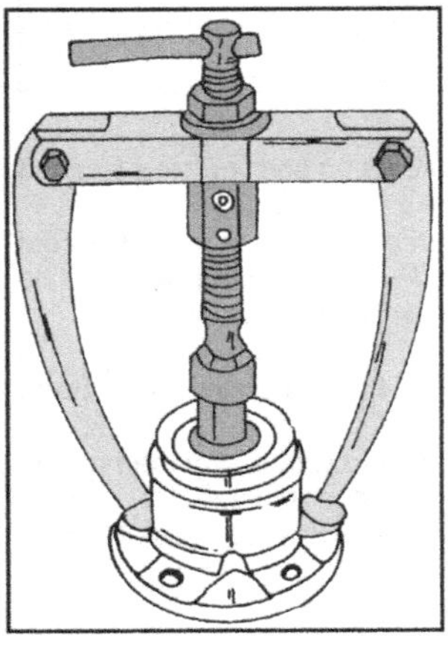

Fig. 4.20 – Inner wheel bearing races can be removed from the inside of the wheel hub in the manner shown. Otherwise, a mandrel will have to do.

- Remove the inner bearing race and the oil seal from the inside of the wheel hub/brake disc, using a two-arm puller. In a similar manner, remove the race of the inner bearing from the wheel hub and the outer race of the outer taper roller bearing from the outside of the wheel hub. Fig. 4.20 shows how the puller is applied.
- If necessary, unscrew the brake disc from the wheel hub. To do this, clamp the brake disc into a vice (soft-metal jaws) and remove the cross-headed bolt.

Bearings cannot be re-used if the following faults can be detected.

- The rollers of the outer bearing cage show signs of pitting, flaking of the chromium or similar defects.
- Bearing rollers show signs of rust (ingress of water or leaking oil seal).
- Coloration of the outer bearing race (overheating).

If a taper roller bearing must be replaced, replace both bearings as a set. Always use bearings of the same manufacturer. If the original bearings are re-used, make sure that bearing races and bearing cages remain together. Refit the bearings as follows:

- If removed, refit the brake disc. Tighten the bolt to 2 kgm (14 ft.lb.). A cross-headed screwdriver is used to tighten the bolt.
- Place the grease washer in position and press or drive the outer bearing race into the front of the wheel hub, using a suitable mandrel. Fig. 4.21 shows this operation. Fit the second grease washer and press the outer race for the inner wheel bearing into the hub. In this case, rest the wheel bearing hub flange onto the press plate in Fig. 4.21.

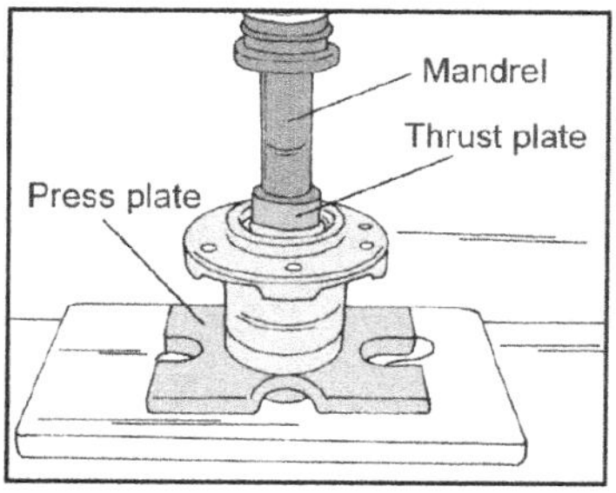

Fig. 4.21 – Pressing the bearing race for the other wheel bearing in position. The grease washer is already installed.

- Smear the bearings with approx. 50 grams of bearing grease. This must be evenly distributed between the bearings and the inside of the hub (per hub). First, fill the spaces between the inner bearing with the grease, insert the bearing and smear some more grease into the end faces of the rollers. Fill the oil seal between the seal lip and the dust lip with grease and carefully drive the oil seal into the rear of the hub until the outer face is flush. Do not damage the seal. *The space between the two grease washers must not be filled.*
- If ABS is fitted and w new wheel hub is fitted, press the rotor for the wheel speed sensor in position. To do this, place the rotor onto the press plate in Fig. 4.21 and press the wheel hub over the rotor.
- Smear the remaining grease into the outer wheel bearing and the inside of the hub (approx. 15 grams) and refit the wheel hub as described in the last section. Never use more grease than specified, as this will be pushed out.
- Adjust the wheel bearing clearance, as described in Section 4.4.0. Make sure that the clearance is adjusted correctly.

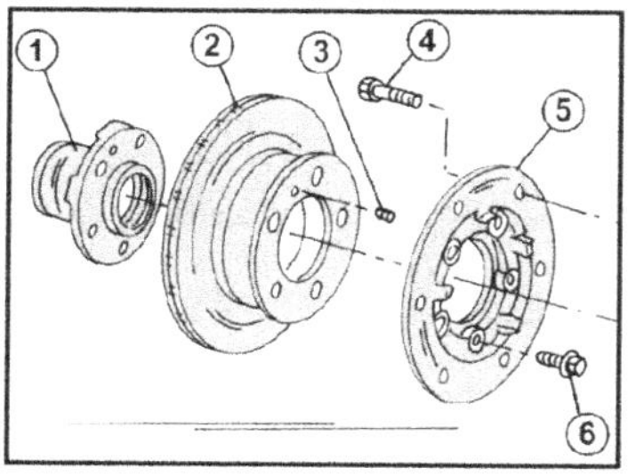

Fig. 4.22 – Arrangement of brake disc and wheel hub (LT46).
1 Wheel hub
2 Brake disc
3 Cross-headed screw, 2.0 kgm
4 Wheel bolt
5 Flange
6 Flange bolt, 18 kgm

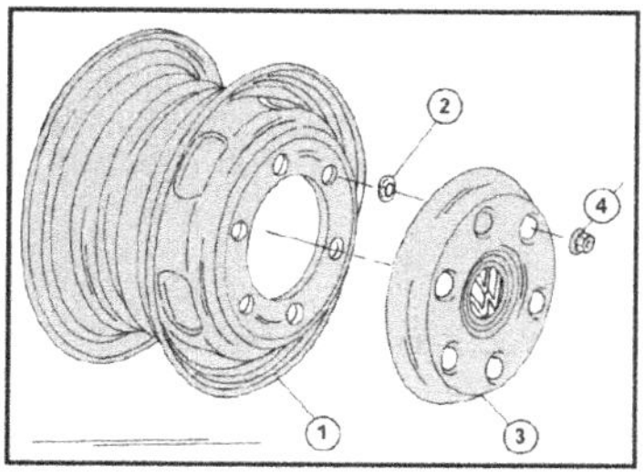

Fig. 4.23 – Wheel and wheel cover (LT46).
1 Wheel
2 Spring washer (over wheel bolt)
3 Wheel cover
4 Wheel nut, 18 kgm

Flange in front of Wheel Hub – LT46

On these models, a flange is fitted in front of the brake disc, as shown in Fig. 4.22, which must be removed before the wheel hub can be taken out. In front of the wheel is a wheel cover that must be removed during the removal of a wheel. After the wheel has been removed, you can unscrew the flange. Brake disc and hub can then be removed in the manner described.

Tighten the flange securing bolts to 18 kgm (130 ft.lb.). Spring washers are inserted between wheel cover and wheel, which must be placed over the studs before the wheel is refitted (18 kgm). The arrangement is shown in Fig. 4.23.

4.5. Front Wheel Alignment

Only the toe-in setting should be checked. Castor and camber setting checks should be carried out at a dealer. The camber and castor cannot be adjusted. The vehicle height at the front and rear must be measured, the height at the front must be deducted from the height at the rear and the resulting value is then read off in a table, which will give the correct settings for castor and camber. This is one reason that you will not be able to adjust the castor or camber under D.I.Y. conditions.

4.5.1. TOE-IN ADJUSTMENT

Toe-in is the difference in the distance between the front of the front wheels and the rear of the front wheels, measured at the wheel rims and in the height of the wheel hubs, when the front wheels are in the straight-ahead position. The dimension should be 0 mm (no toe-in or toe-out), with a tolerance of ±1 mm.

Excessive toe-in of the front wheels can be recognised on wear on the outside of the tyres. Wear on the inside of the tyres points to excessive toe-out, also called negative toe-in.

The toe-in can only be checked properly with the correct equipment, i.e., a good tracking gauge must be available. Otherwise, have the geometry checked by a specialist workshop. Proceed as follows.

- Turn the steering into the centre position.
- Place the tracking gauge against the front of the wheel rims, in height with the wheel hubs and set the gauge to zero. Mark the points where the gauge is applied with chalk.
- Push the vehicle forward by half a turn of the wheels until the two points marked with chalk are at the rear of the wheels, again at the same height as the hubs. Apply the tracking gauge to the rear of the wheel rims and read off the value. This should be within the values given above. If this is the case, there is no need to carry out any adjustments. Otherwise, adjust to 0 mm.
- If the setting is outside the value given, adjust the toe-in by altering the length of the track rods after slackening of the track rod clamps. Before tightening of the track rods, push both track rod ball joints against their stops. After tightening the clamps, check that both track rod ball joints can be moved to and fro.
- Re-check the toe-in setting as described above and tighten the clamps.

4.6. Front Suspension Maintenance Checks

Suspension Ball Joints

Worn suspension ball joints will be spotted during an MOT check. The ball joints are checked, as shown in Figs. 4.24 and 4.25. The suspensions shown are not as fitted to the vehicles covered in this manual, which will, however, be sufficient for the purpose of the check. We must, however, point out that a certain amount of force is required to move the suspension arms and/or the wheel as shown, due to the weight of the parts. Also make sure that the vehicle is well supported on chassis stands to prevent it from "falling" off.

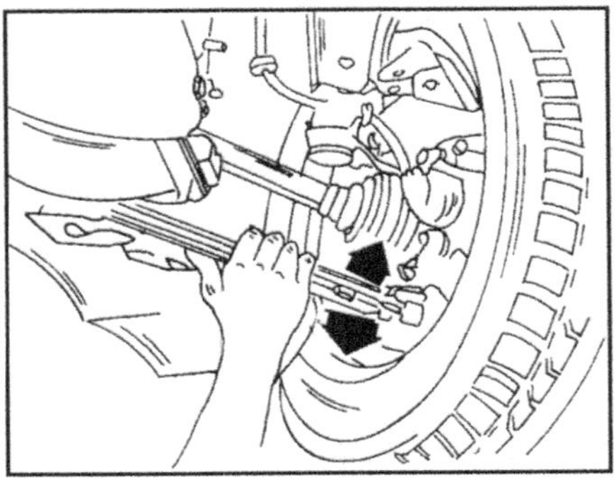

Fig. 4.24 – Checking the up-and-down play of a suspension ball joint.

The first illustration demonstrates axial play, i.e., up-and-down play in the joints. Grip the suspension arm as shown and move it up and down. Excessive movement will indicate a worn ball joint.

The second illustration shows how the side play of the joint is checked. In this case, move the wheel "in" and "out" and check for excessive movement. It should, however, be noted that excessive play in the wheel bearing is also indicated when the wheel is checked as shown.

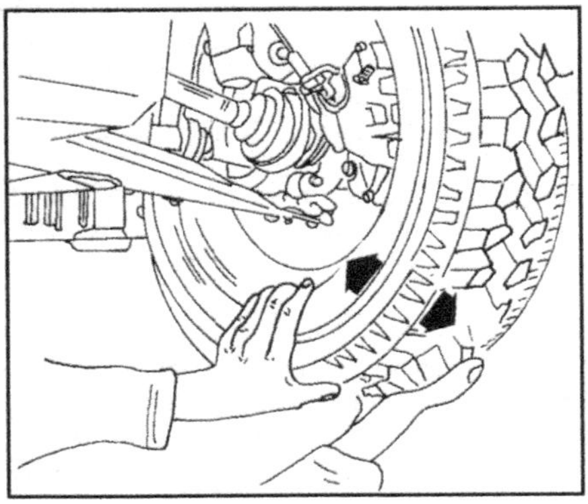

Fig. 4.25 – Checking the side play of a suspension ball joint.

Checking the Suspension Ball Joint Rubber Caps

The suspension ball joints between the lower suspension arms and the steering knuckle are maintenance-free, but the rubber caps should be checked for cuts and other damage in the framework of a good maintenance programme. Damaged caps require the replacement of the ball joint. In this case, you can remove the suspension arm and take the arm to a workshop to have the ball joint replaced, as special mandrels are required.

4.7. Front Axle – Tightening Torques

Suspension arms to front axle	14 kgm (101 ft.lb.)
Upper plate to front axle	4–5 kgm (32.5–36 ft.lb.)
Damper strut to body	9–10 kgm (63–72 ft.lb.)
Front axle to crossmember	9–9.5 kgm (63–47 ft.lb)
Bolt, engine mounting, M12	8–9 kgm (58–63 ft.lb.)
Cable bracket, brake pad wear indicator	1.0 kgm (7 ft.lb.)
Wheels	18 kgm (130 ft.lb.)
Brake caliper to steering knuckle	17.5–20.5 kgm (126–143.5 ft.lb.)
Brake disc to wheel hub	2–2.5 kgm (14.5–18 ft.lb.)
Wheel hub clamp bolt	1.2–1.6 kgm (8.6–11.5 ft.lb.)
Brake disc splash shield	1.0 kgm (7 ft.lb.)
Track rod ball joint to steering lever	12–13 kgm (86.5–94 ft.lb.)
Steering to front axle assembly	5.5–6 kgm (40–43.5 ft.lb.)
Tensioning plate for front spring to front axle assembly, M12 bolts	14–16 kgm (101–116 ft.lb.)
Tensioning plate for front spring to front axle assembly, M10 bolts	6.5–7.5 kgm (47–54 ft.lb.)
Stop plate to suspension arm, M10	4.0–5.0 kgm (30–36 ft.lb.)
Stabiliser bar mounting to front axle assembly, M8	2.2–2.5 kgm (14–18 ft.lb.)
Fluid pressure pipe to steering	3.5–4 kgm (25–30 ft.lb.)
Steering universal joint to steering shaft	3.0–3.5 kgm (22–25 ft.lb.)
Suspension ball joint to suspension arm (nut)	28–30 kgm (202–216 ft.lb.)
Damper strut to steering knuckle (four bolts)	18.5 kgm (133 ft.lb.)
Brake pipe connections	1.2–1.6 kgm (8.6–11.5 ft.lb.)

5. Rear Axle and Propeller Shaft

5.0. Technical Data

Type	Rigid axle with leaf springs, hydraulic telescopic shock absorbers and with or without stabiliser bar (anti-roll bar), auxiliary (helper) springs on certain models.
Oil Capacity of Rear Axle:	1.5 or 1.8 litres, depending on axle
Type of oil	Hypoid gear oil SAE 90 or 85W/90
Rear spring part number	Depending on model version
Shock absorber part number	Depending on model version
Wheel bearing clearance	0.02–0.4 mm (only with twin wheels)

Depending on the vehicle, a different axle is fitted to the various models covered in this manual. The arrangement of the suspension is, however, the same for all axles in models LT28 and LT35, i.e., leaf springs and telescopic shock absorbers are used. A slightly different arrangement is used on LT46 models.

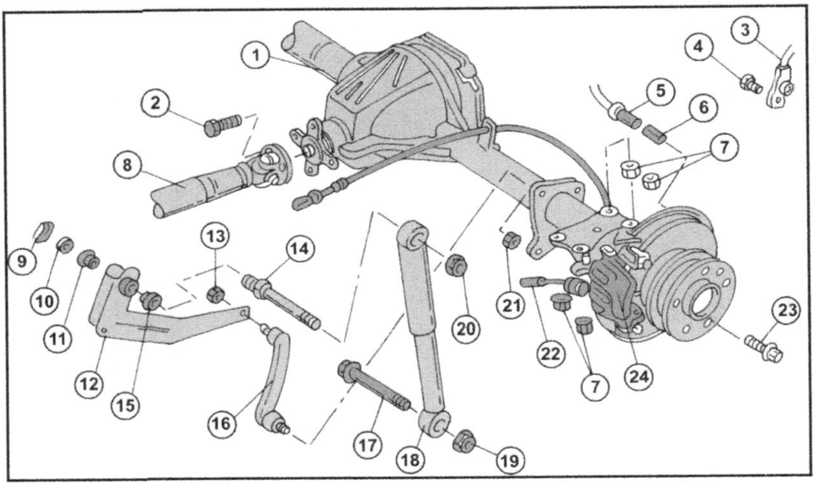

Fig. 5.1 – Details for the removal and installation of the rear axle.

1 Rear axle
2 Bolt, 7.0 kgm
3 Cable, wear indicator
4 Bolt, 1.0 kgm
5 Wheel speed sensor
6 Tensioning sleeve for (5)
7 Nut for leaf spring "U" bolts
8 Propeller shaft
9 Retaining washer
10 Washer
11 Bush
12 Lever, brake pressure regulator
13 Nut, 3 kgm, replace
14 Bolt, replace
15 Bush
16 Lever, regulator
17 Bolt, always replace
18 Shock absorber
19 Nut, 76 Nm + 90°
20 Bolt, 12 kgm, replace
21 Nut, 3 Nm, replace
22 Pad wear indicator
23 Wheel bolt, 18 kgm

5.1. Rear Axle – Removal and Installation

5.1.1. Models without ABS

The attachment of the rear axle is shown in Fig. 5.1. The illustration can be referred to in the following description. A solid strong mobile jack is required to remove the rear axle. A helper is also required to guide the axle during removal and installation. Proceed as follows to remove the axle.

- Remove the wheel covers and slacken the wheel bolts.
- Place the rear end of the vehicle on chassis stands. In this case, the stands must not be placed underneath the axle. Fig. 5.2 shows various positions where the stands can be placed.

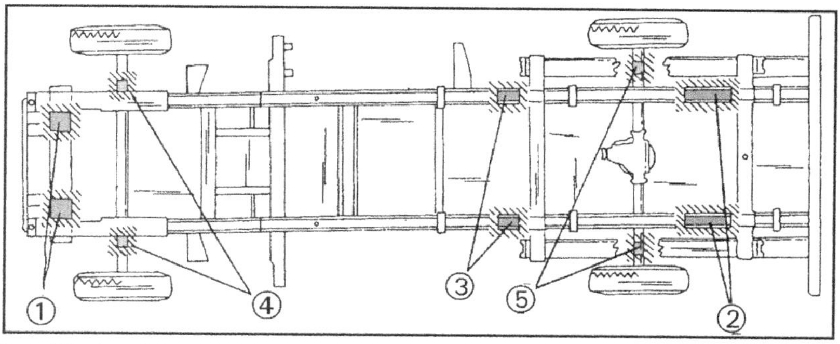

Fig. 5.2 – View of the chassis from underneath. 1–5 show where chassis stands can be placed under the vehicle.

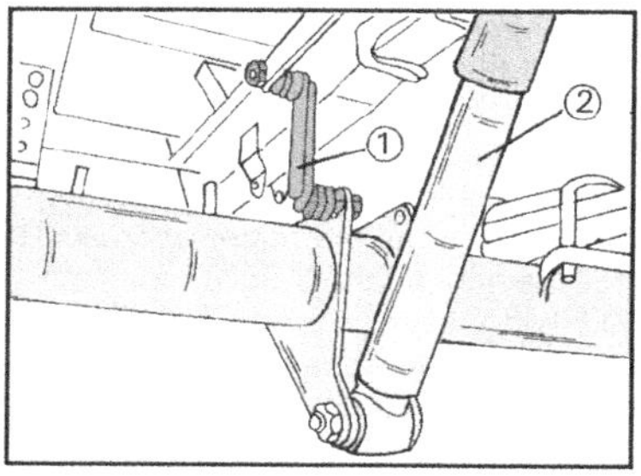

Fig. 5.3 – Lever for brake pressure regulator (1) and shock absorber (2) in the fitted position.

- Withdraw the connector plugs from the brake pad wear indicators, unscrew the indicators and remove (items 3 and 4).
- Below the vehicle remove the handbrake cable brackets.
- Disconnect the front handbrake cable from the lever on the handbrake compensator, slacken the lock bolt for the adjusting cam (eccentric) in the centre of the lever and turn the cam towards the rear until the handbrake cables can be unhooked. Section "Brakes" gives further details.
- Unscrew both shock absorbers from the rear axle. The axle must be lifted until the shock absorbers are slightly compressed (see also Fig. 5.3).
- Above the rear axle, disconnect the brake pressure regulator linkage, using a screwdriver as shown in Fig. 5.4. Both ends of the linkage are secured with nuts which must always be replaced.
- Unscrew the brake hose-to-brake pipe connection. Counterhold the hexagon of the hose with an open-ended spanner when slackening the pipe union nut. Knock out the spring plate and withdraw the hose. This operation must be carried out on both sides of the vehicle. Protect the open connections in a suitable manner.
- Remove the air outlet pipe out of the rear axle.

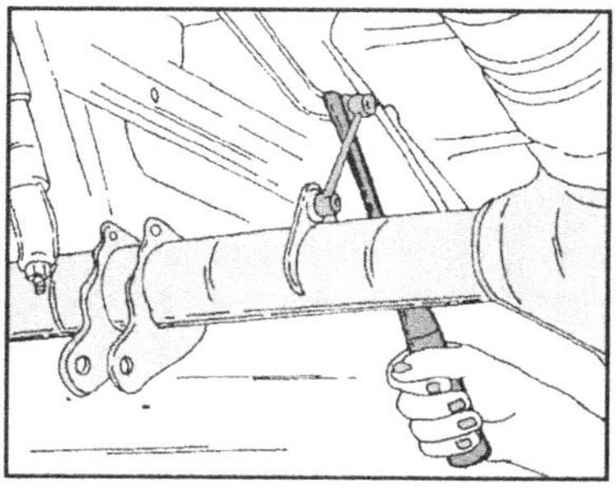

Fig. 5.4 – The brake pressure regulator linkage is separated at the upper end (screwdriver) after removal of the nuts.

- Disconnect the propeller shaft flange from the rear axle flange. Note that the shaft will drop when the last bolt is removed. Take care. Tie the shaft to the underbody after disconnecting it.
- Remove the stabiliser bar if one is fitted (LT46).
- Remove the spring "U" bolt nuts from below the axle (Fig. 5.5) and remove the spring plates. A helper should guide the axle on the jack, to prevent it from tilting to one side.

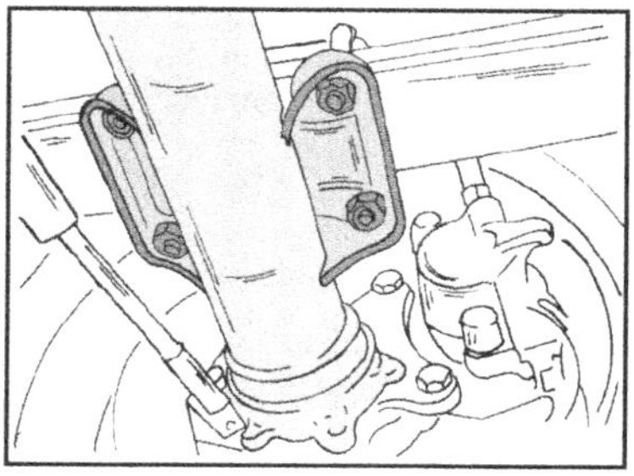

Fig. 5.5 – The "U" bolt nuts for the rear springs are fitted from below. Tighten them to 10–12 kgm (72–86 ft.lb.) during installation.

- Slowly lower the axle on the jack and wheel it out of position.

The installation is carried out as follows, noting that the nuts marked in Fig. 5.1 must be replaced during assembly.

- Wheel the rear axle underneath the vehicle, approximately between the centres of the rear springs.
- Lift the axle in position. The centre bolts of the two springs must engage into the locating holes in the axle plates. When engagement has taken place, lift the axle further until it is in full contact with both springs.
- Fit the spring plates and drive in the "U" bolts without damaging the threads. "U" bolts can be bent in a vice to make them fit, if necessary.

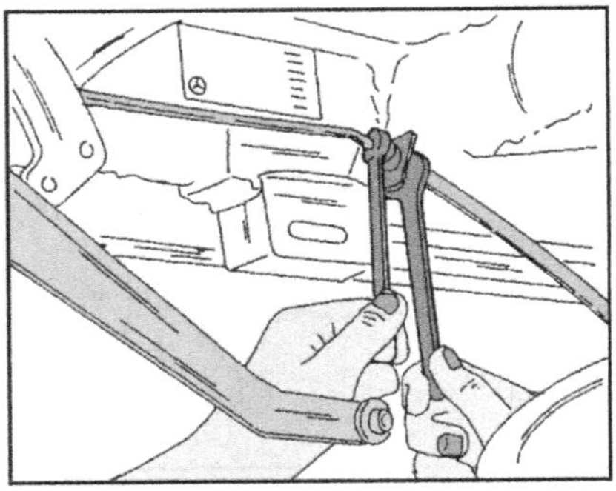

Fig. 5.6 – Disconnecting the brake pipe from the brake hose connections.

- Fit the nuts and evenly tighten them in two stages to 5.0 kgm (36 ft.lb.) and then to 18 kgm (130 ft.lb.). From the end position, tighten the nuts by a further ½ of a turn (refer to Fig. 5.5).
- Refit the brake hose to the retaining bracket, connect the brake pipe to the hose and tighten the union nut. Counterhold the hexagon, as shown in Fig. 5.6.
- Re-connect the brake pressure regulator linkage. New nuts are used and tightened to 3.0 kgm (22 ft.lb.).
- Re-connect both shock absorbers. Tighten the nuts to 12.0 kgm (86.6 ft.lb.) at the upper ends and to 7.5 kgm (54 ft.lb.) at the lower ends. From the final position, tighten the nuts a further 90°.
- Re-connect the propeller shaft to the rear axle flange (7.0 kgm/50.5 ft.lb.).
- Insert the two handbrake cables into the guide plates and secure with the spring clips. Screw the retaining clips to the frame sidemembers.
- Re-connect the handbrake cables to the compensating lever. The cables must be adjusted, as described in the section "Brakes". Finally, bleed the brake system.
- Check the oil level in the rear axle and, if necessary, correct. Refill the axle with oil, if it has been drained.
- Refit the wheels, lower the vehicle to the ground and tighten the wheel nuts. Finally, refit the wheel covers.

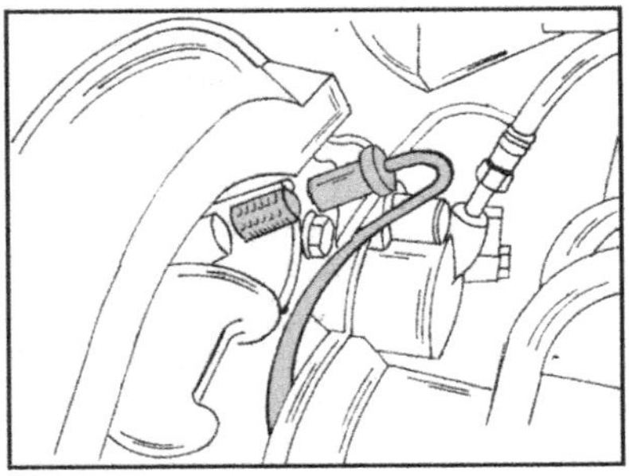

Fig. 5.7 – Removal of a rear wheel speed sensor on models with ABS.

5.1.2. Models with ABS

The removal and installation of the rear axle is carried out in the manner described in the previous section, with the difference that the wheel speed sensor shown in Fig. 5.7 must be removed from the brake backplate. This will free a clamp bush.

5.1.3. Rear Axle Overhaul

Many special tools and experience are necessary to overhaul the rear axle and the differential. To check the axle drive, you can unscrew the rear cover (oil drained) and check the condition of the crown wheel and other parts for major damage. From the condition accessed, you will be able to decide if repairs are necessary. Chipped off metal or other damage will be easily seen. You can remove the rear axle as described above and have it overhauled or fit an exchange unit.

Remember that the rear axle fitted to your particular vehicle is specially allocated and in case of the installation of a second-hand axle into an LT28 or LT35 model, you will have to make sure that the axle is suitable for your vehicle, as axle ratios are not the same for all models. The axle fitted to the LT46 is different in construction, but the above also applies.

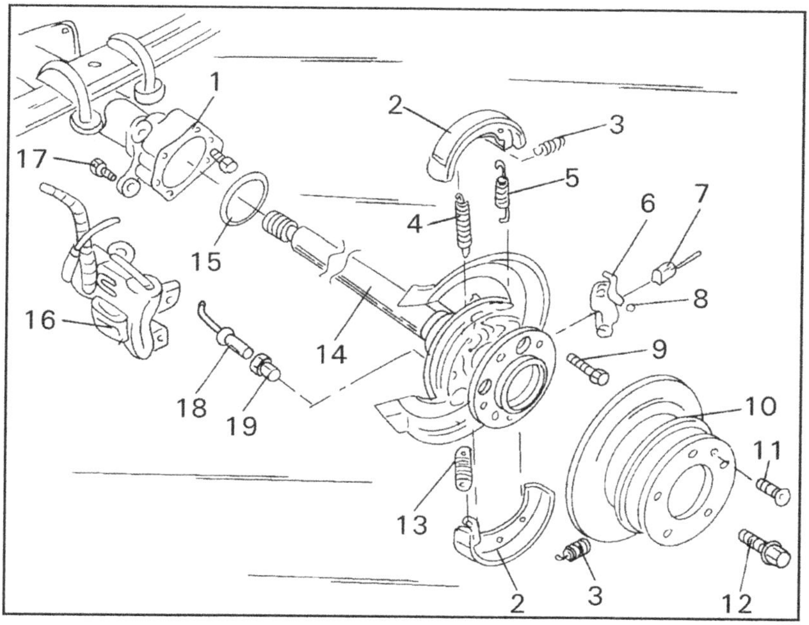

Fig. 5.8 – Details for the removal and installation of a drive shaft (LT28/35).

1	Rear axle housing	8	Fulcrum pin	15	Oil seal, replace
2	Handbrake brake shoes	9	Bolt, 6–7 kgm	16	Brake caliper
3	Return spring	10	Brake disc	17	Bolt, 10.5–11.5 kgm
4	Shoe return spring	11	Bolt, 2 kgm	18	ABS speed sensor
5	Shoe return spring	12	Wheel bolt, 18 kgm	19	Clamp bush for 18
6	Spreader	13	Brake shoe adjuster		
7	Handbrake cable	14	Drive shaft		

5.2. Drive Shafts (Halfshafts)

5.2.0. Removal and Installation

The removal and installation of the drive shafts is different on the LT28/LT35 models and the LT46 model.
The wheel must be removed, the vehicle must be resting on chassis stands.

LT28 and LT36 Models

Fig. 5.8 shows the parts to be removed. It should be noted that it is sufficient to remove the handbrake brake shoes from their springs and leave them in position. Proceed as follows.

- Slacken the lock bolt in the centre of the handbrake cable compensator bracket and turn the adjusting eccentric cam towards the rear (see also the section "Brakes") until the front handbrake cable can be disconnected from the lever on the compensator.
- Remove the rear brake caliper and attach it with a piece of wire to the rear suspension.

The following operations are described in detail in the section "Brakes". These involve the following (see also Fig. 5.8).

- Remove the screw (11) in the outer face of the brake disc (Philips screwdriver) and withdraw the brake disc (10). A tight disc can be freed by turning back the shoe adjuster.
- Disconnect the upper brake shoe return springs (4) and (5) and the horizontally fitted spring (3).
- At the rear of the brake carrier plate, open the lever (6), push out the fulcrum pin (8) and remove the lever.

The remaining operations are carried out as follows.

- Remove the bolts (9) securing the bearing cover to the rear axle. A socket with an extension is inserted through the holes in the drive shaft flange to reach the bolt heads. The shaft must be rotated to reach all bolts (Fig. 5.9).

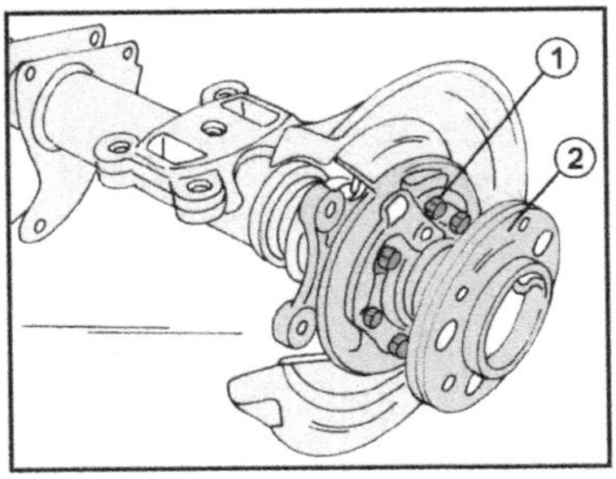

Fig. 5.9 – The bolts (1) behind the drive shaft flange can be removed through the holes (2) in the flange. Turn the shaft to reach all bolts. A socket and an extension are required.

- Withdraw the drive shaft. Mark the drive shafts with L.H. and R.H. if both drive shafts are removed at the same time to make sure that the shafts are refitted to the correct side.
- Remove the oil seal (15) from the brake carrier plate. Sometimes it is possible that the outer bearing race for the axle shaft bearing remains inside the axle tube. In this case, a suitable puller must be used to remove the bearing race.

The installation is carried out as follows.

- Place the brake carrier plate against the rear axle. Insert the gasket between bearing cover and brake carrier plate.
- Insert the axle shaft in accordance with the marking, if both have been removed.
- Align the bearing cover with the holes in the brake carrier plate and the axle tube and fit the bolts. **New bolts must always be fitted.** Tighten the bolts to 7.0 kgm (43 ft.lb.). Again, turn the drive shaft flange to reach the bolt heads, as shown in Fig. 5.9.
- Push the lever for the handbrake cable operation towards the left and insert the fulcrum pin. The lever must be in the position shown in Fig. 5.10.

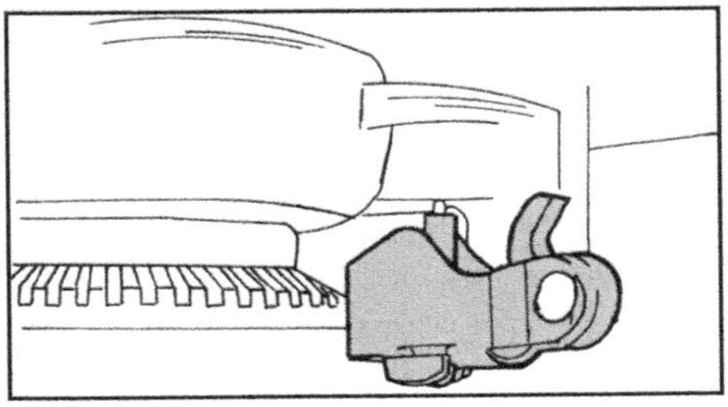

Fig. 5.10 – After connecting the handbrake operating lever, push it towards the left until it is in the position shown.

- Refit the handbrake brake shoes as described in the section "Brakes" and refit the brake drum/disc assembly. Fit the cross-headed screw and tighten it to 2.0 kgm (14.5 ft.lb.).

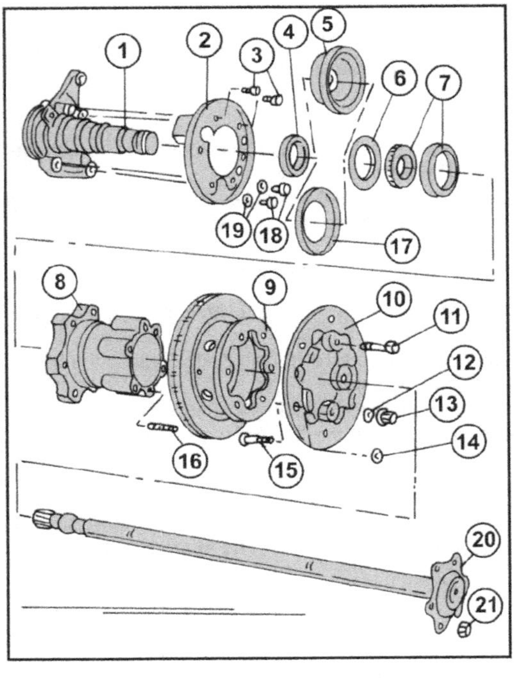

Fig. 5.11 – The component parts of a drive shaft as fitted to the LT46. The various items will be referred to during the replacement of the wheel bearing.

- Refit the wheel speed sensor (ABS only) with a new clamp bush. No adjustments are necessary, as the clearance to the toothed rotor on the drive shaft is self-adjusting. The bush must be fitted with silicone grease. Molycote 44 Light is used in the workshop.
- Finally, check the oil level in the rear axle and correct it as necessary, and adjust the handbrake and tighten the wheel bolts to 18 kgm (130 ft.lb.)

LT46 Model

The arrangement of the drive shaft is completely different, but no problems will be encountered during removal of a shaft. Fig. 5.11 shows the component parts of these shafts.

- Slacken the wheel nuts, place the rear end of the vehicle on chassis stands and remove the wheel.
- Remove the nuts (1) in Fig. 5.12 and withdraw the axle shaft out of the rear axle tube. Attention: oil will be released from the axle tube. Place a drip tray underneath the axle tube.

The installation is a reversal of the removal procedure. The sealing face of the rear axle shaft is coated with sealing compound (AMV 188 200 03). Tighten the nuts evenly across to 6.5 kgm (47 ft.lb.). Finally, check the oil level in the rear axle.

5.2.1. Axle Shaft Repairs

LT28 and LT35

As a slotted nut is used to secure the bearing, it will be possible to replace the bearing, but as a special wrench is required to obtain the high torque setting for the slotted nut (tool number 460 589 01 07 00) and a special puller to withdraw the bearing off the shaft, we advise you to have the bearing replacement carried out in a workshop. The axle shaft will be ready for installation as described above.

LT46

Various special tools are also required to overhaul the drive shafts, but it is considered that the rear wheel bearing can be replaced without them. Section 5.3 describes the replacement of the wheel bearings and should be referred to.

5.3. Rear Wheel Hubs and Rear Wheel Bearings

As mentioned above, remove the drive shaft(s) of an LT28 or LT35 and take them to a dealer to have the wheel bearings replaced. The wheel bearings of the LT46 are located inside the wheel hub. Before removal, read the following instructions, as some special tools or alternatives are required for certain operations. Fig. 5.11 shows an exploded view of a drive shaft. The items mentioned in the text refer to this illustration.

- The rear axle tube (1) is fitted with as splash shield (2), secured by the bolts (3). The bolts are tightened to 2.5 kgm (18 ft.lb.).
- A shrink ring (4) is fitted to the axle stump. A conventional puller can be used to remove it.
- If ABS/EDS is fitted, remove the rotor (5) for the wheel speed sensor from the inside of the hub assembly, also using a two-arm puller. Coat the face of the rotor with sealing compound AMV 188 200 03. Use a plastic mallet and drive the rotor over the hub to its stop.
- Coat the lips of sealing ring (6) with M.P. grease. The seal is pressed into the wheel hub, using a suitable mandrel.
- The inner wheel bearing (7) consists of the bearing cage and the outer race. The outer ring is driven into the wheel hub with a copper hammer. The bearing cage is coated with M.P. grease before installation.
- The removal and installation of the wheel hub (8) is described next. We must, however, point out that a special tool is required to tighten the grooved nut during the adjustment of the wheel bearings.
- The removal of the brake disc (9) and the flange (10) has already been described earlier in this section. The bolts (11) are tightened to18 kgm (130 ft.lb.).
- The flat side of the spring washer (12) is fitted against the wheel nut.
- The wheel nuts (13) are tightened to 18 kgm (130 ft.lb.).
- A ball ring (14) is fitted over the wheel stud, with the flat side against the flange.
- The wheel studs (15) and the studs in the wheel hub (16) can be replaced, if the thread is damaged. Wheel studs are driven out and new ones driven in.
- A deflector (17) is fitted to models with ABS/EDS. Coat the face of the deflector with sealing compound AMV 188 200 03. Use a plastic mallet to drive it over the wheel hub.
- The bolts (18) and the washers (19) are used to secure the deflector (2).
- The rear axle shaft (20) and is secured with nuts (21), already shown in Fig. 5.12.

Wheel Hub – Removal and Installation – LT46

The installation of the wheel hub can be seen in Fig. 5.11. A wheel hub can be removed as follows.

- Slacken the wheel nuts, place the rear end of the vehicle on chassis stands and remove the wheel.
- Remove the brake caliper and tie it to the rear axle, using a piece of wire or string.
- Remove the rear axle drive shaft as described for the LT46,
- Locate the tab of the lock plate in the outer grooved nut and unlock it to free the nut.
- Slacken the outer grooved nut with the special tool already mentioned. This is a peg spanner and is used, as shown in Fig. 5.13. Remove the lock plate after unscrewing the grooved nut.
- Remove the inner grooved nut with the same spanner. Remove the thrust washer and the outer taper roller bearing from the axle tube and withdraw the wheel hub.

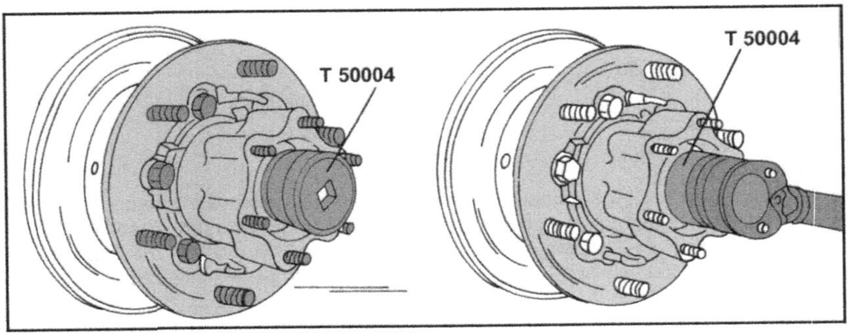

Fig. 5.13 – Removal of the grooved wheel bearing nuts.

The installation is a reversal of the removal procedure. Points to be observed have already been given during the explanation of Fig. 5.11. The wheel bearings must, however, be adjusted as described below, using the peg spanner shown in Fig. 5.13 and a dial gauge that can be attached to the wheel hub. During the following description, we assumed that the inner grooved nut had been tightened finger-tight.

Adjusting the Wheel Bearing Clearance

- Tighten the inner wheel bearing nut as shown in Fig. 5.13 to a torque reading of 30.0 kgm (216 ft.lb.). Rotate the wheel hub continuously during the tightening operation for the bearing to settle.
- Slacken the inner wheel bearing nut by half a turn. Again, rotate the wheel hub.
- Remove the torque wrench from the peg spanner and grip the peg spanner as shown in the R.H. view of Fig. 5.13 and tighten the nut tightly against the thrust washer. This will remove the play in the wheel bearings.
- Fit the outer wheel bearing nut and tighten it to 25 kgm (180 ft.lb.), again using the peg spanner. To allow the lock plate tab to enter a groove of the nut, it is allowed to tighten the nut a little further (do not lock the nut at this stage).

Fig. 5.14 – Check the wheel bearing clearance (play) – LT46.

- Fit a dial gauge in the manner shown in Fig. 5.14 and set the dial to "0". The stylus must be pre-loaded approx. 2 mm.
- Move the wheel hub in and out (make sure that the vehicle is securely supported) and read off the dial gauge. The indication must be between 0.02 and 0.04 mm. Otherwise, tighten the inner nut once more and re-check. Make sure that the outer nut is tightened to 25 kgm, and otherwise, wrong values will be obtained.
- After adjustment, knock the tab of the lockplate into the aligned groove.

5.4. Rear Axle Oil Level Check and Oil Change

An oil level check plug is fitted into the rear cover of the axle drive. A 14-mm Allen head key is required to remove the plug. The surrounding area of the plug must be cleaned before the plug is unscrewed. The oil level in the rear axle is correct when the oil can be seen on the lower edge of the filler hole. If necessary, top up with SAE 90 or 85W90 hypoid oil. The plug is tightened to 8.0–10.0 kgm (57.5–72 ft.lb.).

To change the rear axle oil, proceed as follows.

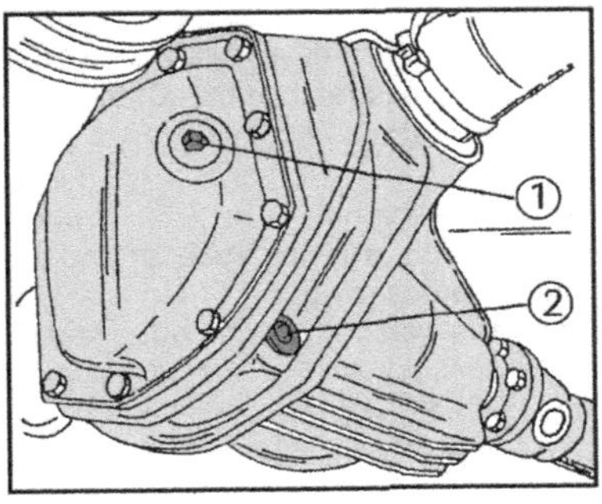

- Drive the vehicle for a few miles to warm up the oil in the axle.
- Place the rear end of the vehicle on chassis stands and unscrew the drain plug at the position shown in Fig. 5.15, using the Allen key mentioned above. Again, the area around the plug must be thoroughly cleaned. Take care if the oil is hot, as it will shoot out immediately.

Fig. 5.15 – The location of the oil level/filler plug (1) and the oil drain plug (2).

- Clean the cover of the axle drive and unscrew the level/filler plug (1) in Fig. 5.15.
- Clean the drain plug and fit it (maybe with a new sealing ring) into the axle. Tighten it to 8.0–10.0 kgm (47.5–72 ft.lb.).

- Fill the rear axle with the specified oil and the quantity given in Section 5.0, depending on the axle type. The oil must be level with the lower edge of the filler bore.
- Clean the filler plug and fit and tighten it to the torque given above.

5.5. Propeller Shaft

The removal and installation of the propeller shaft is straightforward. The shaft is attached at each end to flanges on the gearbox and the rear axle respectively. Depending on the model, a two- or three-part shaft can be fitted. The length of the shaft(s) is also different for the various models. The reason for this is the length of the vehicle or the type of transmission, i.e., manual transmission or automatic transmission. Fig. 5.16 shows the details of a three-part propeller shaft. A two-part shaft only has one centre (support) bearing, and otherwise, the same attachment is used.

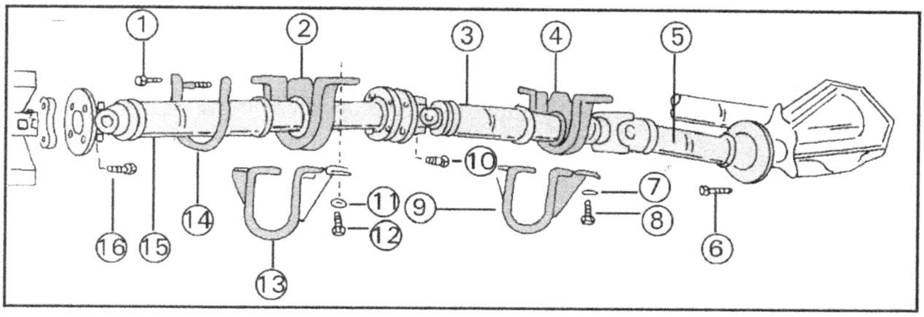

Fig. 5.16 – Exploded view of a three-part propeller shaft.

1 Bolt, 8.5–9-0 kgm
2 Centre bearing, front shaft
3 Centre propeller shaft
4 Centre bearing, rear shaft
5 Rear propeller shaft
6 Bolt, 7.0 kgm
7 Plain washer
8 Bolt, 8.5–9.0 kgm
9 Centre bearing clamp
10 Bolt, 7.0 kgm
11 Plain washer
12 Bolt, 8.5–9.0 kgm
13 Centre bearing bracket
14 Safety clamp, shaft
15 Front propeller shaft
16 Bolt, 7.0 kgm

Propeller shafts are normally not overhauled, but your MB dealer may be able to advise you if the universal joint bearings can be replaced.

5.6. Rear Axle – Tightening Torques

Shock absorbers, upper attachment .... 12.0 kgm (87 ft.lb.)
Shock absorbers, lower attachment, to axle .... 7.5 kgm (54 ft.lb.)
Stabiliser mounting bearings to rear axle .... 2.0–2.5 kgm (14.5–18 ft.lb.)
Stabiliser connecting rod to stabiliser .... 7.5 kgm (54 ft.lb.)

Spring "U" bolt nuts:
- 1st stage .... 5.0 kgm (36 ft.lb.)
- 2nd stage .... 18.0 kgm (130 ft.lb.) + 90°

Nuts for spring shackle plates .... 11.0 kgm (79.5 ft.lb.)
Nut for front spring bolt .... 8.0 kgm (58 ft.lb.) + 90°
Wheel bolts .... 18.0 kgm (130 ft.lb.)
Axle drive shaft cover to axle tube .... 6.0–7.0 kgm (43.5–50.5 ft.lb.)
Slotted nut for axle shaft (LT28/35) .... 50 kgm (360 ft.lb.)
Outer nut for wheel bearing adjustment (LT46) .... 25.0 kgm (180 ft.lb.)
Cover to rear axle housing .... 4.0–4.5 kgm (29.0–32.5 ft.lb.)
Rear brake caliper to axle .... See under "Brakes"
Brake pipe connections .... 1.2–1.6 kgm (9–11.5 ft.lb.)
Oil filler and drain plug .... 8.0–10.0 kgm (58–72 ft.lb.)
Propeller shaft to rear axle flange .... 7.0 kgm (50.5 ft.lb.)
Handbrake cable to rear axle .... 2.0–2.5 kgm (14.5–18 ft.lb.)

6. Front and Rear Suspension

Front axle and rear axle are fitted with leaf springs. Shock absorbers on the rear axle or damper struts on the front axle complete the rear suspension. Both axles have in most cases a stabiliser bar. The attachment of the leaf springs is, however, not the same on the front and the rear axle. The damper struts of the front axle have already been covered in Section 4.

6.1. Replacing a Front Leaf Spring

Fig. 6.1 shows the attachment of the spring. The illustration is to be used in conjunction with the view of the front suspension in Fig. 4.2. Before commencing the operations, you will need a piece of wood, having the length of the spring, as you will need this for the installation of the spring. You will also need a ball joint puller for the separation of the track rod ball joint, a mobile jack and a hydraulic jack. Proceed as follows during the removal of the front spring.

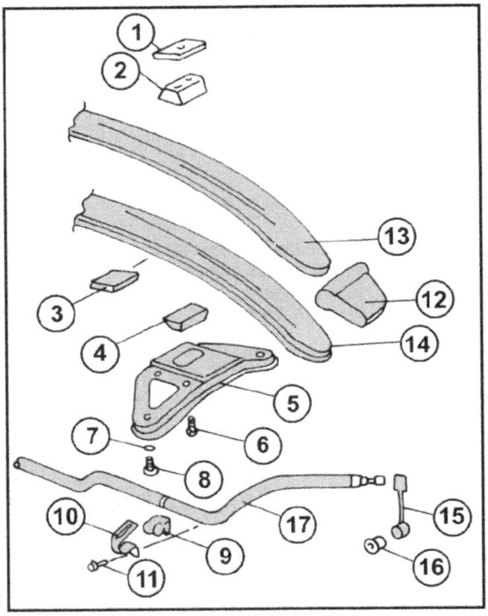

Fig. 6.1 – The component parts and attachment of a front spring.

1 Mounting bracket
2 Upper spring bearing
3 Packing, two-leaf spring
4 Lower spring bearing
5 Spring mounting plate
6 Bolt, 7.0 kgm
7 Bush (sleeve)
8 Bolt, 15 kgm
9 Stabiliser rubber mounting
10 Mounting clamp
11 Bolt, 3.0 kgm
12 End rubber cap
13 One-leaf front spring
14 Two-leaf front spring
15 Connecting linkage
16 Rubber bush in linkage
17 Stabiliser bar

- Jack up the front of the vehicle, place chassis stands underneath the body and unscrew the mounting brackets shown in Fig. 6.2 on both sides. Lower the vehicle back onto its wheel. *Failing to unscrew the bolts will make it impossible to remove them later on.*
- Remove the wheel covers, slacken the wheel bolts and jack up the vehicle once more.

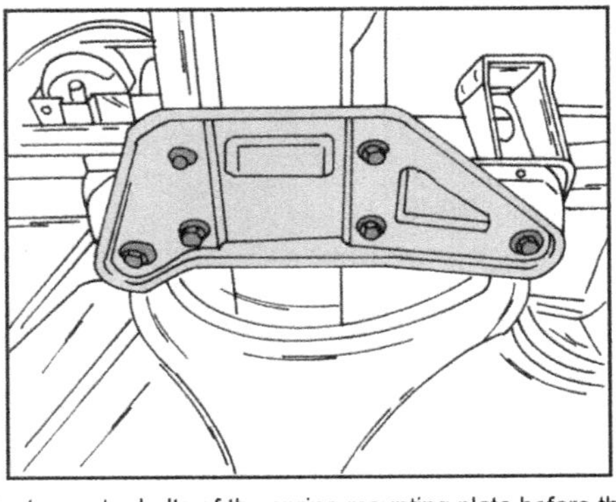

Fig. 6.2 – Slacken and remove the two outer bolts of the spring mounting plate before the front of the vehicle is jacked up for the removal of the spring.

- Place stands underneath the chassis in accordance with Fig. 5.2. Remove the wheel.
- Remove the steering knuckle on the R.H. side from the damper strut (remove the four bolts) as described earlier and pull the steering knuckle towards the outside (Fig. 6.3) to separate the connection.

Fig. 6.3 –Pull the steering knuckle towards the outside to remove the spring from the R.H. side.

- Remove the brake caliper and tie it to the front suspension with a piece of wire. The brake hose is not disconnected.
- On the L.H. and R.H. sides, remove the spring stop plates by removing the bolts shown by the arrows in Fig. 6.4 and take out the plates.

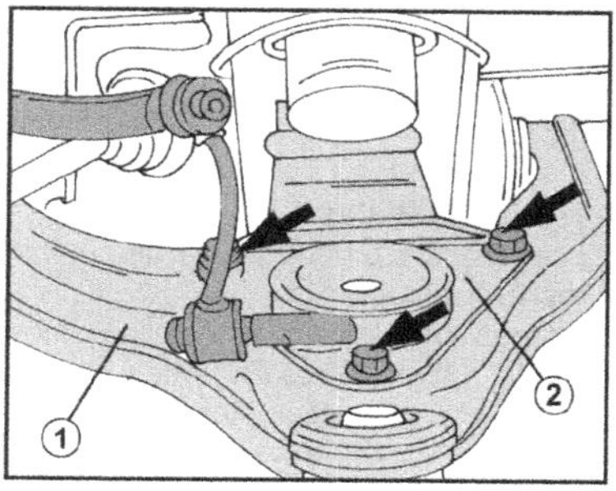

Fig. 6.4 – The bolts shown by the arrows secure the spring stop plates (2) to the suspension arms (1). The illustration shows a vehicle with stabiliser bar.

- If a stabiliser bar is fitted, remove the spring stop plates in the manner described, but leave the connecting linkage in position. Push the spring stop plate together with the linkage to one side into the wheel housing and tie it up with a piece of wire.
- Place a mobile or hydraulic jack underneath the lower suspension arm and lift the arm approx. 4 in., to compress the damper strut. This will release the load.
- *Slacken* the bolt securing the rear and front end of the suspension arm to release the tension in the rubber bearings.
- Place a jack underneath the centre of the front spring (hydraulic jack if possible).
- Remove the suspension arm on the L.H. side completely as already described earlier. Slowly lower the jack. A helper should hold the spring on one side. Remove the spring as soon as it is free.

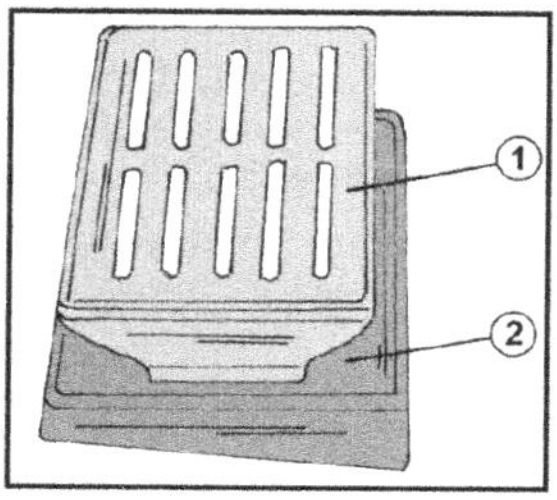

The installation of the front spring is carried out as follows. The installation of the spring mounting plates is described under a separate heading following the description below.

Fig. 6.5 – After gluing the upper spring rubber stops (1) to the retainers (2), you will have the parts assembled as shown.

- Mark the centre of the front spring and the front axle with a felt tip pen.
- Coat the upper spring rubber stops (2) in Fig. 6.1 and the bearing bracket with the special glue VW No. D 002 100 and join them together, as shown in Fig. 6.5. Insert the parts into the openings of the axle carrier. Note the correct direction of installation and the colour markings. The black stop is fitted to the driver's side, the red stop to the passenger's side. The two bearing brackets are, however, identical. Coat the bottom of the upper spring bearing with some liquid soap.

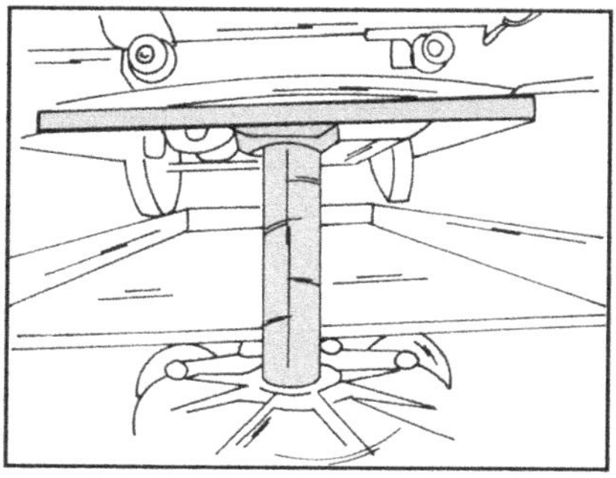

- Coat the two rubber caps (12) with liquid soap and push them over the ends of the spring as far as they will go. Fit the rubbers in the direction shown in the case of rubber cap (12).
- Slide the two ends of the spring with the rubber caps into the locating openings of the suspension arms. You now must make sure that the marked centres of spring and front axle are aligned. To facilitate the fitting of the rubber caps, coat them with vaseline (petroleum jelly). Do not use standard grease, as this will attack the rubber.

Fig. 6.6 – Lift the front spring with a wooden plank as shown until the upper rubber stops are fully compressed.

- Place the wooden plank, mentioned above on the jack. The plank must be underneath the two spring ends, as shown in Fig. 6.6. Operate the jack until the two upper rubber stops on the spring are fully compressed.

- Place a mobile jack underneath the suspension arm, lift it in position and insert the front end of the arm into the mounting bracket. Drive in the fulcrum pin (from the inside) and fit the washer and the nut. Tighten the nut, without tightening it fully. Lower the jack.
- Pull the suspension arm towards the rear and insert the rear end of the arm into the mounting bracket on the inside. Align the holes in arm and mounting bracket, drive in the fulcrum pin and fit the washer and the nut to the other side. Tighten the nut, without tightening to the specified torque setting.
- Fit the upper suspension carrier plate (three bolts) and tighten to the torque values given in Fig. 6.1.
- Connect the damper strut with the steering knuckle and tighten the four bolts in Fig. 4.13 to the values given in the illustration).
- Refit the brake caliper.
- Refit the wheels and lower the vehicle to the ground. Tighten the wheel bolts (18.0 kgm).

6.1.1. Spring Mounting Plates – Installation

The following operations are normally carried out on a vehicle lift or over a pit. Otherwise, you will find that the operating room is restricted.

- Fit a spring mounting plate (5) in Fig. 6.1 with the lower rubber stop (4) and attach it at the position shown in Fig. 6.7. The sleeve (7) is fitted into the locating bore. Note that the flat rubber stop is fitted to the driver's side, and the higher stop is fitted to the passenger's side.

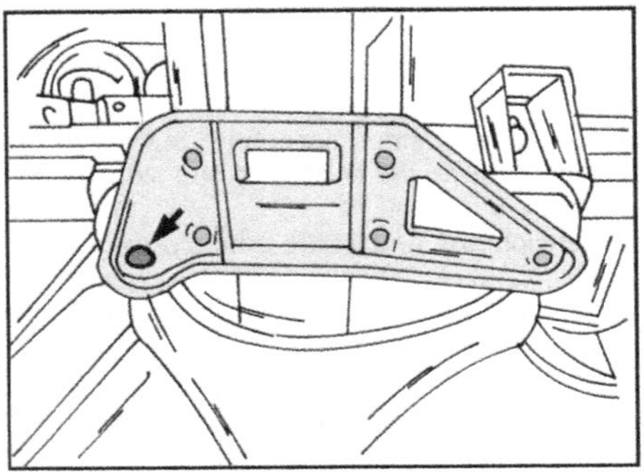

Fig. 6.7 – Fit the spring mounting plate together with the rubber stop as shown and tighten the bolt shown by the arrow with the inserted sleeve finger-tight.

- Tighten the four centre bolts in Fig. 6.8 hand-tight, without applying the applicable torque. The bolt in Fig. 6.7 is now tightened to 15 kgm (108 ft.lb.).
- Insert a drift or rod of suitable diameter as shown in Fig. 6.8 to align the bores in the mounting plate with the threaded bores. The sleeve must not be inserted. The four bolts in the centre of Fig. 6.8 can now be tightened to 7.0 kgm (50.5 ft.lb.). After tightening, remove the drift, insert the bolt with the sleeve and tighten the bolt to 15 kgm (108 ft.lb.).
- Move the vehicle approx. 3–4 feet forwards and then backwards and "swing" the front end a few times up and down. The mountings of the suspension arms on the inside, shown in Fig. 6.9, are now tightened in this position to 14 kgm (101 ft.lb.). This will pre-tension the rubber bushes/bearings in the suspension arms to their correct value. As already mentioned, the working space is limited.

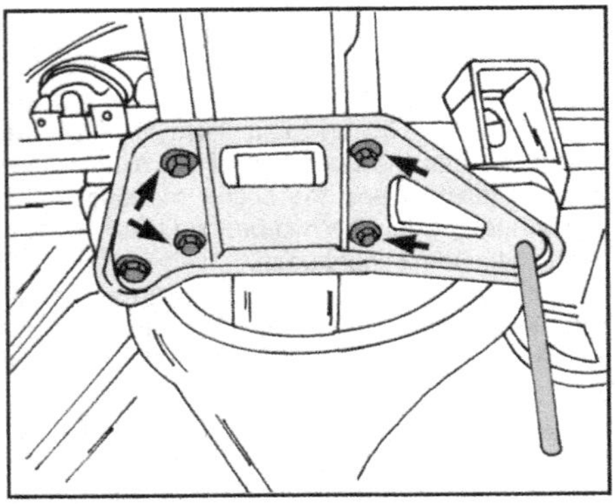

Fig. 6.8 – Tighten the four bolts in the centre (arrows) after the spring mounting plate has been aligned with an inserted drift (R.H. side).

- After completing installation, check once more that the wheel bolts have been tightened. Also check the pressure of the brakes before you drive off.

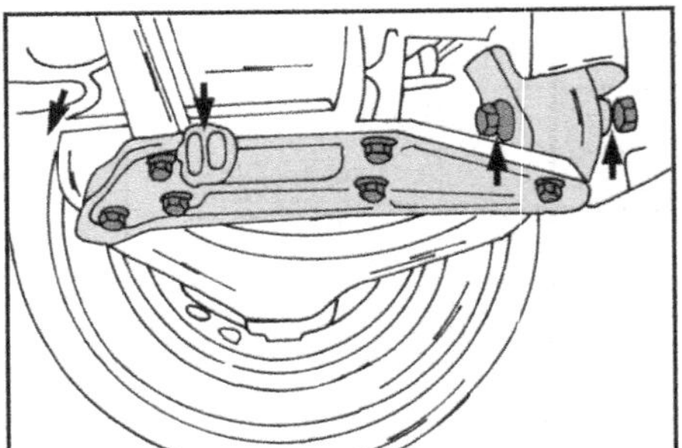

Fig. 6.9 – Suspension arm bolts.

Notes when fitting front springs/bearings: As one-leaf or two-leaf springs are used, there is a difference in the shape of the upper and low spring rubber stops. If one of the stops is to be replaced, refer to Fig. 6.10 and note the differences.

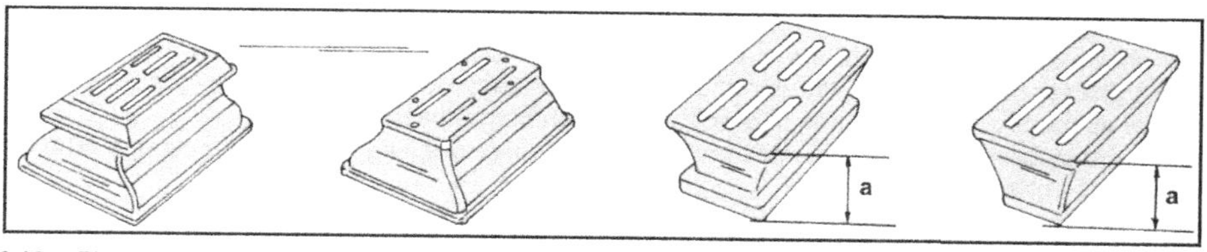

Fig. 6.10 – The various upper and lower spring rubber stops. Compare the shape in accordance with the explanation below.

From left to right: Upper stop for one-leaf spring, upper stop for two-leaf spring, lower stop for one-leaf spring (a = 36 mm, L.H. side, and 41 mm, R.H. side) and lower stop for two-leaf spring (a = 29 mm, L.H. side, and 34 mm, R.H. side).

The fitted position of the spring stops can be seen in Fig. 6.11, i.e., items 1–5 are fitted in the manner shown between the front axle beam "A", the leaf spring "C" and the spring mounting plate "B".

Take the old part to your parts supplier when parts are replaced.

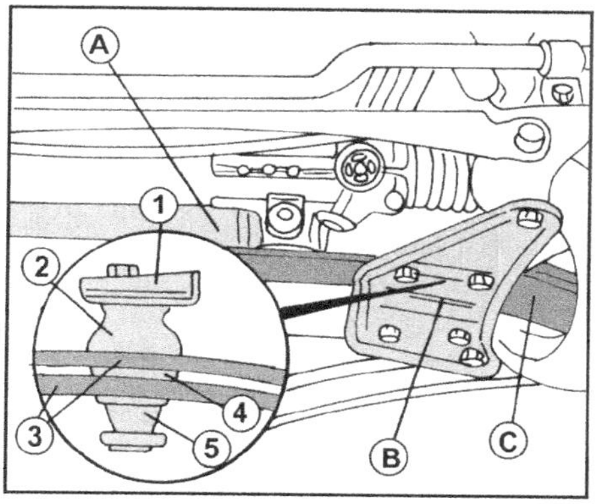

Fig. 6.11 – Fitted position of spring rubber stops (A, B and C are explained in the text).

1 Stop retainer
2 Upper rubber stop
3 Two-leaf spring
4 Packing (only two-leaf spring)
5 Lower rubber stop

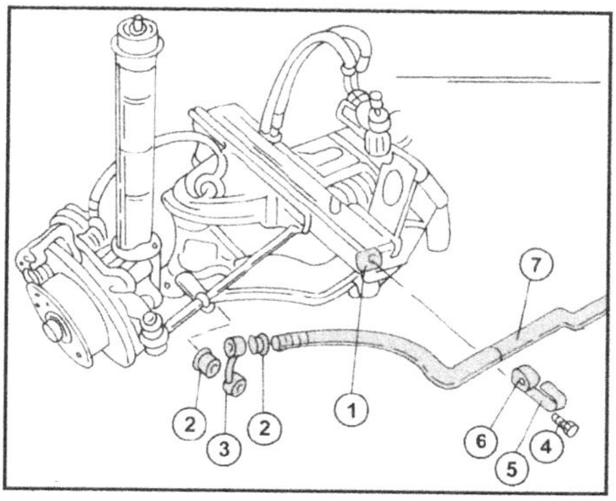

Fig. 6.12 – View of the front suspension on one side (without leaf spring). Rubber bushes (2) and (6) must be coated with oil before installation.

1 Nut, 2.5–2.8 kgm
2 Rubber/metal bearing for connecting linkage
3 Connecting linkage
4 Bolt
5 Mounting clamp
6 Rubber/metal bearing
7 Stabiliser bar

6.2. Front Stabiliser Bar

Fig. 6.12 shows the attachment of the stabiliser bar on one side. The same parts are used on the other side. The stabiliser mountings are fitted to the sidemembers and connected with linkages on bearing pins at the upper suspension carrier plates. A stabiliser bar can be removed as follows.

- Slacken the stabiliser mounting clamps on the sidemembers and slightly open the clamps. The clamps remain on the chassis.
- Push the rubber bushes out of the mounting clamps, using a suitable drift.
- Again, using a drift, knock the linkages on both sides off the bearing pins on the upper suspension carrier plate. The attachment is shown in Fig. 6.13. Part of the mounting bush remains on the pin and must be removed with a pair of pliers, by turning it to and fro.
- Clamp the linkage into a vice and knock the linkages off the stabiliser ends. Note how the linkages are fitted. Again, the inner part of the bush will remain on the linkage and must be removed with a pair of pliers, as explained above.

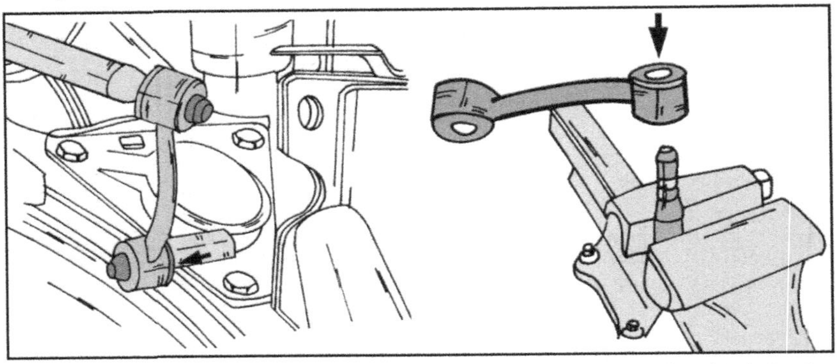

Fig. 6.13 – Drive the stabiliser linkage with a drift off the bearing pin on the upper mounting plate in the direction of the arrow (left). The R.H. view shows the fitting of the stabiliser linkage over the ends of the stabiliser bar.

- If the rubber bushes for the linkages must be replaced, place the end(s) over a vice and drive out the sleeves of the rubber bushes. Coat new rubber bushes with soap and press them in position.
- Clamp the ends of the stabiliser bar into a vice and fit the linkages as shown in Fig. 6.13 in the R.H. view. Coat the inside of the bush with soap to facilitate the fitting. Once in position, use a hammer and drive the linkage fully home.

The installation of the stabiliser bar is carried out as follows. The connecting rods are "curved" and must be fitted so that the "curved" side is facing towards the inside. This applies to both sides.

- Fit the stabiliser bar to the chassis and insert it into the mounting clamps. Tighten the bolts hand-tight.
- Coat the inside of the bush in the lower ends with soap and push them over the bearing pins on the upper carrier plate. Use a hammer and drive the linkages fully home.
- Coat the rubber bushes for the mounting clamps with soap and push them over the stabiliser bar. Push the bushes underneath the mounting clamps until both sides protrude by the same amount.
- Align the stabiliser bar exactly in the centre position and tighten the mounting clamp bolts to 2.5–2.8 kgm (18–20 ft.lb.).

6.3. Rear Leaf Springs

Details for the removal and installation of a rear spring are shown in Fig. 6.14 for the LT28/LT35 models. The arrangement on the LT46 is similar, with the difference that the spring shackle (11) is facing with its ends towards the bottom. As can be seen, the spring is attached at the front end by a single spring eye bolt and at the rear with a spring shackle arrangement. A spring can be replaced as follows.

- Remove the wheel covers, slacken the wheel bolts, jack up the rear end of the vehicle and remove the wheels.
- Place a jack underneath the centre of the rear axle and jack up the axle until the axle tube is just under tension.
- Remove the "U" bolt nuts and carefully drive the "U" bolts out of their locations. Remove the spring plate above the spring.
- Slowly lower the axle until the spring is suspended by its mountings at the front and rear.
- Remove the nut securing the front spring bolt and drive out the bolt. A helper should hold the spring in position until the bolt is removed.

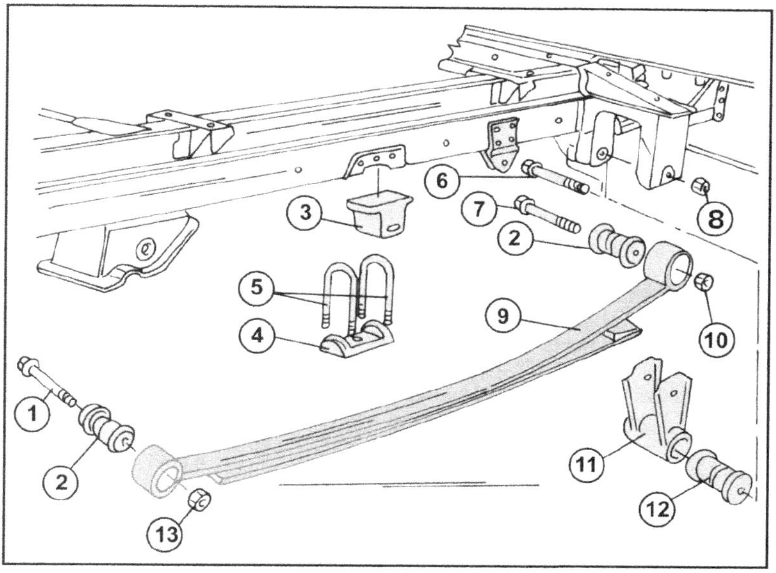

Fig. 6.14 – Attachment of a rear spring.

1 Bolt, always replace, note different versions
2 Rubber/metal bush, front and rear different
3 Rebound rubber
4 Upper spring plate
5 Spring "U" bolt
6 Bolt, replace
7 Bolt, replace
8 Self-locking nut, replace, 8 kgm + 90°
9 Leaf spring
10 Nut, replace, 11 kgm
11 Spring shackle
12 Rubber/metal bush, can be replaced
13 Self-locking nut, always replace, 8.0 kgm + 90°

- Remove the spring in the same manner from the spring shackle mounting. Again, a helper must hold the spring – this time it will drop on the floor – **prevent accidents.**
- If required, remove the spring shackle from the spring. The spring must be fitted the correct way round. The hole for the spring pin is larger in diameter and this side of the spring is fitted nearer the front.

If you find a broken spring leaf, enquire in an MB workshop if it can be replaced. Otherwise, a new spring must be fitted. The rubber bushes in the spring eye and the shackle mounting can be replaced by pressing them out and pressing new ones in. Coat the outside of the bushes with liquid soap to facilitate the installation.

The installation is a reversal of the removal procedure, noting the following points. Refer to Fig.6.14 for various information regarding replacement of bolts/nuts and other points to be observed:

- Fit the shackle mounting to the spring and push it at an angle towards the outside until the dimension between the centre of the spring eye bore at the front and the centre of the spring shackle bore is exactly 1470 mm. Ask a helper to hold the spring shackle mounting in this position and tighten the bolt and nut to 11.0 kgm (80 ft.lb.).
- Fit the spring to the sidemember. The front spring bolt and the spring shackle-to-sidemember bolt are tightened when the vehicle is with its wheels on the ground. The tightening torques are given in Fig. 6.14.

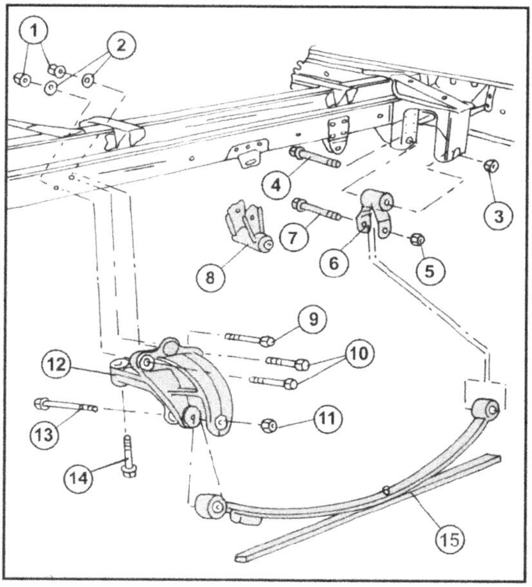

Fig. 6.15 – Arrangement of a rear spring in the case of a platform chassis or a vehicle with double cabin. Refer to the description below.

Notes for spring mounting on vehicles with platform chassis and double cabin:
The following description refers to Fig. 6.15. Differences are given in the following description.

- The self-locking nuts (1) must always be replaced once slackened. Washers (2) are under the nuts. Tightening torque is 18.5 kgm (133 ft.lb.).
- The self-locking nut (3) must be replaced if slackened. Is tightened to 9.0 kgm (75 ft.lb.) – LT28/LT35.
- The bolt (4) is not the same on all models. Always replace. Tighten to 18.0 kgm (130 ft.lb.) – LT46.
- The self-locking nut (5) must be replaced if slackened or removed. Tighten to 11.0 kgm (80 ft.lb.) – LT28/35 or 24.0 kgm (173 ft.lb.).
- The spring shackle plate (6) is only fitted to the LT46. The rubber bearing is at the upper end.
- The bolt (7) is not the same on all models. The spring shackle (8) is only fitted to LT28/LT35 models. The rubber bearing is at the lower end.
- The bolt (9) is only fitted to some models. Tightening torque is 18.5 kgm (133 ft.lb.).
- The two bolts (10) are M14 bolts. Identical on all models.
- The self-locking nut (11) is tightened to 9.0 kgm (65 ft.lb) + 90° (LT28/LT35) or 18.0 kgm (130 ft.lb.) + 90° (LT44).
- The rear spring bracket (12) is the same for all models.
- The bolt (13) and the bolt (14) are different, depending on the model. Bolt (14) is tightened to 18.5 kgm (133 ft.lb.).
- The rear spring (15) is attached to the bracket and the shackle mounting in accordance with the interrupted lines.

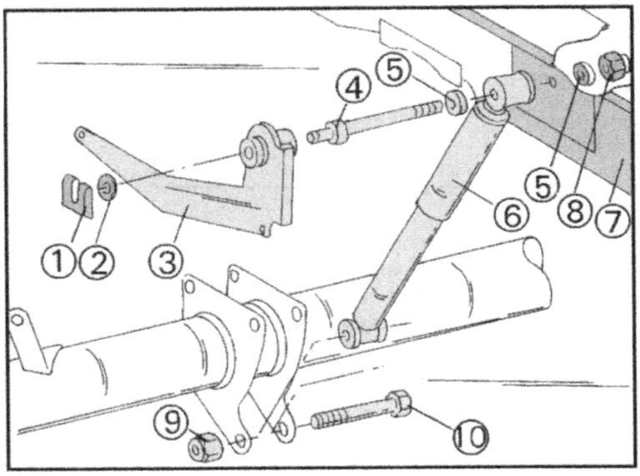

Fig. 6.16 – Attachment of a rear shock absorber.
1 Spring clip
2 Plain washer
3 Operating arm
4 Connecting bolt
5 Plain washer
6 Shock absorber
7 Sidemember
8 Nut, 12.0 kgm
9 Nut, 7.5 kgm + 90°
10 Bolt

6.4. Rear Shock Absorbers

Shock absorbers are secured at the upper end with a nut and at the lower end with bolts and nuts. The upper securing bolt carries on the other side the lever for the brake pressure regulator.

Fig. 6.16 shows details of the attachment. Also refer to Fig. 5.1. The shock absorbers of the LT46 models is secured with a bolt and nut at the upper end lower ends. A washer is inserted between the shock absorber and the nut or washer. The axle must be lifted until the shock absorber is slightly compressed when a shock absorber is removed. The following instructions refer to the LT28/35 models. In the case of the LT46 model, simply unscrew the nut and bolt at top and bottom, after the damper has been compressed.

- Remove the shock absorber at the lower end from the rear axle tube.
- Remove the spring clip (1) in Fig. 6.16 to free the operating lever (3) and take off the washer (2). Withdraw the lever from the mounting bolt (4).
- Counterhold the bolt (4) on the hexagon and slacken the nut (8) on the other side. Remove the nut and the washer and drive out the bolt.

To check a shock absorber, clamp it vertically into a vice and "pump" it up and down. The shock absorber should show even resistance throughout the travel. "Dead" travel requires the replacement of the damper unit.

The installation is a reversal of the removal procedure. The tightening torques are already given in Section 5.6. The nut/bolt of an LT46 model is tightened to 11.0 kgm (80 ft.lb.).

6.5. Rear Stabiliser Bar

Fig. 6.17 shows the mountings of the stabiliser bar on the rear axle. Linkages are used to connect the ends of the bar with the chassis. The vehicle must be resting on chassis stands to remove the stabiliser bar.

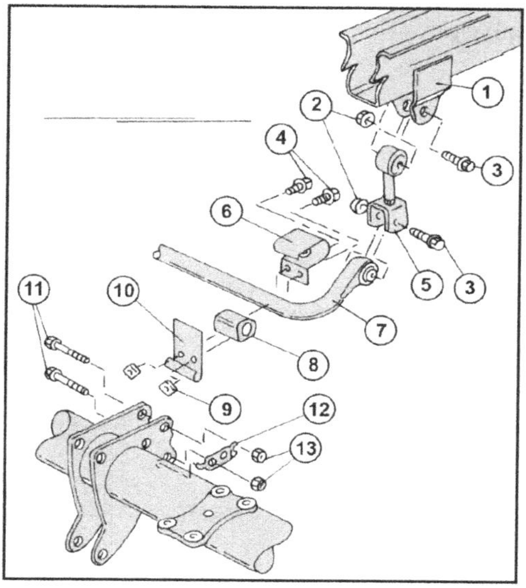

Fig. 6.12 – Stabiliser bar – Rear axle.
1 Welded nut
2 Nut, 7.5 kgm
3 Bolt, M12 thread
4 Bolt, M8, 2.5 kgm
5 Connecting linkage
6 Mounting clamp
7 Stabiliser bar
8 Rubber bearing
9 Square nut
10 Stabiliser bracket
11 Bolt, M8 thread
12 Bracket (cable)
13 Nut, 2.5 kgm

- On each side of the stabiliser bar, remove the linkages (5) at the upper end. Fig. 6.18 shows the connection. Remove nut (2) and bolt (3).
- Remove the nuts (9) securing the mounting clamps (6).

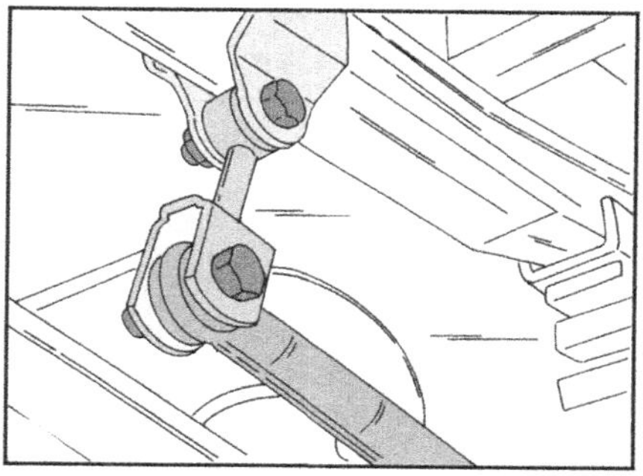

Fig. 6.17 – Fitted linkage between stabiliser bar and chassis on one side.

- Pull out the bolts (4) and remove the bracket (10) on the other side.
- Remove the stabiliser bar (7) together with the mounting clamps.
- Remove the nuts (13) securing the retaining bracket (10) and remove the bolts (11). Remove the bracket out of the mounting clamp.
- Remove the mounting clamp (6) from the rubber mounting (8) and slide both parts off the stabiliser bar.
- From the outside of the stabiliser, unscrew the linkages (5).
- If the rubber bushes are to be replaced, use a press to press out the old bushes and press in new bushes. Bushes in the linkages and the stabiliser bar can be replaced. Coat the new bushes with liquid soap before installation.

Refit the stabiliser bar as follows.

- Fit the linkages with bolts and nuts to the end of the stabiliser bar.
- Push the rubber bushes (8) over the stabiliser bar and slide the mounting clamp over the bushes. Talcum powder will facilitate this operation (coat both parts).
- Slide the retaining bracket (10) on one side into the mounting clamp and tighten the connection with the bolt.
- Fit the stabiliser bar with the linkages to the chassis.
- Align the stabiliser bar in the centre and tighten the bolts and nuts securing the linkages to 7.5 kgm (54 ft.lb.).
- Tighten the square nuts (9) to 2.5 kgm (18 ft.lb.). The flat side of the nuts must be placed against the bracket (10).

6.6. Tightening Torques – Suspension

A summary of the tightening torques for the front suspension are listed in Section 4.7 on page , and the values for the rear suspension are given in Section 5.6 on page 103.

Note the differences between LT28/35 models and the LT46, concerning mainly the rear shock absorbers and rear springs.

7. Steering

7.0. Technical Data

Type	Rack and pinion steering with servo assistance, fitted behind the engine. Progressive ratio.
Type of fluid used	As in automatic transmissions
Fluid used	VW G 052 162 steering fluid

7.1. Steering – Removal and Installation

The following precautions must be taken when work is carried out on the steering system.

- Before removing any connections, clean the immediate surrounding and the connection.
- All removed parts must be placed onto a clean workbench and covered up in a suitable manner if not refitted immediately.
- Never use fluffy rags to clean component parts. New parts must be unwrapped from their packing immediately before they are fitted. Only use original spare parts.
- Never re-use fluid that has been drained from the system.

Remove the steering as follows. It is assumed that the front end of the vehicle is resting on chassis stands:

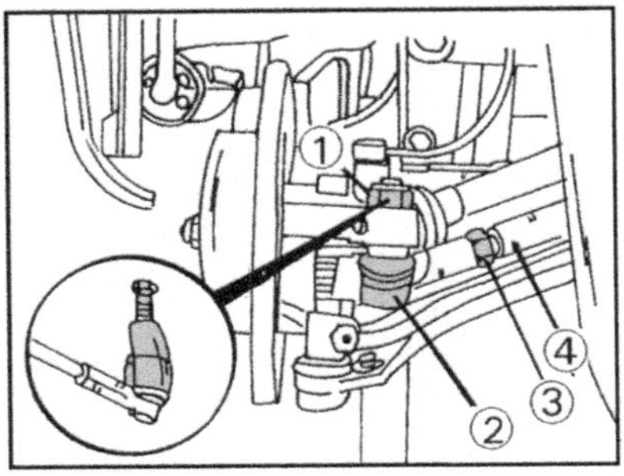

Fig. 7.1 – View of the track rod connection on the steering lever. The ball joint is separated (inset) after removal of the nut (1).

1 Nut
2 Track rod ball joint
3 Locknut
4 Track rod

- Remove the nut (1) in Fig. 7.1 and separate the track rod ball joint with a suitable puller from the lever on the steering knuckle. The illustration shows a suitable puller. Both track rod ball joints must be removed. Remove the ball joint end of the L.H. track rod (4) after slackening the locknut (3). Unscrew the ball joint end, at the same time counting the turns (and half turns). Write it onto a piece of paper. During installation, screw the track rod end by the same number of rotations onto the track rod. This will bring the toe-in to the approximate value.
- Turn the steering into the centre position, i.e., both front wheels must be in the straight-ahead position.
- Remove the fluid out of the fluid reservoir, using a syringe.
- Disconnect the steering hose and pipe. The hose is the pressure hose, the pipe is the return line. The fluid seals must always be replaced. Fluid can be collected in a suitable container. Tighten the high-pressure hose and the return pipe to 3.5 kgm (25 ft.lb.).
- Remove the clamp bolt securing the universal joint to the steering shaft (torx-head bolt) and slide the universal joint off the shaft.
- Unscrew the steering from the front axle and remove it. Tilt the steering towards the front and take it out towards the right.

The installation is a reversal of the removal procedure, but the following points should be noted.

- Before the steering is fitted, turn the steering pinion to one stop and from this position turn the pinion to the other lock, counting the numbers of turns. Now, turn the steering pinion the other way by half the number of turns. Check that the alignment marks near the entry of the steering pinion are opposite each other. This is the centre position.

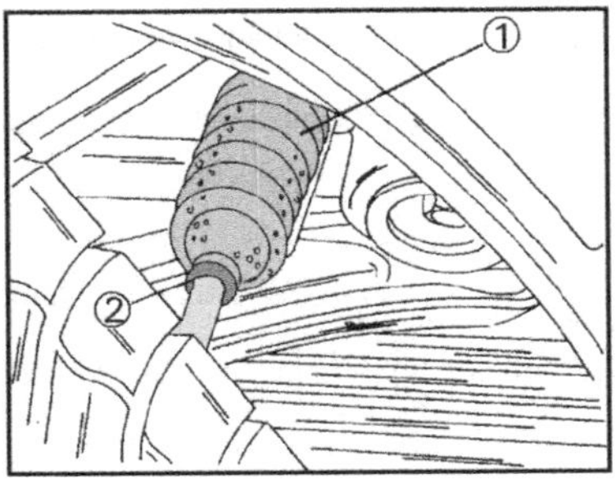

- Insert the steering from the left and tilt upwards. Set the steering into its mounting position and tighten the bolts evenly across to 5.0 kgm (36 43.5 ft.lb.).
- Slide the universal joint over the shaft, align the bore in the joint with the groove in the steering pinion and insert the bolt. Tighten the bolt to 2.5 kgm (18 ft.lb.), again using a torx-head insert.
- Fit the track rod ball joint end over the track rod (in the original position) and connect the ball joint stud to the lever on the steering knuckle. Tighten the nut to 13 kgm (94 ft.lb.).

Fig. 7.2 – Slacken the gaiter clamp (2) on the outside and remove the gaiter (1) from the steering box. The large clamp on the inside is already removed.

- Refit the two fluid lines to the steering (tightening torque given above). Fill the system with the recommended fluid and bleed it as described later on.
- Lower the vehicle to the ground and tighten the wheels.
- Check that the steering is in the centre position (wheels in the straight-ahead position and steering wheel in the correct position). If this is not the case, remove the universal joint once more and turn it by one tooth. This will reset the steering wheel by 6°.
- Finally, check the steering system for leaks and re-check the fluid level in the steering fluid reservoir.

7.2. Track rod Ball Joint Ends and Gaiters

The replacement of a ball joint follows the description during the removal and installation of the steering, given in this case for the L.H. track rod. Slacken the gaiter clamp on the inside and the small one on the outside, at the position shown in Fig. 7.2, and remove the gaiter. Gaiter clamps are tightened to 1–1.2 kgm (7–9 ft.lb.). Observe the tightening torques given during the installation of the steering, if applicable.

7.3. Operations on the Hydraulic System

7.3.0. FILLING THE HYDRAULIC SYSTEM

After the installation of the steering or after draining the system, it must be filled and bled in accordance with the instructions below.

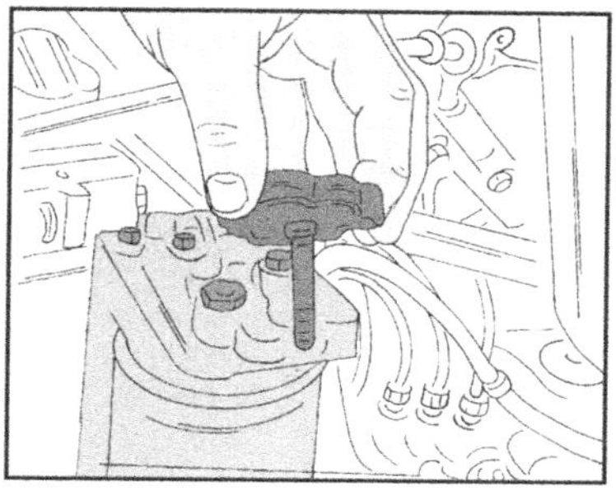

Fig. 7.3 – After unscrewing the filler cap, check the fluid level on the dipstick.

Fluid drained during removal operations must not be re-used. The fluid level is checked after unscrewing the filler cap as shown in Fig. 7.3 and checking if the fluid level is between the "Min" and "Max" mark on the dipstick. The fluid level will be higher or lower, depending on the fluid temperature. To make sure you obtained the correct reading, clean the dipstick and re-insert it, screwing the cap fully down.

Check the filler cap seal before refitting it. The fluid level in the reservoir should be checked during each engine oil change. Take care that no foreign matter can enter the fluid reservoir.

7.3.1. Bleeding the Hydraulic System

After the fluid level in the reservoir remains steady after the engine has been started, turn the steering wheel several times from one lock to the other to expel the air. Ask a helper to observe the fluid level during this operation. If necessary, correct the fluid level to keep it on the "Max" mark. No air bubbles must be visible when the steering wheel is turned as described above.

7.4. Steering Pump Drive Belt

The steering pump is driven by the same belt as the other units at the front of the engine. To replace the belt, refer to Section 1.4.10, commencing on page 50.

8. Brake System

8.1. Technical Data

Type of system	See description below
Front Brakes	
Caliper piston diameter, front brakes, except LT46:	60.00 mm
Caliper piston diameter, front brakes, LT46:	44.00 mm
Thickness of brake pads, incl. back plate:	
- Except LT46:	13.0 mm
- LT46:	13.5 mm
Min. thickness of linings:	2.0 mm
Min. thickness including metal plate:	
- Except LT46:	7.5 mm
- LT46:	7.0 mm
Brake disc diameter, all models:	276.0 mm
Brake disc thickness:	22.0 mm
Min. thickness of brake disc:	19.0 mm
Max. run-out of brake discs:	0.10 mm
Wear limit of brake discs, per side:	max. 0.05 mm
Rear Disc Brakes	
Caliper piston diameter:	
- LT28:	45.0 mm
- LT35:	52.0 mm
- LT46:	44.0 mm
Brake Disc Diameter:	
- LT28:	258.0 mm
- LT35:	272.0 mm
- LT46:	278.0 mm
Brake Disc Thickness:	
- LT28/LT35 (Teves/ATE):	12.0 mm
- LT35 (Bosch):	16.0 mm
- LT46:	22.0 mm
Min. Brake Disc Thickness:	
- LT28/LT35 (Teves/ATE):	11.0 mm
- LT35 (Bosch):	14.0 mm
- LT46:	19.0 mm
Thickness of brake pads, incl. metal plate:	
- LT28:	11.0 mm
- LT35:	12.0 mm
- LT46:	13.5 mm
Min. thickness of pad linings	2.0 mm
Handbrake	
Width of brake shoes, all models:	40.0 mm
Brake Master Cylinder	
Cylinder diameter:	23.81 mm
Diameter of pressure piston:	20.0 mm
Diameter of intermediate piston:	22.5 mm

8.2. Short Description

All models covered in this manual are fitted with a hydraulic dual-circuit brake with vacuum-operated brake servo unit. The brake servo unit is supplied with vacuum from a separate exhauster pump. Sliding brake calipers with one or two pistons, depending on the model, are used for the front wheels. LT28 and LT35 models have brake calipers with one piston, and the LT46 model has calipers with two pistons.

Different makes of disc brakes are fitted. Disc brakes, working on the same principle as the front brakes, are used on the rear wheels.

The brake calipers are not the same on all models, and again, you can consult the technical data section for details.

A simple brake pressure regulator is fitted to the rear brake circuit. Fig. 8.1 shows a layout of the system.

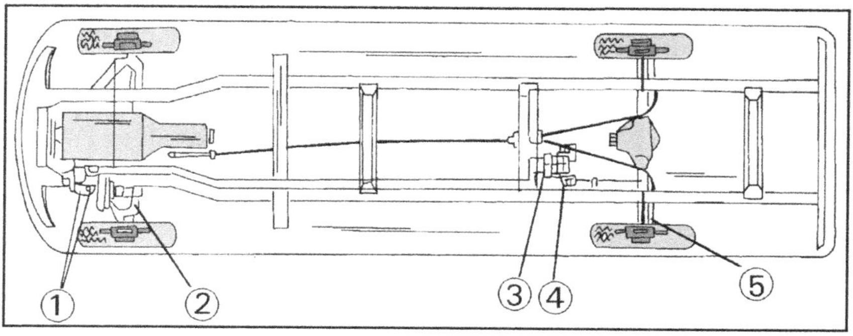

Fig. 8.1 – Brake circuit with brake pressure compensating valve.

1 Brake pressure compensating valve
2 Front brake caliper
3 Dual brake pressure regulator, adjustable
4 Adjustable brake pressure regulator
5 Rear brake caliper

- The brake system is diagonally split, i.e., one circuit serves one of the front brake calipers and the diagonally opposed rear brake caliper. The other circuit operates the other two brake assemblies accordingly.
- If one brake circuits fails, the brakes will operate as normal, but more brake force will be required.
- Depending on the model, ABS and an automatic differential lock can be fitted to your vehicle.

8.3. Front Disc Brakes

As already mentioned, sliding brake calipers with one or two pistons are fitted to the front wheels. The assemblies consist of a caliper mounting bracket bolted rigidly to the front axle steering knuckle and a separate caliper cylinder, irrespective if the cylinder contains one or two pistons. When the brake is operated, the piston pressed first with its brake pad against the brake disc. The caliper cylinder then slides on glide bolts and moves against the direction of the pressure until the other brake pad is pressed against the brake disc on the other side. Only the caliper cylinder must be removed to replace the brake pads, the mounting bracket remains on the steering knuckle.

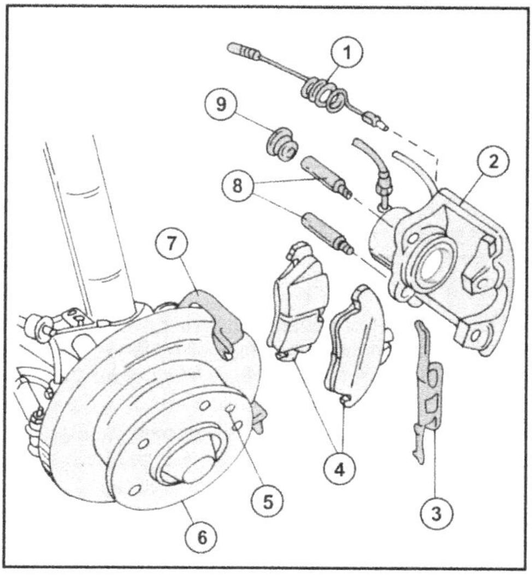

Fig. 8.2 – Exploded view of the front brake caliper assembly and pads.

1 Brake pad wear indicator
2 Brake caliper cylinder
3 Elastic spring
4 Brake pad set
5 Brake disc bolt, 2.0 kgm
6 Brake disc
7 Brake caliper bracket
8 Guide bolt, 2.5 kgm
9 Rubber dust cap

8.3.0. BRAKE PAD REPLACEMENT

Models LT28 and LT35

If the remaining brake pad material thickness is less than 2 mm (0.08 in), a warning light will light up in the dashboard, indicating that the brake pads must be replaced. The thickness can also be checked through the openings in the brake caliper after the wheels have been removed. Fig. 8.2 shows the parts belonging to the brake pads. Refer to the illustration during the following instructions, if you are not familiar with this type of arrangement. Proceed as follows.

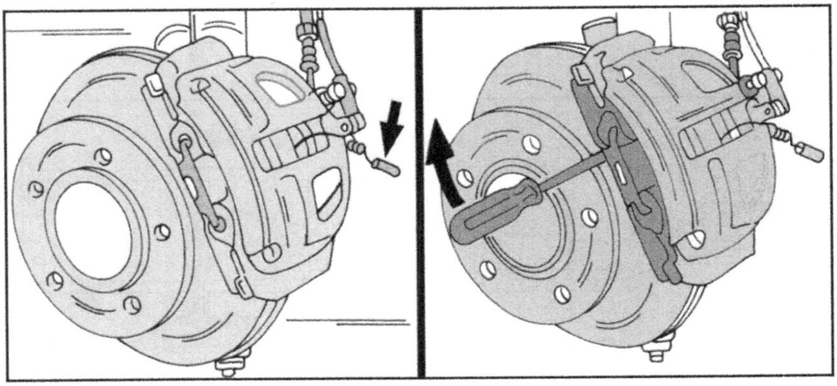

Fig. 8.3 – View of the fitted caliper on the L.H. side. The wear indicator cable is shown by the arrow. The R.H. view shows the removal of the elastic spring.

- Place the front end of the vehicle on chassis stands (underneath the front axle beam). The wheels must be removed. The caliper is then exposed, as shown in Fig. 8.3.
- Disconnect the lead for the brake pad wear indicator. To do this, remove the small cover and push off the connector, using a small screwdriver. Fig. 8.3 (left) shows where the connector is located.

Fig. 8.4 – After removing the small caps, unscrew the two bolts securing the caliper cylinder to the mounting bracket.

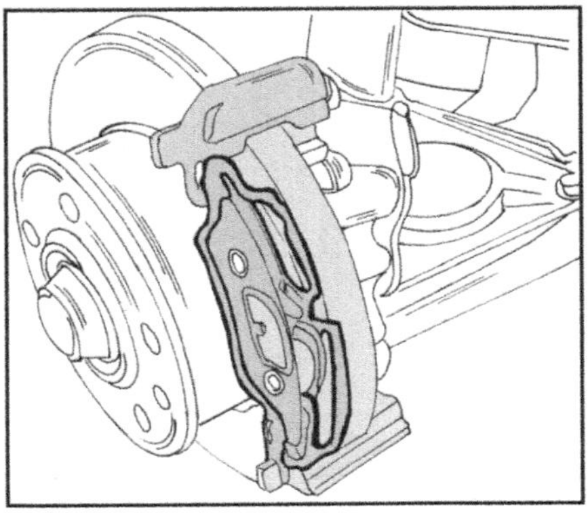

Fig. 8.5 – After removal of the brake caliper cylinder, the outer brake pad will remain on the side of the brake disc and must be removed. The inner pad is secured with a retaining spring to the brake caliper.

- Again, using the screwdriver, as shown in Fig. 8.3, remove the elastic spring (3) in Fig. 8.2 between brake caliper cylinder and caliper mounting bracket.
- At the rear of the brake caliper, remove the two caps fitted over the caliper housing securing bolts (screwdriver) and unscrew the Allen-headed bolts below it. A straight Allen key with a socket should be used. The caliper cylinder can be lifted off after both bolts have been removed. Only remove the two bolts shown by the arrows in Fig. 8.4. Use a piece of wire and tie the caliper to the front suspension.
- Remove the brake pads from the side of the brake caliper. The second pad remains on the mounting bracket and is removed separately.

If the brake pad material has worn to a thickness of 3.5 mm, fit a new set of pads. Although the min. permissible material thickness is considered to be 2.0 mm, you will find that the brake pad wear indicator warning light will light up when the thickness has been reduced to 3.5 mm. Never replace one brake pad only, even though if the other one appears to be in good condition.

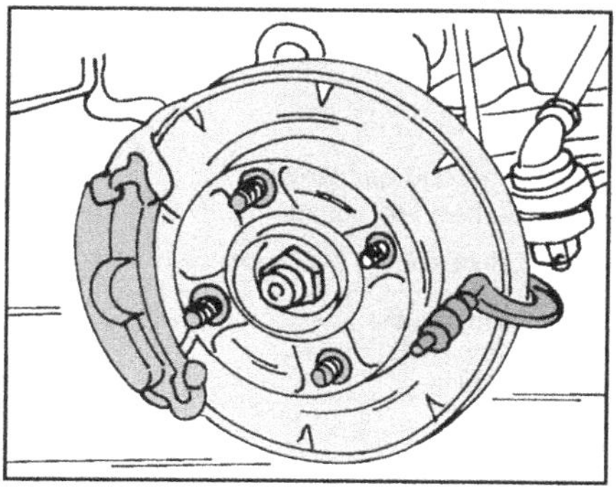

If brake pads have worn more than expected, there is a possibility that the piston(s) have seized inside the bores. In this case, it may be necessary to overhaul the brake calipers or have them overhauled. The same applies if the rubber dust boots on the cylinder housing are damaged.

Check the brake disc faces and clean them thoroughly before new pads are fitted. If possible, measure the brake disc thickness (as shown in Fig. 8.6) and compare the result with the dimensions given in Section 8.0. If the minimum thickness has been reached, replace the disc(s). Disc with uneven faces can perhaps be re-faced. Check with an MB workshop.

Fig. 8.6 – Brake discs can be measured with a micrometre as shown.

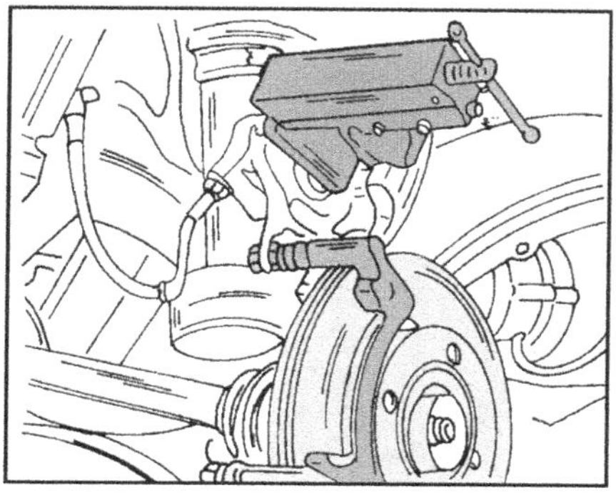

Refit the brake pads as follows.

- Clean the contact areas for the brake pads inside the caliper housing.
- Remove some brake fluid from the fluid reservoir, if it is fairly full, to prevent fluid overflow during the next operation. You will have to push the piston(s) into the caliper bore(s), to fit the new brake pads. Normally, a special pair of pliers is used, which is employed in the manner shown in Fig.8.7.

Fig. 8.7 – Piston reset pliers fitted to the caliper cylinder. Push the piston (or both) to the bottom of the bore. A piece of wood can also be used.

- Fit the brake pad with the retaining spring into the caliper cylinder as shown in Fig. 8.8 and push it into the direction of the arrow until it snaps into place. The outer brake pad is fitted into the opening of the caliper mounting bracket, into the position shown in Fig. 8.5.

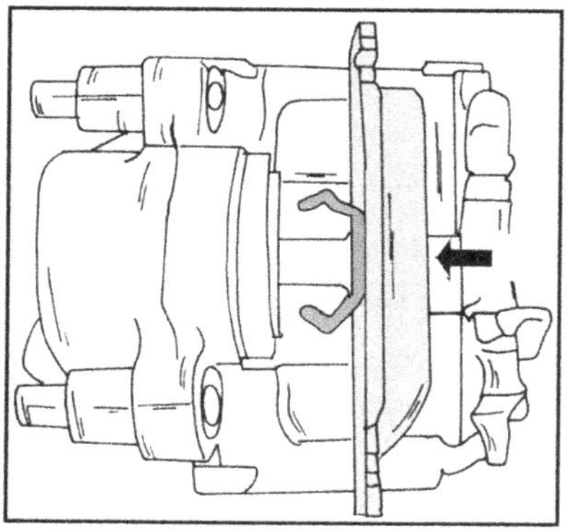

Fig. 8.8 – Fitting the inner brake pad with the retaining spring into the caliper cylinder.

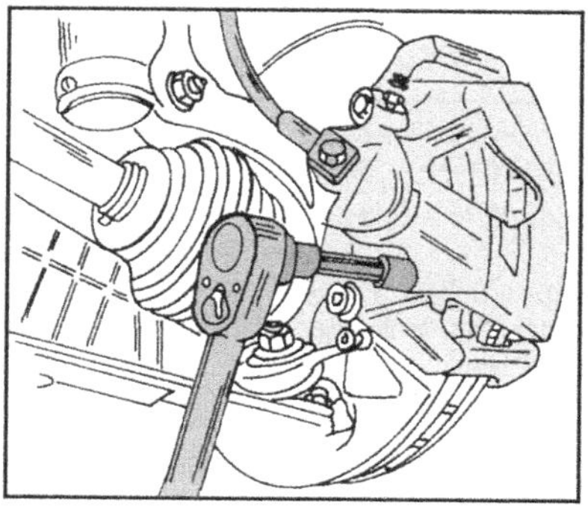

Fig. 8.9 – Fitting the caliper cylinder to the mounting bracket.

- Place the cylinder caliper carefully over the brake pads and insert the bolts through the cylinder housing into the mounting bracket. Tighten the bolts with an Allen key to 2.5 kgm (18 ft.lb.). Only use an Allen key of the shape shown in Fig. 8.9 in order to obtain the correct tightening torque. Push the two caps over the bolt openings after tightening the bolts.
- Refit the elastic spring into brake caliper cylinder and brake caliper mounting bracket.
- Roll the brake pad wear indicator cable to a spiral and connect to the connector spade terminal of the brake pad. Push the cover over the connector. Operate the brake pedal a few times to set the brake pads against the brake disc.

- Check the fluid level in the fluid reservoir and correct if necessary. Finally, fit the wheels and tighten the wheel bolts to the correct torque setting.
- Treat the new brake pads with care during the first 100 miles or, so, remember that the new material must "bed in" before they reach their full performance. To speed up the process, brake the vehicle several times from a speed of approx. 50 mph to 25 mph. Allow the brakes to cool between each braking operation.

Brake Pad Replacement – LT46

Fig. 8.10 shows the component parts of a caliper (made by Perrot). Proceed as follows.

- Proceed as described for the other type until the front wheels have been removed.

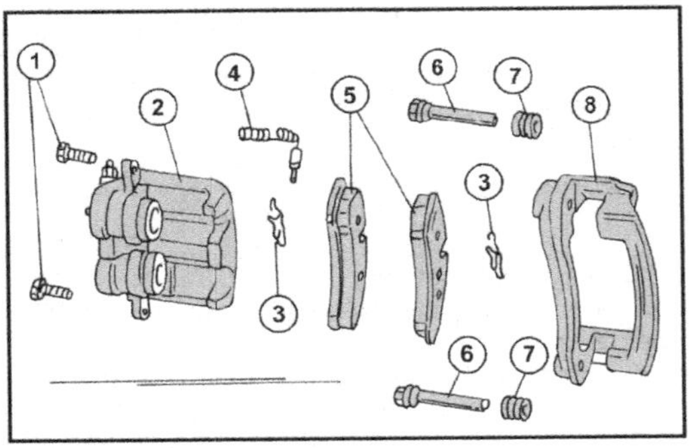

Fig. 8.10 – Brake caliper and brake pads of an LT45 model.

1 Bolts, 35 Nm, replace
2 Caliper (two pistons)
3 Retaining spring
4 Wear indicator
5 Brake pad set
6 Guide bolt
7 Rubber dust sleeve
8 Brake caliper bracket

- Refer to Fig. 8.11 and disconnect the wear indicator cable plug at the position shown by the arrow. Hold the guide bolt (2) at the inside by means of an open-ended spanner and unscrew the bolt (1). A socket with straight extension is required. It may be necessary to ground the spanner in order to insert it.

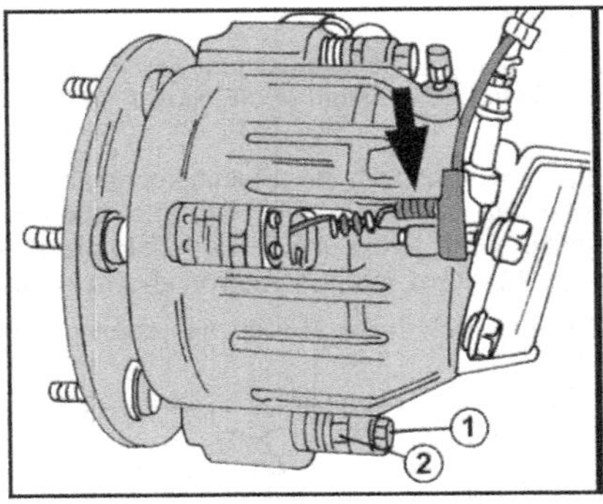

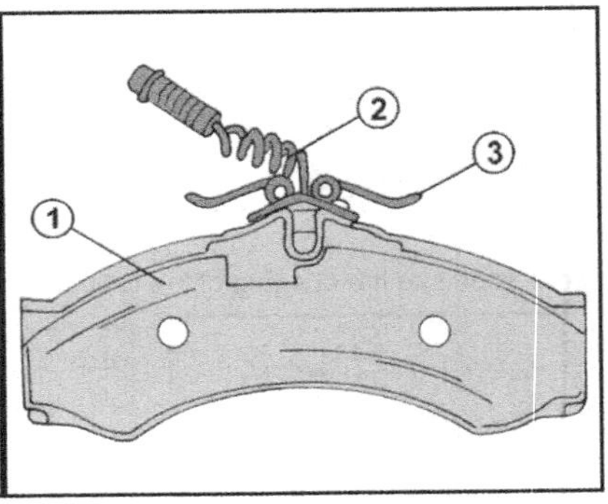

Fig. 8.11 – L.H. side: the connector for the wear indicator cable (arrow). Remove bolt (1) and (2) as described. R.H. side: attachment of the brake pad wear indicator cable (2) on the brake pad (1). The spring (3) must be removed.

- Grip the caliper at the bottom and swivel it towards the top. Tie the caliper in this position to the damper strut.
- Release the spring (3) in Fig. 8.11 from the brake pad and carefully disengage the brake pad wear indicator (2) from the brake pad (1), using a screwdriver. All removed parts can now be checked as described for the other brake caliper.

The installation is a reversal of the removal procedure, noting the following points.

- Swing the caliper towards the bottom and insert one of the new bolts in the pad repair kit.
- Hold the hexagon of the bolt on the inside (open-ended spanner) and tighten the bolt to 3.5 kgm (18 ft.lb.). Then, remove the upper guide bolt, replace it with the second bolt in the repair kit and tighten it as described.
- All other instructions given for the calipers in LT28/LT35 models also apply to these calipers.

8.3.1. Brake Caliper – Removal and Installation

Fig. 8.12 shows the attachment details of a brake caliper. The brake pedal must be operated a few times to expel the vacuum in the system and the front of the vehicle must be on chassis stands, wheel(s) removed.

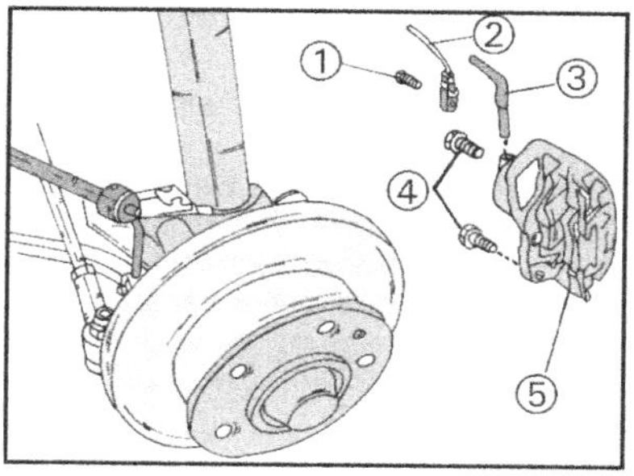

Fig. 8.12 – Details for the removal and installation of a front brake caliper cylinder.
1 Screw, 1.0 kgm
2 Cable for brake pad wear indicator
3 Brake hose, 1.2–1.6 kgm
4 Bolts, caliper cylinder
5 Brake caliper cylinder

- Push a bleeder hose over one of the bleed screws in the brake calipers and insert the other end of the hose into a jar. Open the bleed screw and operate the brake pedal until the brake fluid has been drained.
- Unscrew the union nut from the brake pipe connection on the caliper. Close the open connections in a suitable manner to prevent entry of foreign matter.
- Remove the screw (1) for the brake pad wear indicator connection and withdraw the cable.
- Remove the caliper cylinder in the manner described above and lift off the caliper. The caliper can now be unscrewed from the brake hose. The brake pads can be removed if required.
- If necessary, unscrew the two bolts (4) and remove the mounting bracket from the steering knuckle.

The installation is carried out in reverse order, noting the following points.

- Use new bolts for the mounting bracket. These are specially designed for the mounting bracket. Tighten the bolts to 11 kgm (80 ft.lb) – LT28/35 or 19 kgm 137 ft.lb.) – LT46.
- Screw the brake caliper onto the brake hose until nearly tight. The final tightening takes place when the cylinder is in position.
- Refit the brake pads and the cylinder as described above.
- Tighten the brake hose connection to 1.2–1.8 kgm (9–13 ft.lb.). Make sure that the hose cannot twist during this operation. If this is the case, you can slacken the connection between hose and pipe and reset the hose accordingly. Move the steering from lock to lock and check that the hose cannot tub against other parts of the front suspension. The brake system must be bled of air after installation.

8.3.2. Brake Calipers – Overhaul

We do not recommend the dismantling and/or overhaul of the brake calipers. Either fit a new caliper or have the cylinder overhauled in a workshop dealing with brake cylinder units.

8.4. Rear Disc Brakes – Brake Pads

The brake pads must be replaced if the thickness of the brake pad material is approx. 2 mm (0.08 in.). The thickness can be checked with the calipers fitted to the rear axle, as has been described for the front calipers. It should be noted that different calipers are fitted to the various models. Fig. 8.13 shows the attachment of a rear caliper and brake pads as fitted to the LT28 and LT35 models and made by Teves/ATE. A caliper made by Bosch can be used in LT35 models. The LT46 caliper is made by Perrot, again with two pistons. As the attachment is not the same in all cases, refer to the relevant heading.

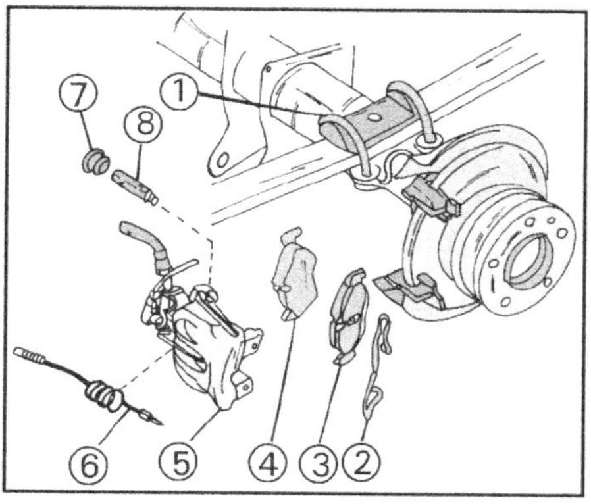

Fig. 8.13 – Details for the removal and installation of the rear disc brake pads (LT28/LT35).
1 Rear axle
2 Retaining spring
3 Outer brake pad
4 Inner brake pad
5 Brake caliper cylinder
6 Pad wear indicator cable
7 Rubber cap
8 Guide bolt

Teves/ATE Calipers

- As in the case of the front brakes, always replace the brake pads as a set.
- Operate the brake pedal a few times to expel the vacuum from the system, jack up the rear end of the vehicle and place chassis stands underneath the axle tube. Remove the rear wheels.
- On both sides of the vehicle, disconnect the brake pad wear indicator cables from the connection shown in Fig. 8.14.

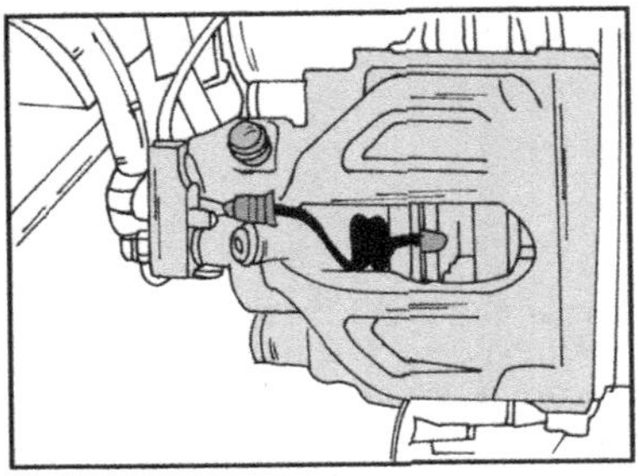

Fig. 8.14 – Disconnect the cable connection for the brake pad wear indicator from the connector on the L.H. side.

- Remove the two retaining springs (1) with a small screwdriver and remove the spring.
- At the rear of the brake caliper, remove the two rubber dust caps (Fig. 8.15) over the guide bolts and using a small screwdriver and unscrew the bolts, using a straight Allen key, as shown in the R.H. view. Lift off the cylinder and tie it to the rear axle with a piece of wire, as shown in Fig. 8.16 (L.H. view). Do not suspend the caliper on its hose. Remove the brake pad from the caliper.

Fig. 8.15 – Remove the two socket head screws at the top and bottom after removal of the rubber dust caps. An Allen key is required.

- Remove the second brake pad from the brake caliper mounting bracket.

Refit the brake pads as follows.

- Clean the contact areas for the brake pads inside the caliper housing.
- Remove some brake fluid from the fluid reservoir, if it is fairly full, to prevent fluid overflow during the next operation. You will have the push the piston(s) into the caliper bore(s), to fit the new brake pads. Normally, a special pair of pliers is used, which is employed in the manner shown in Fig. 8.15 (R.H. view).
- Fit the brake pad with the retaining spring into the caliper cylinder as shown in Fig. 8.9 for the front brake pads and push it into the direction of the arrow until it snaps into place. The outer brake pad is fitted into the opening of the caliper mounting bracket, into the position shown in Fig. 8.16.
- Place the cylinder caliper carefully over the brake pads and insert the bolts through the cylinder housing into the mounting bracket. Tighten the bolts with an Allen key to 2.5 kgm (18 ft.lb.). Only use an Allen key of the shape shown in Fig. 8.15 in order to obtain the correct tightening torque. Push the two caps over the bolt openings after tightening the bolts.

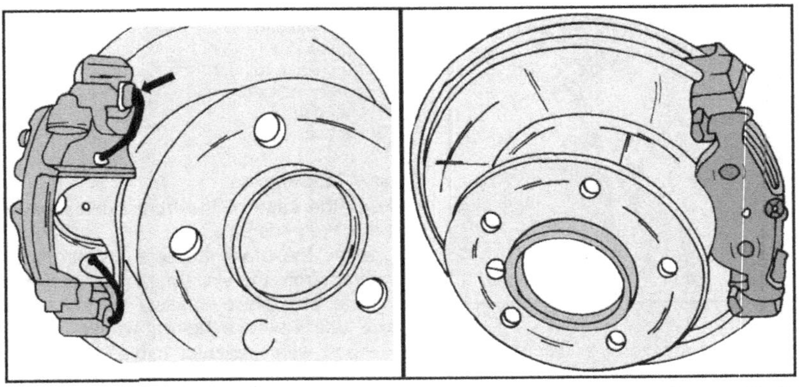

Fig. 8.17 – Fit the outer brake pad as shown against the brake caliper mounting bracket, as shown on the R.H. side. The L.H. view shows where the retaining spring is inserted at the top and bottom.

- Refit the retaining spring into brake caliper cylinder and brake caliper mounting bracket. Check with Fig. 8.17 to make sure that the spring is fitted correctly.
- Roll the brake pad wear indicator cable to a spiral and connect to the connector spade terminal of the brake pad. Push the cover over the connector. Operate the brake pedal a few times to set the brake pads against the brake disc.
- Check the fluid level in the fluid reservoir and correct if necessary. Finally, fit the wheels and tighten the wheel bolts to the correct torque setting.
- Treat the new brake pads with care during the first 100 miles or, so, remember that the new material must "bed in" before they reach their full performance. To speed up the process, brake the vehicle several times from a speed of approx. 50 mph to 25 mph. Allow the brakes to cool between each braking operation.

Bosch Brake Calipers – LT35

The calipers are secured with simple guide bolts. The removal is carried out in a similar manner as described for the Teves/ATE calipers. Disconnect the connector plug from the pad wear indicator, unscrew the bolt at the upper and lower end and lift off the cylinder, guiding the cable through the "window".

The installation is a reversal of the removal procedure. The bolt threads are coated with "Loctite" to secure them in position. Tighten the bolts to 3.0 kgm (22 ft.lb.). All instructions for the previous caliper also apply to the Bosch version.

Perrot Calipers – LT46

The attachment of these calipers is similar to the front calipers, as shown in Fig. 8.10. Removal and installation is carried out by referring to the instructions given earlier on. Fig. 8.11 also applies to these calipers.

8.4.0. Brake Caliper – Removal and Installation

The removal and installation of the brake calipers follows the same pattern as described for the front calipers. Remove the brake pads as described above and remove the caliper. To remove the caliper mounting bracket unscrew the two bolts. The bolts are tightened to 11 kgm (80 ft.lb.) in the case of the LT28/35 model or 18 kgm (130 ft.lb.) during installation. Fit the brake hose in the manner described, making sure that it is not twisted. Tighten the connection on the caliper to 1.2–1.6 kgm (9–13 ft.lb.)

Make sure that the hose cannot rub against other parts of the rear axle.

8.4.1. Brake Calipers – Repairs and Overhaul

Rear brake calipers should not be overhauled. Note, however, that different calipers are fitted to the various models, depending on the model and on the series. Model LT46 has calipers with two pistons. If a new caliper is purchased, always quote the model, the chassis number and the series number.

8.5. Brake Discs

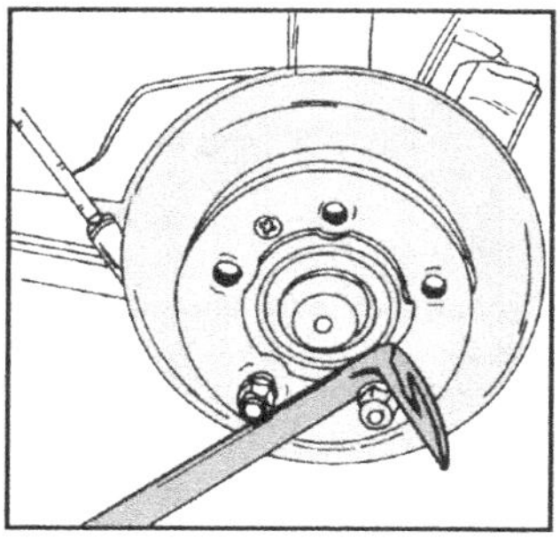

8.5.0. Front Discs – Removal and Installation

Removal of a brake disc requires the removal of the front brake pads as already described.

Fig. 8.18 – A front wheel hub can be held against rotation as shown.

Unscrew the caliper mounting bracket and take off the bracket. The disc is attached to the outside of the wheel hub with a single bolt (2.0 kgm/15 ft.lb.). To prevent the disc/hub from rotating, insert a tyre lever, as shown in Fig. 8.18. Take care not to damage the wheel bolt threads. Free a sticking disc with a rubber or plastic mallet. Before fitting the original brake disc, read the information in Section 8.5.2. Tighten the brake caliper mounting bracket bolts as already given.

8.5.1. Rear Discs – Removal and Installation

The rear brake discs of the LT28/LT35 models are attached to the wheel hub with a single bolt (2.0 kgm/15 ft.lb.). To remove the bolt, apply the handbrake to lock the disc/hub. Release the handbrake after removal of the bolt. To remove a brake disc, remove the brake pads and caliper mounting bracket as described during the removal and installation of the rear brake pads/caliper.

On LT46, there is a flange fitted in front of the brake disc. The flange must be removed in order to remove the brake disc.

The brake shoes in the inside of the handbrake must be reset in order to remove the disc. To do this, insert a screwdriver at the position shown in Fig. 8.19 and reset the adjuster wheel for the brake shoes. The brake shoes will now be away from the brake disc and the disc can be removed. A sticking disc can be freed with a rubber or plastic mallet.

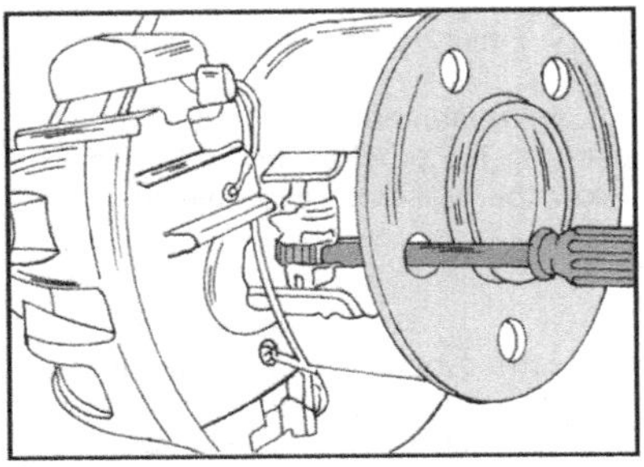

The installation is a reversal of the removal procedure. Tighten the caliper mounting bracket to the value already given. The flange securing bolts on the LT46 are tightened to 20.0 kgm (145 ft.lb.).

The handbrake brake shoes must now be set to their basic position. Insert the screwdriver as shown in Fig. 8.19 and rotate the adjuster wheel until the brake disc is tight and can no longer be rotated. From this position, turn the adjuster wheel in the opposite direction by five "clicks".

Fig. 8.19 – Resetting the brake shoes inside the brake disc/drum.

After installation, check the operation of the handbrake and adjust if necessary.

8.5.2. Re-Facing of Brake Discs

Brake discs can be re-faced, provided that their minimum thickness has not been reached. The thickness of the front discs is the same on all models (22 mm), as is the min. permissible thickness (19 mm). Discs with a different thickness are used on the rear wheels and have therefore a different min. permissible thickness. Refer to Section 8.0. for the values. If you intend to have the discs re-faced, we advise you to have it done at a VW workshop, as all necessary data will be available.

Discs should be checked for run-out after installation. To do this, attach a dial indicator with the stylus placed against the outer edge of the disc. Slowly turn the disc and read off the dial indicator reading. If the reading exceeds 0.2 mm (0.008 in.), remove the disc and check the disc and hub faces for foreign matter. Otherwise, the disc may be out-of-true.

8.6. Master Brake Cylinder

All models are fitted with a tandem master cylinder with a double reservoir, supplying both circuits of the dual circuit with brake fluid. The brake lines are diagonally split.

The brake fluid level is indicated by means of a warning light in the dashboard when it is below the "Min" mark.

8.6.0. Removal and Installation

The master cylinder is fitted to the front face of the brake servo unit and is removed from the engine compartment.

- Push a bleeder hose over one of the bleed screws in a front brake caliper (remove the rubber dust cap first), suspend the other end of the hose into a container, open the screw and pump the brake pedal until the system has been drained.
- Withdraw the cable plug from the brake fluid level contact switch on the side of the reservoir cap.
- Clamp off the hose leading to the fluid reservoir with a suitable clamp. Cut the hose clamp with a wire cutter and withdraw the hose.
- Move the two retainers for the fluid reservoir towards the outside as shown in Fig. 8.20 and remove the reservoir.

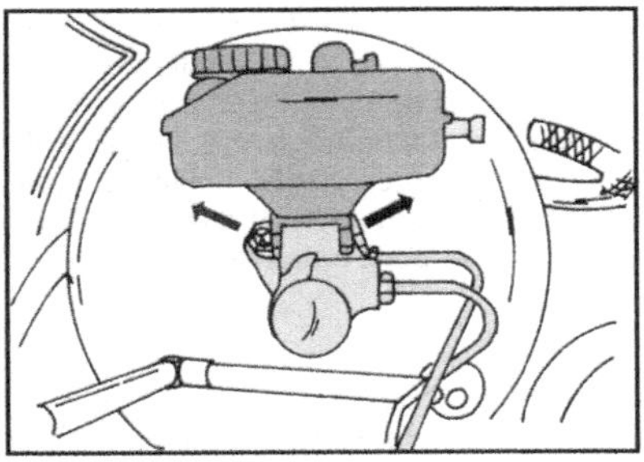

Fig. 8.20 – Removal of the master brake cylinder.

- Disconnect the brake pipes from the cylinder and push them carefully to one side. Close the pipe ends in a suitable manner to prevent entry of foreign matter. Tighten the union nuts during installation to 1.2–1.6 kgm (9–13 ft.lb.).
- Unscrew the cylinder and lift it off in horizontal position. Take care not to tilt the cylinder, as this will bring the push rod out of alignment. A rubber sealing ring is inserted between cylinder and brake servo unit. Take care to collect it. Fig. 8.20 shows the cylinder as it is fitted to the brake servo unit.

The installation is a reversal of the removal procedure. Always replace the sealing ring. Tighten the cylinder to 2.5 kgm (18 ft.lb.). There is no need to adjust the brake servo push rod.

Finally, fill the reservoir and bleed the brake system as described later on.

8.6.1. Master Cylinder – Overhaul

Master brake cylinders are no longer overhauled. A new cylinder must be fitted if the old one fails. Always quote the model and the chassis number when ordering a new cylinder, as cylinders are sometimes changed and a different cylinder may have been allocated to your particular vehicle.

8.7. The Handbrake

The handbrake is known as duo-servo-brake. "Duo" indicates that the braking effect is the same in both rotating directions of the brake disc.

"Servo" indicates that the operation of one brake shoe is transmitted to the other brake shoe. The brake shoes for the handbrake are located inside the rear brake discs. The remaining component parts of the handbrake system are a normal handbrake lever, a compensating lever assembly underneath the vehicle, the necessary handbrake cables and the operating levers on the two rear brake calipers.

8.7.0. Brake Shoes – Removal and Installation

A special tool is required to remove the brake shoes with the rear axle shafts fitted, in order to remove the shoe return springs. Otherwise, you will have to remove the axle shafts to expose the brake shoes. The arrangement of the brake shoes is different for the LT28/35 model and the LT46 model. The brake shoes are situated in the case of the former models at the top and bottom, whereas the LT46 model has the shoes on the L.H. and R.H. sides. The removal and installation is therefore not the same on all models. The following description assumes that the latter recommendation is followed, as it will not be possible to reach the springs. The operations are described for one side, but shoes must always be replaced as a set. The rear of the vehicle must be on chassis stands, with the wheels removed.

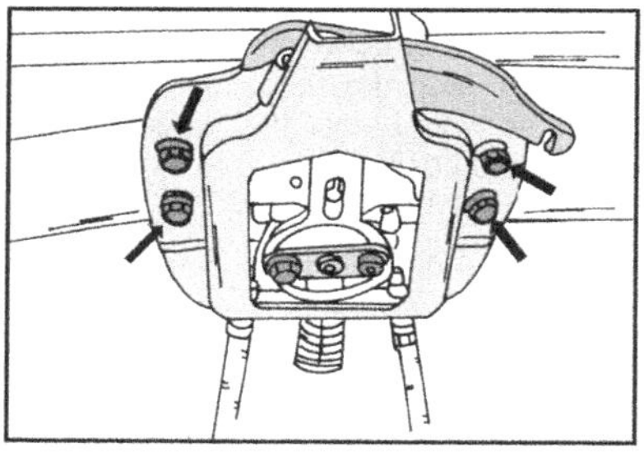

Fig. 8.21 – View of the compensating lever. The front cable has already been disconnected from the lever on the right.

LT28/LT35

- Underneath the vehicle, disconnect the front handbrake cable from the compensating lever by unscrewing the adjuster until the cable can be unhooked.
- Refer to Fig. 8.21 and slacken the handbrake compensator lever to disconnect the rear handbrake cables.
- Remove the rear brake disc as already described. If necessary, slacken the handbrake cable adjusting nuts until the disc is free.
- On the outside of the brake assembly, disconnect the handbrake cable from the spreader lock.
- Remove the brake shoe return springs. One of the springs must be compressed until the spring seat can be rotated by 90°. Remove the other springs with a pair of pliers.
- Expand the two brake shoes until they are free to be removed. The adjuster wheel at the lower end will come free during this operation.

The installation is carried out as follows.

- Connect the handbrake cable with the pin to the spreader lock and push the spreader lock into the brake back plate.
- Coat the thread of the adjuster push rod and the cylindrical part with long-term grease and turn back the adjuster into the body.
- Fit the adjuster mechanism into the brake shoes with the adjuster wheel at the lower end. Fit the upper return spring from the front into the brake shoe.
- Expand the brake shoes until one shoe can be engaged with the spreader lock. Fit the spring from the side into the brake shoe.

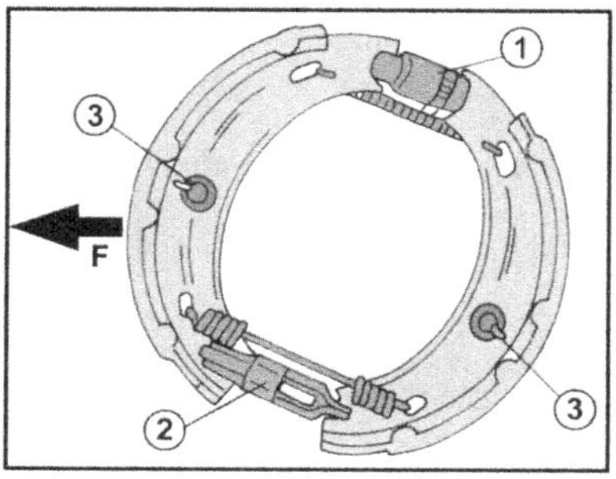

- Compress the other spring from the front until the inner end can be engaged into the brake back plate and turn it by 90° to secure it.
- Fit the remaining spring between the brake shoes. A wire hook can be used to stretch and insert the spring.
- Refit the brake disc and the brake caliper assembly and adjust the handbrake as described below.

Fig. 8.22 – Correctly assembled handbrake brake shows (LT46).
1 Adjuster wheel
2 Spreader lock
3 Brake shoe holding pins

LT46

The removal and installation is carried out in a similar manner, but the brake shoes are fitted to the L.H. and R.H. sides. The flange in front of the brake disc must be removed before the brake disc can be withdrawn. Make a note of the position of the individual parts. Also note which side is facing towards the front and mark the brake shoe accordingly.

Brake shoes and adjuster and the upper shoe return spring must be pre-assembled before the shoes are placed in position. Make sure that the side marked "Front" is facing in the correct direction. Compare the assembled brake shoes with Fig. 8.22.

8.7.1. Handbrake Adjustment

The handbrake must be adjusted if the handbrake lever can be pulled by more than four "clicks" before the braking effect takes place. The handbrake is adjusted inside the brake discs, but it may also be necessary to adjust the handbrake cables (see the following section). Proceed as follows to adjust the brake shoes, noting the difference as applicable.

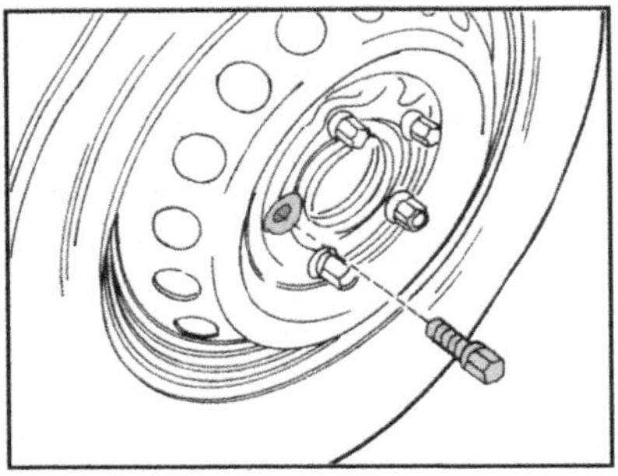

Fig. 8.23 – Turn the wheel into the position shown, remove the wheel bolt and insert a screwdriver into the hole to rotate the brake shoe adjuster inside the brake disc.

LT28 and LT35

- Place the rear end of the vehicle on chassis stands and slacken the handbrake cables at the compensating lever.
- Turn the wheel into the position shown in Fig. 8.23 and remove the wheel bolt. Insert a screwdriver into the hole and engage it by "feeling" with the teeth of the adjuster wheel. Turn the adjuster wheel until the wheel can no longer be rotated. From this position, turn the wheel in the reverse direction by five "notches" until the wheel can be rotated. The screwdriver must be moved as follows during the adjustment.
- **On the L.H.** side from the bottom towards the top.
- **On the R.H.** side from the top towards the bottom.
- Operate the handbrake a few times. Check that the brake shoes release the brake disc/drum. Otherwise, re-adjust.
- Fit the removed wheel bolt and tighten it to the correct torque (18 kgm).
- Adjust the handbrake cables as described below.

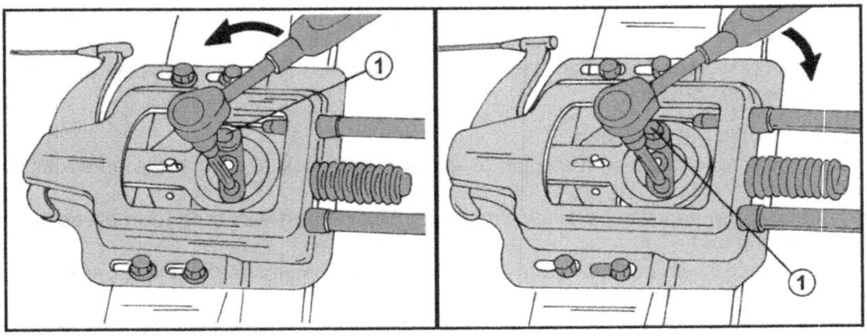

Fig. 8.24 – The two stages to adjust the handbrake on LT46.

LT46

- Place the vehicle on chassis stands and remove the wheels.
- Slacken the bolt (1) in Fig. 8.24 and turn the adjusting eccentric in the direction of the arrow as shown in the L.H. view.
- Refer to Fig. 8.21 and slacken the four bolts securing the handbrake compensating lever and push the assembly towards the front.
- Rotate the brake disc until the adjuster wheel can be seen through one of the hole in the brake disc. Insert a screwdriver and rotate the adjuster until the brake disc can no longer be rotated. From this position, turn the adjuster wheel in the reverse direction by five "notches" until the brake disc can be rotated freely. Adjust the handbrake cable as described below and then pull the handbrake lever by four "notches".

- Refer to Fig. 8.24 and pre-load the adjusting eccentric by 1.5 kgm and tighten the bolt (1). Release the handbrake lever.
- Refit the wheels and rotate them. No resistance must be felt. Re-tighten the wheels after approx. 30 miles.

8.7.2. Handbrake Cable Adjustment

- Release the handbrake, slacken the bolt (1) in Fig. 8.24 and rotate the adjusting eccentric in the direction of the arrow to its stop.
- Slacken the four bolts shown in Fig. 8.21 and push the compensating lever towards the front.
- Insert an Allen key of 6-mm A/F between the handbrake compensating lever and the front lever, as shown in Fig. 8.25, and push the compensating lever towards the rear, as indicated by the arrow, until the handbrake cable is just tight (no play and no tension). In this position, tighten the bolts in Fig. 8.21 and remove the Allen key. The rear end of the vehicle must, of course, be on chassis stands.

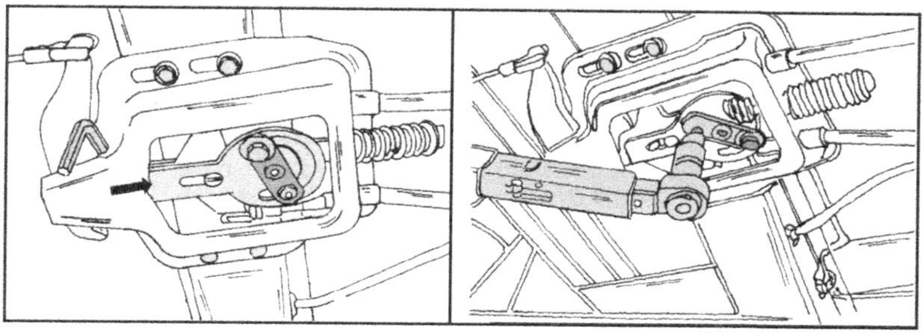

Fig. 8.25 – Push the handbrake compensating lever in the direction of the arrow to adjust the handbrake cables. The R.H. view shows the tightening of the adjusting eccentric on the compensating lever as specified in the description.

- Pull the handbrake lever in the driver's cabin by two notches.
- Slacken the lock bolt in the centre of the adjusting eccentric. Apply a socket to the eccentric as shown in Fig. 8.25 and tighten it to 1.5 kgm (11 ft.lb.). Tighten the lock bolt in the centre.
- Apply the handbrake lever by four "notches".
- Check that the two rear wheels cannot be rotated. Then, release the handbrake lever and check that the wheels can be rotate freely. Lower the vehicle onto its wheels.

Note: If the brake shoes have been replaced, "brake in" the new shoe linings before using the vehicle in day-to-day traffic. To do this, drive with approx. 50 MP/h and apply the handbrake to the first notch, but not longer than 10 seconds. Release the handbrake lever and stop the vehicle. Apply the handbrake lever 2–3 times and jack up the rear end of the vehicle. Check that the wheels can be rotated freely.

8.7.3. Replacing Handbrake Cables

The front handbrake cable must be replaced by your VW workshop, as a special tool is required to remove a circlip which secures the handbrake cable to the body underfloor. The rear cable can be replaced more easily. After the cable is detached from the compensating lever, remove the rear disc and disconnect the cable from the brake shoe. Again, we would like to point out that the operations are fairly complicated.

8.8. Brake Servo Unit

Brake servo units should not be dismantled, as special tools are required to dismantle, assemble and test the unit. Different servo units are fitted to the range covered in this manual. Always make sure to fit the correct unit of the part is replaced. Remember that a failure of the servo unit to act will not affect the efficiency of the brake system, but, of course, additional effort will be required for the same braking distance to be maintained.

Note: If you coast downhill without the engine running, for whatever reason, remember that the vacuum in the unit will be used up after a few applications of the brake pedal and the brake system will from then onwards operate without the power assistance. Be prepared for his.

The brake servo can be removed as follows.

- Remove the master cylinder as already described.
- Disconnect the vacuum hose from the brake servo unit.
- Disconnect the brake servo push rod from the brake pedal. To do this, remove the lock and pull out the pin.
- Remove the brake servo securing nuts from the inside and lift it out from the engine compartment side.

The installation is a reversal of the removal procedure. Tighten the securing nuts to 2.5–2.8 kgm (18–20 ft.lb.).

8.9. Bleeding of the Brake System

Bleeding of the brake system should be carried out at any time that any part of the system has been disconnected, for whatever reason. If only one of the brake circuits (caliper or wheel brake cylinder) has been opened, either bleed the front left and rear right or front right and rear left. If both circuits have been opened, first bleed the primary circuit and then the secondary circuit. The procedure given below should be followed and it should be noted that an assistant will be required, unless a so-called "one-man" bleeding kit is available.

Always use clean fresh brake fluid of the recommended specification and never re-use fluid bled from the system. Be ready to top up the reservoir with fluid (a brake bleeding kit will do this automatically) as the operations proceed. If the level is allowed to fall below the minimum, the operations will have to be re-started.

- Obtain a length of plastic tube, preferably clear, and a clean container. Put in an inch or two of brake fluid into the container and then go to the first bleed point. Take off the dust cap and attach the tube to the screw, immersing the other end of the tube into the fluid in the container.
- Open the bleed screw about three quarters of a turn and have your assistant depress the brake pedal firmly to its full extent, while you keep the end of the tube well below the fluid level in the container. Watch the bubbles emerging from the tube and repeat the operation until no more are seen. Depress the brake pedal once more, hold it down and tighten the bleed screw firmly.
- Check the fluid level, go to the next point and repeat the operations in the same way. Install all dust caps, depress the brake pedal several times and finally top up the reservoirs.

Note: A pressure-bleeding apparatus is used in the workshop to bleed the brake system. If you are not 100% happy with the performance of the brakes, have the brakes bled at a dealership. Do not take chances. If you do the bleeding operations yourself, only use the brake fluids supplied by Mercedes-Benz. Although other fluids are suitable, we advise against it. Never use fluid that has been standing for a while, as fluid will deteriorate over the years. Also do not re-use fluid drained from the system.

Changing the Brake Fluid

The brake fluid should be changed every two years. In general, this is normally not carried out, but we highly recommend it if heavy loads or passengers are carried. Empty the brake system by opening all bleeding screws and pumping the pedal. Old fluid should be collected into a jar as described above.

After filling the fluid reservoir, bleed the system in the manner described above. In this case, it is more important to observe the fluid level in the reservoir, as it will sink rapidly. It is of advantage to have an additional helper, one to operate the brake pedal and another one to top up the reservoir.

Make sure to fit the rubber dust caps over the bleeder screws after the system has been bled. Operate the brake pedal a few times to check the pedal pressure. If the pedal "lifts" during the pumping operation, re-bleed the system.

8.10. Brakes – Tightening Torques

Brake servo unit:................................2.1–2.9 kgm (15–21 ft.lb.)
Master brake cylinder:................................2.1–2.9 kgm (15–21 ft.lb.)
Brake pipe union nuts:................................1.2–1.6 kgm (9–11.5 ft.lb.)
Bleeder screws:................................1.2–1.6 kgm (9–11.5 ft.lb.)

Front brake caliper to steering knuckle:
- LT28/LT35:................................11.0 kgm (80 ft.lb.)
- LT46:................................19.0 kgm (137 ft.lb.)
Rear caliper/rear axle:................ 11.0 kgm (77 ft.lb.), LT28/35, 18.0 kgm (130 ft.lb.), LT46

Brake caliper cylinders to mounting bracket:
- LT28/LT35:................................2.5 kgm (18 ft.lb.)
- LT46:................................3.5 kgm (25 ft.lb.)

Brake caliper to mounting bracket:
- LT28/LT35:................2.5 kgm (18 ft.lb.), Teves, 3.0 kgm (22 ft.lb.), Bosch
- LT46 (Perrot):................................3.5 kgm (25 ft.lb.)
Brake disc to wheel hub (LT28/35):................................2.0–2.5 kgm (15–18 ft.lb.)
Brake disc to wheel hub (LT46), with flange:................................20.0 kgm (144 ft.lb.)
Handbrake cable to rear axle:................................2.0–2.5 kgm (15–18 ft.lb.)
Brake pressure regulator to bracket:................................2.3–2.5 kgm (16–18 ft.lb.)

9. Electrical System

9.0. Battery

Voltage	12 volts
Polarity	Negative earth
Capacity	Depending on equipment
Condition of Charge:	
- Fully charged	1.280
- Half charged	1.200
- Discharged	1.120

To check the voltage of the battery, use an ordinary voltmeter and apply between the two battery terminals. A voltage of 12.5 volts or more should be obtained.

If a hydrometer is available, the specific gravity of the electrolyte can be checked. The readings of all cells must be approximately the same. A cell with a low reading indicates a short circuit in that particular cell. Two adjacent cells with a low reading indicate a leak between these two cells.

A battery can be re-charged, but the charging rate must not exceed 10% of the battery capacity, for example, 7.2 amperes. The battery must be disconnected from the electrical system. Charge the battery until the specific gravity and the charging voltage are no longer increasing within 2 hours. Add distilled water only. Never add acid to the battery.

The level of the battery electrolyte should always be kept above the top of the plates.

9.1. Alternator

9.1.0. ROUTINE PRECAUTIONS

The vehicles covered in this manual employ an alternator and control unit. This equipment contains polarity-sensitive components and the precautions below **must** be observed to avoid damage.

- Check the battery polarity **before** connecting the terminals. Immediate damage will result to the silicon diodes from a wrong connection – even if only momentarily.
- Never disconnect the battery or alternator terminals **whilst the engine is running.**
- Never allow the alternator to be rotated by the engine unless **ALL** connections are made.
- Disconnect the alternator multi-pin connector **before** using electric welding equipment anywhere on the vehicle.
- Disconnect the battery leads if a rapid battery charger is to be used.
- If an auxiliary battery is used to start the engine, take care that the polarity is correct. **Do not** disconnect the cables from the vehicle battery.

9.1.1. REMOVAL AND INSTALLATION

The alternator is rigidly attached to the engine. Remove as follows.

- Disconnect the battery and jack up the front end of the vehicle. Remove the protective panel underneath the engine compartment to improve the access to some of the parts.
- Disconnect the cable harness by withdrawing the plug at the rear end of the alternator. Also disconnect the single cable.
- Slacken the poly drive belt as described in the section dealing with the engine and lift off the belt.
- Remove the two alternator mounting bolts and lift out the alternator from below the vehicle.

The installation of the alternator is a reversal of the removal procedure. Take care to make the correct connections. Adjust the drive belt tension.

Tighten the alternator bolts to 4.2 kgm (30 ft.lb.), the nut of the small terminal to 0.5 kgm and the nut of the large terminal to 1.8 kgm (13 ft.lb.).

9.1.2. SERVICING

A Bosch alternator is used on the engines dealt with in this manual, having an output of 90 or 120 amperes. We do not recommend that the alternator or control unit should be adjusted or serviced by the owner. Special equipment is required in the way of test instruments and the incorrect application of meters could result in damage to the circuits.

The alternator is fitted with sealed-for-life bearings and no routine attention is required for lubrication. Keep the outside of the alternator clean and do not allow it to be sprayed with water or any solvent.

The alternator brush gear runs in plain slip rings and the brushes have a long life, requiring inspection only after a high mileage has been covered. To inspect the brushes, we recommend the removal of the alternator. Take out the two screws from the brush holder assembly and withdraw for inspection. Measure the protruding length of the brushes. If the protruding length is less than 5.0 mm (0.2 in.) or approaching this length, replace the brushes. New brushes will have to be soldered in position. We would like to point out that it is not an easy operation to guide the brushes over the slip rings when the slip ring cover is being fitted.

9.2. Starter Motor

9.2.0. REMOVAL AND INSTALLATION

- Disconnect the battery earth (ground) cable.
- Remove the starter motor mounting bolts.
- Disconnect the starter motor cables from the starter motor solenoid switch. Withdraw the unit from the vehicle. The unit must be guided past the engine crossmember. Tilt the motor so that the solenoid switch is facing downwards and then remove it sideways.

Install in the reverse sequence to removal. Tighten the starter motor bolts to 4.5 kgm (32.5 ft.lb.). Make sure that the mating faces are clean before bolting up. Re-connect the wires and the battery terminals. Tighten the nut to 2.5 kgm (18 ft.lb.).

9.2.1. SERVICING

It may be of advantage to fit an exchange starter motor if the old one has shown faults. Exchange starter motors carry the same warranty as new units and are therefore a far better proposition.

The following repairs can, however, be attempted without any problems:

Solenoid Switch

- Disconnect the cable between the solenoid switch and the starter motor terminal.
- From the front of the solenoid switch, remove the bolts and withdraw the switch from the starter motor at the same time disengaging the engagement lever from the starter motor drive.
- Refit the new switch in reverse order.

Starter Motor Brush Replacement

- Unscrew and remove the cover from the rear of the starter motor.
- Remove the two screws securing the commutator bearing cover.
- Remove the "C" clip from the rear end of the armature shaft (screwdriver), remove the sundry parts and withdraw the cover from the starter motor yoke. If necessary, insert a screwdriver to get the parts apart. Remove the shims.
- Remove the brushes from their holders. The brush leads are either screwed-on or soldered. If they are soldered, you will have to be familiar with a soldering iron. Otherwise, take the unit to an auto electrical repair shop to have the bushes replaced.
- Replace the brushes as applicable and re-assemble the starter motor in reverse order.

9.2.2. Starter Motor Faults

Following are a few tips to locate faults in the starting system before the starter motor is replaced.

- *When the key is turned to the starting position, the starter motor turns very slowly, nothing happens and starts to crank the engine and stops:* If the warning lights in the instrument panel are nearly or fully extinguished, check the battery. Loose or corroded cables on the starter motor, bad earth connection are other causes. If the battery has been connected previously, check the terminals for good contact. Use a jumper lead or get a tow-start to start the engine.
- *If the warning lights in the instrument panel remain "on" and a faint click is heard from the area of the starter motor,* it is possible that the engagement plunger of the solenoid switch is sticking. Use a heavy spanner and knock against the outside of the switch. Sometimes this will free the plunger. Other causes are worn starter motor brushes or defective contacts in the solenoid switch. In both cases, remove the starter motor to replace the bushes and/or the solenoid switch. If the warning lights remain "on" and no click can be heard, check the cable connections between battery and starter motor.
- *If the starter motor is heard to operate without cranking the engine,* suspect a damaged solenoid switch, or the engagement mechanism, the starter motor pinion of the flywheel teeth is damaged. Try to engage a gear, push the vehicle backwards and forwards a few times and try the starter motor again. Otherwise, fit a new starter motor.
- *If the starter motor "spins" after the engine has started and the key has been turned to the normal position,* a sticking solenoid switch is mostly the cause for the fault (contacts fail to open). The ignition switch can also be damaged. *Immediately switch off the engine.*

If the starter motor pinion "spins" after starting the engine (howling noise), switch off the engine immediately. In most cases, a new starter motor will be necessary.

9.3. Headlamps

9.3.0. Headlamp/Side Light Bulb – Replacement

All front bulbs are fitted to the rear of a light unit. Fig. 9.1 shows the location of the various bulbs in a headlamp, as seen from the inside. Replace the bulbs as follows.

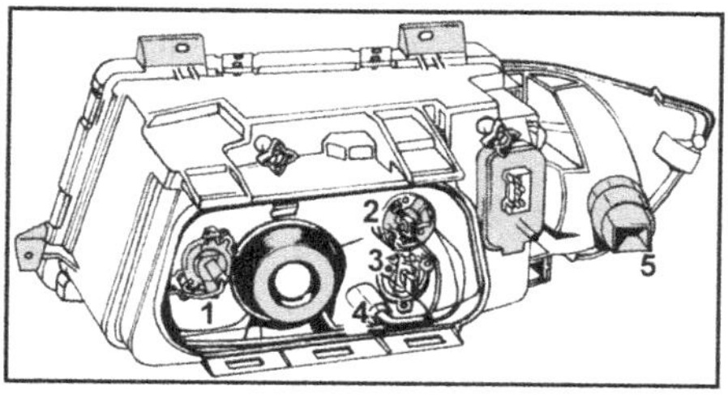

Fig. 9.1 – View of a headlamp from the rear, showing the location of the various bulbs.
1 Bulb for main beam, H1, 55 W
2 Bulb for dipped beam, H1, 55 W
3 Bulb for fog lamp, H1, 55 W
4 Bulb for side lamp, 21 W
5 Bulb for flasher lamp, 21 W

- Open the bonnet, and from the rear of the headlight unit, remove the plastic cover by releasing the spring clips (arrows, Fig. 9.2) and removing it in the direction of the white arrow.
- Disconnect the battery.

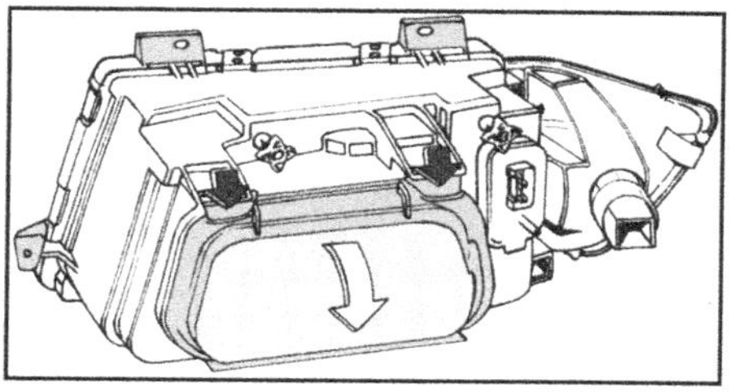

Fig. 9.2 – Removal of the headlamp cover at the rear. Unclip the retainers (black arrows) and tilt the cover in the direction of the white arrow.

Replacing the Bulb for the Dipped Beam:

Snap back the spring clip, remove the bulb (3) and disconnect the connector plug. The installation is carried out in reverse order. Do not touch the glass part of the bulb with bare fingers.

Replacing the Bulb for the Side Light:

The bulb holder (4) with the bulb is inserted into the rear of the headlamp. To remove the bulb holder, push it towards the inside and turn it until it can be removed. The installation is carried out in reverse order. Do not touch the glass part of the bulb with bare fingers.

Replacing the Bulb for the Fog Lamp:

Snap back the retaining clip, remove the bulb (3) and disconnect the cable harness connector. The installation is carried out in reverse order. Do not touch the glass part of the bulb with bare fingers.

Replacing the Bulb for the Main Beam:

The headlamp must be removed to gain access to the main beam bulb. Snap back the retaining clip, remove the bulb (1) and disconnect the cable harness connector. The installation is carried out in reverse order. Do not touch the glass part of the bulb with bare fingers. There should be no need to adjust the headlamps if only the bulb has been replaced.

9.3.1. HEADLAMP REPLACEMENT

As the headlamps must be adjusted after replacement of a headlight unit and the adjustment can only be carried out with optical equipment, we expect that you will have the beam alignment checked at your dealer or a service station dealing with headlamp work.

9.3.2. Headlamp Adjustment

The adjustment of the headlamps is carried out from the rear of the headlamp units. You will find two knurled bolts on each headlamp.

The headlamps should be adjusted with optical equipment and only in emergency should you reset the headlamps.

- Start the engine and, if fitted, turn the light beam height regulation to position "0".
- Drive the vehicle in front of a dark wall or garage door and switch on the headlamps. If only one headlamp requires adjustment, set the beam to the same height as the other headlamp.

- If both headlamps must be adjusted, transfer the centre point of each headlamp to the wall or door by measuring from the ground to the headlamp, the dimension between the headlamps, etc. The vehicle must be 10 m away from the adjusting area.
- Switch the headlamps to high beam, place a blanket over one headlamp and check that the light beam is identical with the centre of the marked cross for the headlamp in question. If this is not the case, adjust the headlamp with the adjusting screws shown.
- Check the high beam adjustment of the other headlamp in the same manner.
- Switch the headlamps to dipped beam and check that the light beam is identical on each headlamp.

We instruct you again: Have the headlamp adjustment checked professionally at the earliest opportunity. Only consider the above instructions in an emergency.

10. SERVICING AND MAINTENANCE

Most of the maintenance operations can be carried out without much difficulties. In many cases, it is, however, better to have certain maintenance operations carried out in a workshop as experience and special equipment, for example, test instruments are required to carry out a certain job. Most important are the regular inspections and checks which are described below. Operations to be carried out after a certain mileage are described later on in this section and the text will advise when specific jobs should be left to a VW Dealer.

10.0. Regular Maintenance

Oil Level Check: Check the engine oil level every 500 miles. With the vehicle standing on level ground, remove the oil dipstick and wipe it clean with a clean rag or a piece of tissue paper. Re-insert the oil dipstick and remove once more. The oil level must be visible between the upper and the lower mark on the dipstick. If the oil level is below the lower mark, top up with engine oil of the correct viscosity. From the actual level indicated, you will be able to tell how much oil is missing. Never overfill the engine – the level must never be above the upper dipstick mark.

Checking the Brake Fluid Level: The brake fluid reservoir is in the engine compartment on the driver's side. The reservoir is transparent and it is easy to check whether the fluid level is between the "Min" and "Max" mark. If necessary, top up to the "Max" mark with the correct brake fluid.

Checking the Brake Lights: The operation of the brake lights can either be checked with the help of another person or you can check it by yourself by driving the vehicle backwards near the garage door. Operate the brake pedal and check if the reflection of the brake lights can be seen on the garage door by looking through the rear view mirror.

Checking the Vehicle Lights: In turn, check every vehicle light, including the horn and the hazard warning light system. Rear lights and reversing lights can be checked in the dark in front of a garage door, without leaving the vehicle.

Checking the Tyre Pressures: Check the tyre pressures at a petrol station. Pressures are different for the various models. Either your operator's manual or tyre charts will give you the correct pressures.

If continuous speeds of more than 100 mph are anticipated, increase the tyre pressure by 0.2 kg/sq.cm. (3 psi.).

Checking the Coolant Level: See Section 1.8.1. Never open the radiator filler cap when the engine is hot.

10.1. Service every 9000 Miles or Once a Year

Changing the Engine Oil and Oil Filter: Some petrol stations will carry out an oil change free of charge – you only pay for the oil. The same applies to the oil filter (there may be a small extra charge), but not every petrol station will be able to obtain a VW filter. To change the filter yourself, refer to the relevant page.

Lubrication Jobs: Apart from the engine lubrication, there are further lubrication points that should be attended to. These include the throttle linkage and shafts (only grease the swivel points), the engine bonnet catch and the hinges (use a drop of engine oil) and perhaps the door mechanism.

10.2. Additional Service Every 20,000 Miles

Checking the Idle Speed: If the engine no longer idles as expected, have the idle speed checked and, if necessary, adjusted at your dealer.

Air Filter Service: Remove the air filter element for cleaning.

Checking the Brake System: If no trouble has been experienced with the brake system, there is little need to carry out extensive checks. To safeguard for the next 9000 miles, however, follow the brake pipes underneath the vehicle. No rust or corrosion must be visible. Dark deposits near the pipe endpoint to leaking joints. Brake hoses must show no signs of chafing or breaks. All rubber dust caps must be in position on the bleeder valves of the calipers. Insert a finger underneath the master cylinder, where it is fitted to the brake servo unit. Moisture indicates a slightly leaking cylinder.

The brake pads must be checked for the remaining material thickness as has been described in the section "Brakes" for the front and rear brakes.

Adjusting the Parking Brake: Adjust the parking brake as described in the section "Brakes" under the relevant heading.

Brake Test: A brake test is recommended at this interval. Your will decide yourself if the brakes perform as you expect them to. Otherwise, have the brakes tested on a dynamometer. The read-out of the meter will show you the efficiency of the brake system on all four wheels.

Checking the Wheel Suspension and Steering: In the case of the front suspension, remove both wheels and check the shock absorbers for signs of moisture, indicating fluid leaks.

Check the free play of the steering wheel. If the steering wheel can be moved by more than 25 mm (1 in.) before the front wheels respond, have the steering checked professionally.

Check the rubber dust boots of the track rod and suspension ball joints. Although rubber boots can be replaced individually, dirt may have entered the joints already. In this case, replace the ball joint end piece or the suspension ball joint.

Check the fluid level in the reservoir for the power-assisted steering. Refer to the "Steering" section for details. If the steering fluid is always missing after the 20,000 miles check, suspect a leak somewhere in the system – see your dealer.

Tyre Check: Jack up the vehicle and check all tyres for uneven wear. Tyres should be evenly worn on the entire surface. Uneven wear at the inner or outer edge of front tyres points to misalignment of the front wheel geometry. Have the geometry measured at your dealer. Make sure that a tread depth of 1.6 mm is still visible to remain within the legal requirements. Make sure to fit tyres suitable for your model, mainly if you buy them from an independent tyre company.

Re-Tighten Wheel Bolts: Re-tighten the wheel bolts to 18.0 kgm (130 ft.lb.). Tighten every second bolt in turn until all bolts have been re-tightened.

Checking the Cooling System: Check all coolant hoses for cuts, chafing and other damage. Check the radiator for leaks, normally indicated by a deposit, left by the leaking anti-freeze. Slight radiator leaks can be stopped with one of the proprietary sealants available for this purpose.

Checking the Clutch: Check the clutch operation. The fluid reservoir should be full. If it is suspected that the clutch linings are worn near their limit, take the vehicle to a dealer.

Checking the Anti-Freeze: The strength of the anti-freeze should be checked. Petrol stations normally have a hydrometer to carry out this check. Make sure that only anti-freeze suitable for VW engines is used.

Checking the Manual Transmission Fluid Level: Refer to Section 3.3.

Checking the Rear Axle Oil Level: Refer to Section 5.3.

10.3. Additional Service Every 38,000 Miles

Air Cleaner Element Change: Refer to the relevant section.

Diesel Oil Filter – Change: Replace the fuel filter as described in the section "Diesel Injection System". Bleed the fuel system.

Timing Belt (2.5-Litre Engine): The belt should be replaced. Refer to your dealer as the intervals are sometimes changed.

10.4. Once Every Year

Brake Fluid Change: We recommend to have the brake fluid changed at your dealer. Road safety is involved and the job should be carried out professionally. If you are experienced with brake systems, follow the instructions in the "Brakes" section to drain, fill and bleed the brake system.

10.5. Once Every Three Years

Cooling System: The anti-freeze must be changed. Refer to Section 1.8.1. to drain and refill the cooling system.

FAULT FINDING SECTION

The following section lists some of the more common faults that can develop in a motor car. For the purpose of this manual, references to diesel engines are of course, first and foremost. The section is divided into various categories and it should be possible to locale faults or damage by referring to the assembly group of the vehicle in question.

The faults are listed in no particular order and their causes are given a number. By referring to this number, it is possible to read off the possible cause and to carry out the necessary remedies, if this is within the scope of your facilities.

ENGINE FAULTS

Engine will not crank:	1, 2, 3, 4
Engine cranks, but will not start:	5, 6, 7, 8
Engine cranks very slowly:	1, 2, 3
Engine starts, but cuts out:	5, 6, 9, 10
Engine misfires in the lower speed ranges:	5, 6, 9, 11
Engine misfires in the higher speed ranges:	5, 6, 11, 12
Continuous misfiring:	5, 6, 7, 10–15, 21, 22
Max. revs not obtained:	5, 6, 12, 22
Faulty idling:	5, 6, 8–11, 13, 15, 16, 21 and 22
Lack of power:	3, 5–11, 13–15, 22
Lack of acceleration:	5–8, 12, 14 to 16
Lack of max. speed:	5–8, 10, 12, 13–15, 22
Excessive fuel consumption:	3, 5, 6, 15, 16
Excessive oil consumption:	16–19
Low compression:	7, 11–13, 16, 20–22

CAUSES AND REMEDIES

1. Fault in the starter motor or its connection. Refer to “Electrical Faults”.
2. Engine oil too thick. This can be caused by using the wrong oil, low temperatures or using oil not suitable for the prevailing climates. Depress the clutch whilst starting (models with manual transmission). Otherwise, refill the engine with the correct oil grade, suitable for diesel engines.
3. Moveable parts of the engine not run-in. This fault may be noticed when the engine has been overhauled. It may be possible to free the engine by adding oil to the fuel for a while.
4. Mechanical fault. This may be due to seizure of the piston(s), broken crankshaft, connecting rods, clutch or other moveable parts of the engine. The engine must be stripped for inspection.
5. Faults in the glow plug system. Refer to “Glow Plug Faults”.
6. Faults in the fuel system. Refer to “Fuel Faults”.
7. Incorrect valve timing. This will only be noticed after the engine has been reassembled after overhaul and the timing belt has been replaced incorrectly. Re-dismantle the engine and check the timing marks on the timing gear wheels.
8. Compression leak due to faulty closing of valves. See also under (7) or leakage past worn piston rings or pistons. Cylinder head gasket blown.
9. Entry of air at inlet manifold, due to split manifold or damaged gasket.
10. Restriction in exhaust system, due to damaged exhaust pipes, dirt in end of exhaust pipe(s), kinked pipe(s), or collapsed silencer. Repair as necessary.
11. Worn valves or valve seats, no longer closing the valves properly. Top overhaul of engine is asked for.
12. Sticking valves due to excessive carbon deposits or weak valve springs. Top overhaul is asked for.
13. Cylinder head gasket blown. Replace gasket and check block and head surfaces for distortion.
14. Camshaft worn, not opening or closing one of the valves properly, preventing proper combustion. Check and, if necessary, fit new camshaft.
15. Incorrect valve (tappet) clearance. There could be a fault in the hydraulic tappets.
16. Cylinder bores, pistons or piston rings worn. Overhaul is the only cure. Fault may be corrected for a while by adding “Piston Seal Liquid” into the cylinders, but will re-develop.
17. Worn valve guides and/or valve stems. Top overhaul is asked for.
18. Damaged valve stem seals. Top overhaul is asked for.
19. Leaking crankshaft oil seal, worn piston rings or pistons, worn cylinders. Correct as necessary.
20. Loose glow plugs, gas escaping past thread or plug sealing washer damaged. Correct.
21. Cracked cylinder or cylinder block. Dismantle, investigate and replace block, if necessary.
22. Broken, weak or collapsed valve spring(s). Top overhaul is asked for.

GLOW PLUG FAULTS

Check a suspect glow plug as follows.

- Remove the glow plug lead from the rear glow plug and from the remaining plugs the bus bars.
- Connect a 12-volt test lamp to the plus terminal of the battery and with the other lead of the lamp touch in turn the connecting threads of each glow plug. The faulty plug is detected when the test lamp does not light up.

Further faults in the glow plug system can develop in the glow plug relay. Check as follows.

- Disconnect the electrical lead from the glow plug on the flywheel end and connect a test lamp between the lead and a good earthing point.
- Disconnect the electrical lead from the coolant temperature sender unit and move away from the earth.
- Turn the ignition key to the "glowing" position and observe the test lamp. The lamp should light up for approx. 25–30 seconds and then switch off.
- Turn the ignition switch "off" and then again to the "glowing" position. The test lamp must light up once more.
- Hold the disconnected lead from the temperature sender unit against earth. The test lamp must switch off.
- Turn the ignition switch "off" and then again to the "glowing" position. The test lamp should light up.
- Operate the starter motor and check that the test lamp remains "on".
- If the above tests cannot be carried out satisfactory, see your dealer.

LUBRICATION SYSTEM FAULTS

The only problem the lubrication system should give is excessive oil consumption or low oil pressure, or the oil warning light not going off.

Excessive oil consumption can be caused by worn cylinder bores, pistons and/or piston rings, worn valve guides, worn valves stem seals or a damaged crankshaft oil seal or leaking gasket on any of the engine parts. In most cases, the engine must be dismantled to locate the fault.

Low oil pressure can be caused by a faulty oil pressure gauge, sender unit or wiring, a defective relief valve, low oil level, blocked oil pick-up pipe for the oil pump, worn oil pump or damaged main or big end bearings. In most cases, it is logical to check the oil level first. All other causes require the dismantling and repair of the engine. If the oil warning light stays on, switch off the engine immediately, as delay could cause complete seizure within minutes.

COOLING SYSTEM FAULTS

Common Faults Are Given as Follows: Overheating, loss of coolant and slow warming-up of the engine.

Overheating:

1. ***Lack of Coolant:*** Open the radiator cap with care to avoid injuries. Never pour cold water into an overheated engine. Wait until the engine cools down and pour in coolant whilst the engine is running.
2. ***Radiator Core Obstructed by Leaves, Insects, etc.***: Blow with air line from the back of the radiator or with a water hose to clean.
3. ***Cooling Fan Not Operating:*** Check the fan for proper cut-in and cut-out temperature. If necessary, change the temperature switch or see your dealer.
4. ***Thermostat Sticking:*** If sticking in the closed position, coolant can only circulate within the cylinder head or block. Remove thermostat and check as described in the section "Cooling".
5. ***Water Hose Split:*** Identified by rising steam from the engine compartment or the front of the vehicle. Slight splits can be repaired with insulation tape. Drive without expansion tank cap to keep the pressure in the system down, to the nearest service station.
6. ***Water Pump Belt Torn:*** Replace and tension belt.
7. ***Water Pump Inoperative:*** Replace water pump.
8. ***Cylinder Head Gasket Blown:*** Replace the cylinder head gasket.

Loss of Coolant:

1. ***Radiator Leaks:*** Slight leaks may be stopped by using a radiator sealing compound (follow the instructions of the manufacturer. In emergency, an egg can be cracked open and poured into the radiator filler neck.
2. **Hose Leaks:** See under 5, "Overheating".
3. **Water Pump Leaks:** Check the gasket for proper sealing or replace the pump.

Long Warming-Up Periods:

1. Thermostat sticking in the open position: Remove thermostat, check and if necessary replace.

DIESEL FUEL SYSTEM FAULTS

Engine is difficult to start or does not start	1 to 13
Engine starts, but stop soon afterwards:	14 to 20
Engine misfires continuously:	1 to 13
Bad idling:	14 to 20
Black, white or blue exhaust smoke:	21 to 29
Lack of power:	30 to 39
Excessive fuel consumption:	40 to 47

CAUSES AND REMEDIES

1. Fuel tank empty. Refuel.
2. Pre-glowing time too short. Operate until warning light goes "off".
3. Cold starting device not operated. Pull cable and push in after approx. 1 mm.
4. Glow plug system inoperative. Refer to "Glow Plug Faults".
5. Electro-magnetic cut-off device, loose or no current. Check cable to cut-off at top of injection pump. Ask a second person to operate ignition key and check if a "click" is heard. Either interrupted current supply or defective cut-off device.
6. Air in fuel system. Operate starter motor until fuel is delivered.
7. Fuel supply faulty. Slacken the injection pipes at injectors, and check if fuel is running out. Other faults: kinked, blocked or leaking injection pipes blocked fuel filter, tank breathing system blocked. Wrong fuel for cold temperatures.

& Injection pipes refitted in wrong order after repair.

9. Injection timing of pump out of phase: have the adjustment checked and corrected.
10. One or more injectors faulty, dirty or incorrect injection pressure. Have injectors repaired or replace them.
11. Injection pump not operating properly. Fit an exchange pump or have it repaired.
12. Valves not opening properly.
13. Compression pressures too low. See item "8" under "Engine Faults".
14. Idle speed not properly adjusted. Adjust.
15. Throttle cable not properly adjusted or sticking. Re-adjust or free-off.
16. Fuel hose between filter and pump not tightened properly. Tighten connections.
17. Rear mounting of injection pump loose or cracked. Tighten or replace.
18. See items 6, 7, 9, 11, 12 and 13
19. Engine mounting not tightened properly or worn. Tighten or replace.
20. Sticking accelerator pedal. Free-off pedal.
21. Engine not at operating temperature. Check exhaust smoke colour again when engine is warm.
22. Too much acceleration at low revs. Use individual gears in accordance with acceleration.
23. Air cleaner contaminated. Clean or replace.
24. Fuel filter contaminated. Replace.
25. Max. speed adjustment incorrect. Re-adjust.
26. Injectors are dripping. Have them checked or replace faulty ones.
27. Injector nozzles sticking or broken. Replace injector.
28. Injection pressure too low. Have injectors checked and adjusted.
29. See items 9, 11, 12 and 13
30. Throttle cable travel restricted. Re-adjust. Check that floor mats cannot obstruct pedal movement.
31. Throttle cable not correctly adjusted. Re-adjust.
32. Operating lever loose on pump. Re-tighten.
33. Max. speed not obtained. Re-adjust max. speed or have it adjusted.
34. Injector pipes restricted in diameter (near connections). Disconnect pipes and check that diameter is at least 2.0 mm (0.08 in.).
35. Heat protection sealing gaskets under injectors not sealing or damaged. Remove injectors and check. Replace if necessary. Fit the washers correctly.
36. Injection pressure of injectors wrong. Have them re-adjusted.
37. See items 6, 7, 9, 11 and 13
38. See item 20.
39. See items 23, 24, 26 and 27.
40. Road wheels dragging. Brakes seized or wheel bearings not running freely.
41. Engine not running "free". Refers to new or overhauled engine.
42. Fuel system leaking. Check hoses, pipes, filter, injection pump, etc. for leaks.
43. Fuel return line blocked. Clean with compressed air if possible.
44. Idle speed too high. Re-adjust.
45. Max. speed too high. Re-adjust.
46. See items 10, 11, 12 and 13.
47. See items 24, 26, 27 and 28.

CLUTCH FAULTS

Clutch slipping:	1, 2, 3, 4, 5
Clutch will not disengage fully:	4, 6–12, 14
Whining from clutch when pedal is depressed:	13
Clutch judder:	1, 2, 7, 10–13
Clutch noise when idling:	2, 3
Clutch noise during engagement:	2

CAUSES AND REMEDIES

1. Insufficient clutch free play at pedal.
2. Clutch disc linings worn, hardened, oiled-up, loose or broken. Disc distorted or hub loose. Clutch disc must be replaced.
3. Pressure plate faulty. Replace clutch.
4. Air in hydraulic system. Low fluid level in clutch cylinder reservoir.
5. Insufficient play at clutch pedal and clutch release linkage. Rectify as described.
6. Excessive free play in release linkage (only for cable operated clutch, not applicable). Adjust or replace worn parts.
7. Misalignment of clutch housing. Very rare fault, but possible on transmissions with separate clutch housings. Re-align to correct.
8. Clutch disc hub binding on splines of main drive shaft (clutch shaft) due to dirt or burrs on splines. Remove clutch and clean and check splines.
9. Clutch disc linings loose or broken. Replace disc.
10. Pressure plate distorted. Replace clutch.
11. Clutch cover distorted. Replace clutch.
12. Fault in transmission or loose engine mountings.
13. Release bearing defective. Remove clutch and replace bearing.
14. A bent clutch release lever. Check lever and replace or straighten, if possible.

- The above faults and remedies are for hydraulic and mechanical clutch operation and should be read as applicable to the model in question, as the clutch fault finding section is written for all types of clutch operation.

STEERING FAULTS

Steering very heavy:	1–6
Steering very loose:	5, 7–9, 11–13
Steering wheel wobbles:	4, 5, 7–9, 11–16
Vehicle pulls to one side:	1, 4, 8, 10, 14–18
Steering wheel does not return to centre position:	1–6, 18
Abnormal tyre wear:	1, 4, 7–9, 14–19
Knocking noise in column:	6, 7, 11, 12

CAUSES AND REMEDIES

1. Tyre pressures not correct or uneven. Correct.
2. Lack of lubricant in steering.
3. Stiff steering linkage ball joints. Replace ball joints in question.
4. Incorrect steering wheel alignment. Correct as necessary.
5. Steering needs adjustment. See your dealer for advice.
6. Steering column bearings too tight or seized or steering column bent. Correct as necessary.
7. Steering linkage joints loose or worn. Check and replace joints as necessary.
8. Front wheel bearings worn, damaged or loose. Replace bearing.
9. Front suspension parts loose. Check and correct.
10. Wheel nuts loose. Re-tighten.
11. Steering wheel loose. Re-tighten nut.
12. Steering gear mounting loose. Check and tighten.
13. Steering gear worn. Replace the steering gear.
14. Steering track rods defective or loose.
15. Wheels not properly balanced or tyre pressures uneven. Correct pressures or balance wheels.
16. Suspension springs weak or broken. Replace spring in question or both.
17. Brakes are pulling to one side. See under "Brake Faults".
18. Suspension out of alignment. Have the complete suspension checked by a dealer.
19. Improper driving. We do not intend to tell you how to drive and are quite sure that this is not the cause of the fault.

BRAKE FAULTS

Brake Failure: Brake shoe linings or pads excessively worn, incorrect brake fluid (after overhaul), insufficient brake fluid, fluid leak, master cylinder defective, wheel cylinder or caliper failure. Remedies are obvious in each instance.

Brakes Ineffective: Shoe linings or pads worn, incorrect lining material or brake fluid, linings contaminated, fluid level low, air in brake system (bleed brakes), leak in pipes or cylinders and master cylinder defective. Remedies are obvious in each instance.

Brakes Pull to One Side: Shoes or linings worn, incorrect linings or pads, contaminated linings, drums or discs scored, fluid pipe blocked, unequal tyre pressures, brake back plate or caliper mounting loose, wheel bearings not properly adjusted and wheel cylinder seized. Rectify as necessary.

Brake Pedal Spongy: Air in hydraulic system. System must be bled of air.

Pedal Travel Too Far: Linings or pads worn, drums or discs scored, master cylinder or wheel cylinders defective and system needs bleeding. Rectify as necessary.

Loss of Brake Pressure: Fluid leak, air in system, leak in master or wheel cylinders, brake servo not operating (vacuum hose disconnected or exhauster pump not operating). Place vehicle on dry ground and depress brake pedal. Check where fluid runs out and rectify as necessary.

Brakes Binding: Incorrect brake fluid (boiling), weak shoe return springs, basic brake adjustment incorrect (after fitting new rear shoes), piston in caliper of wheel cylinder seized, push rod play on master cylinder insufficient (compensation port obstructed), handbrake adjusted too tightly. Rectify as necessary. Swelling of cylinder cups through use of incorrect brake fluid could be another reason.

Handbrake Ineffective: Brake shoe linings worn, linings contaminated, operating lever on brake shoe seized, brake shoes or handbrake need adjustment. Rectify as necessary.

Excessive Pedal Pressure Required: Brake shoe linings or pads worn, linings or pads contaminated, brake servo vacuum hose (for brake servo) disconnected or wheel cylinders seized. Exhauster pump not operating. Rectify as necessary.

Brakes Squealing: Brake shoe linings or pads worn so far that metal is grinding against drum or disc. Inside of drum is full of lining dust. Remove and replace, or clean out the drum(s). Do not inhale brake dust.

Note: Any operation on the steering and brake systems must be carried out with the necessary care and attention. Always think of your safety and the safety of other mad users. Make sure to use the correct fluid for the power-assisted steering and the correct brake fluid.

Faults in an ABS system should be investigated by a dealer.
Change the brake fluid in regular Intervals (approx. every two years).

ELECTRICAL FAULTS

Starter motor failure:	2–5, 8, 9
No starter motor drive:	1–3, 5–7
Slow cranking speed:	1 to 3
Charge warning light remains on:	3, 10, 12
Charge warning light does not come on:	2, 3, 9, 11, 13
Headlamp failure:	2, 3, 11, 13, 14
Battery needs frequent topping-up:	11
Direction indicators not working properly:	2, 3, 9, 13, 14
Battery frequently discharged:	3, 10, 11, 12

CAUSES AND REMEDIES

1. Tight engine. Check and rectify.
2. Battery discharged or defective. Re-charge battery or replace if older than approx. two years.
3. Interrupted connection in circuit. Trace and rectify.
4. Starter motor pinion jammed in flywheel. Release.
5. Also 6, 7 and 8. Starter motor defective, no engagement in flywheel, pinion or flywheel worn or solenoid switch defective. Correct as necessary.
9. Ignition/starter switch inoperative. Replace.
10. Drive belt loose or broken. Adjust or replace.
11. Regulator defective. Adjust or replace.

12. Generator inoperative. Overhaul or replace.
13. Bulb burnt out. Replace bulb.
14. Flasher unit defective. Replace unit.

Wiring Diagram Index

Cable Colour Code

ws	=	white
sw	=	black
ro	=	red
br	=	brown
gn	=	green
bl	=	blue
gr	=	grey
li	=	violet
ge	=	yellow
or	=	orange

Battery, Ignition/Starter Switch

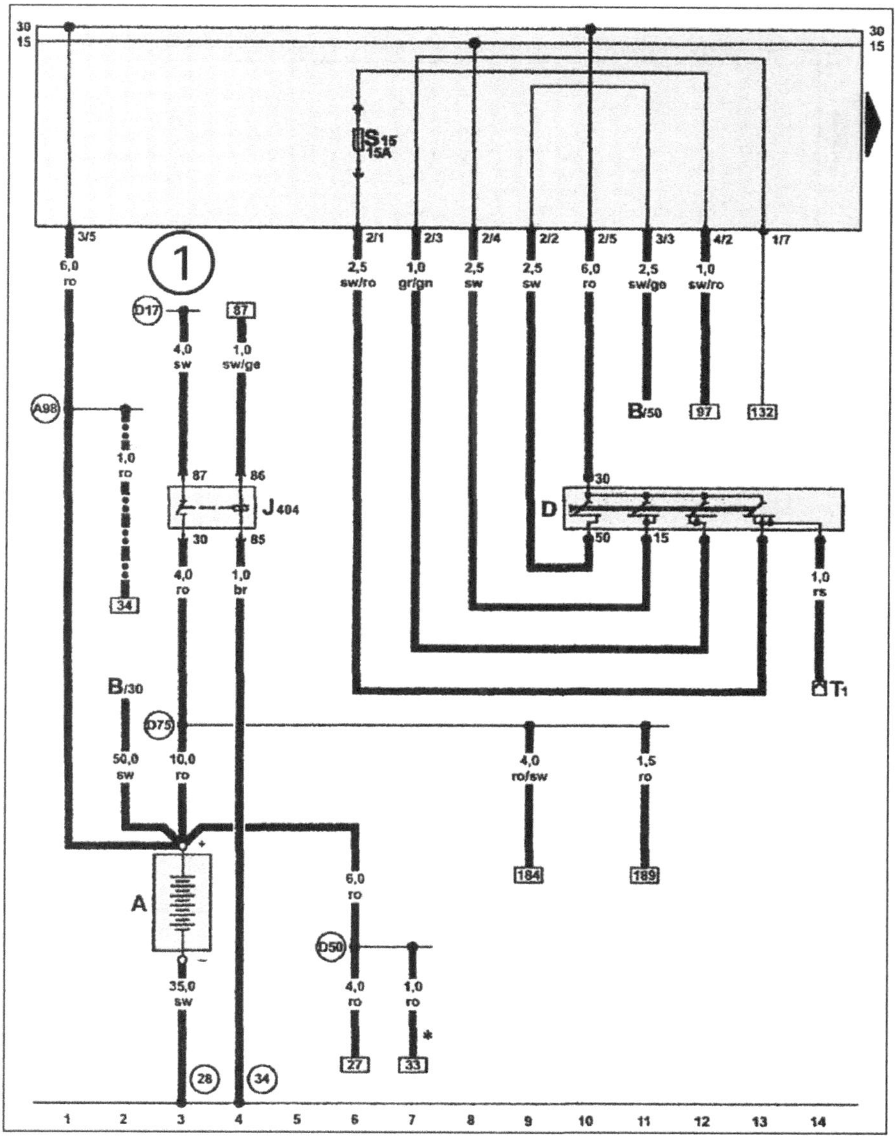

A Battery

B Starter motor

D Ignition/starter switch

J404 Relief relay, under driver's seat

T1 Single connector, behind steering column, cover, with airbag

In Circle:

28 Earth point, bulkhead, left

A88 Plus connection (30), in instrument panel wiring loom. to January 1999

D17 Plus connection (15), engine cable harness

D60 Plus connection (39), engine compartment cable harness

D75 Plus connection (30), engine

* Vehicles from February 1999

** Vehicles to January 1999

34 Earth point, below driver's seat

Fresh Air Blower, Two-way Valve for Fresh/Recirculation Air

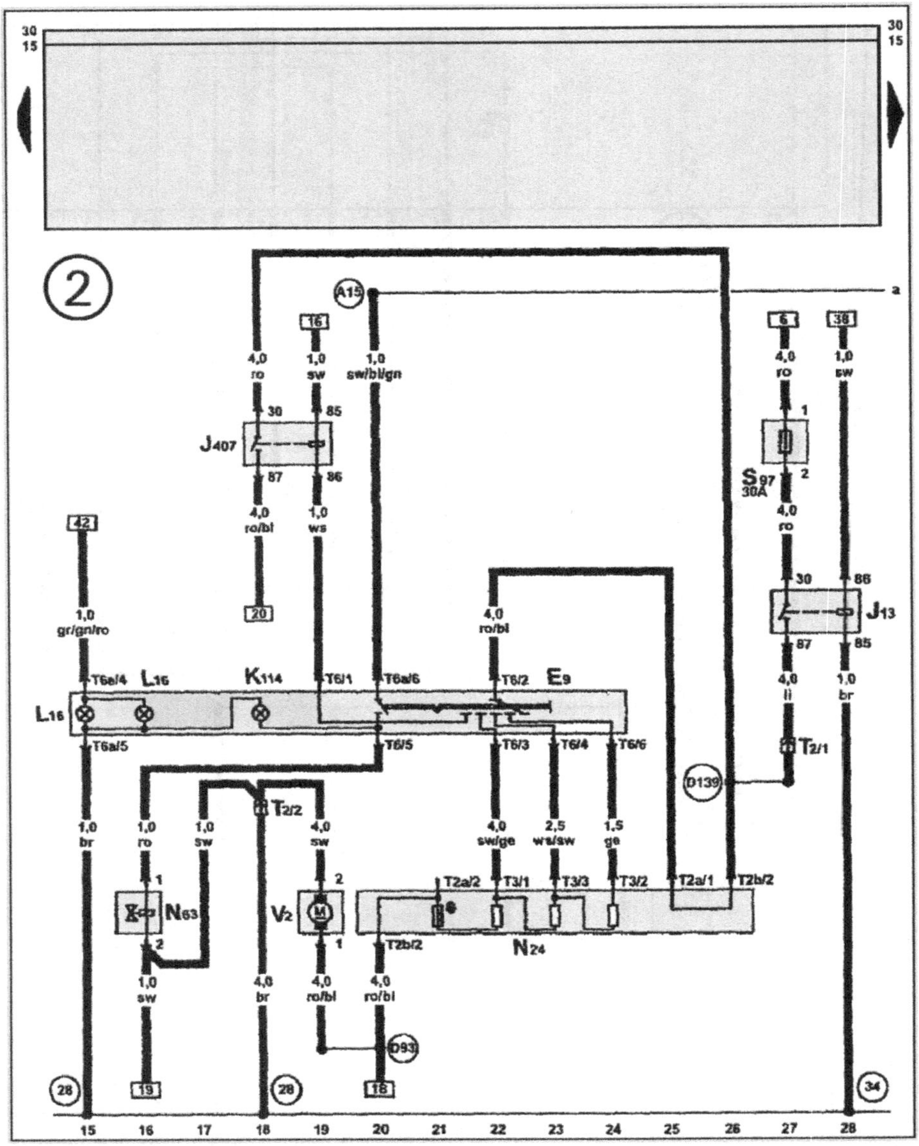

Switch, fresh air blower
Relay, fresh air blower, below seat
Relay, blower, 4th stage
Warning light, fresh air/rec. air operation
Bulb, fresh air regulator
Resistor, fresh air blower
Two-way valve, fresh air/rec. air flap
Fuse 1, fresh air blower, below seat
2-pin connector
2-pin connector
2-pin connector

T3 3-pin connector
T6 6-pin connector
T6a 6-pin connector
V2 Fresh air blower
In Circle:
28 Earth point, bulkhead, left
34 Earth point, driver's seat
A15 Plus connection (15), dashboard harness
D93 Connection (1), engine comp. harness
D139 Connection (2), engine comp. harness

Instrument Insert, Speedometer

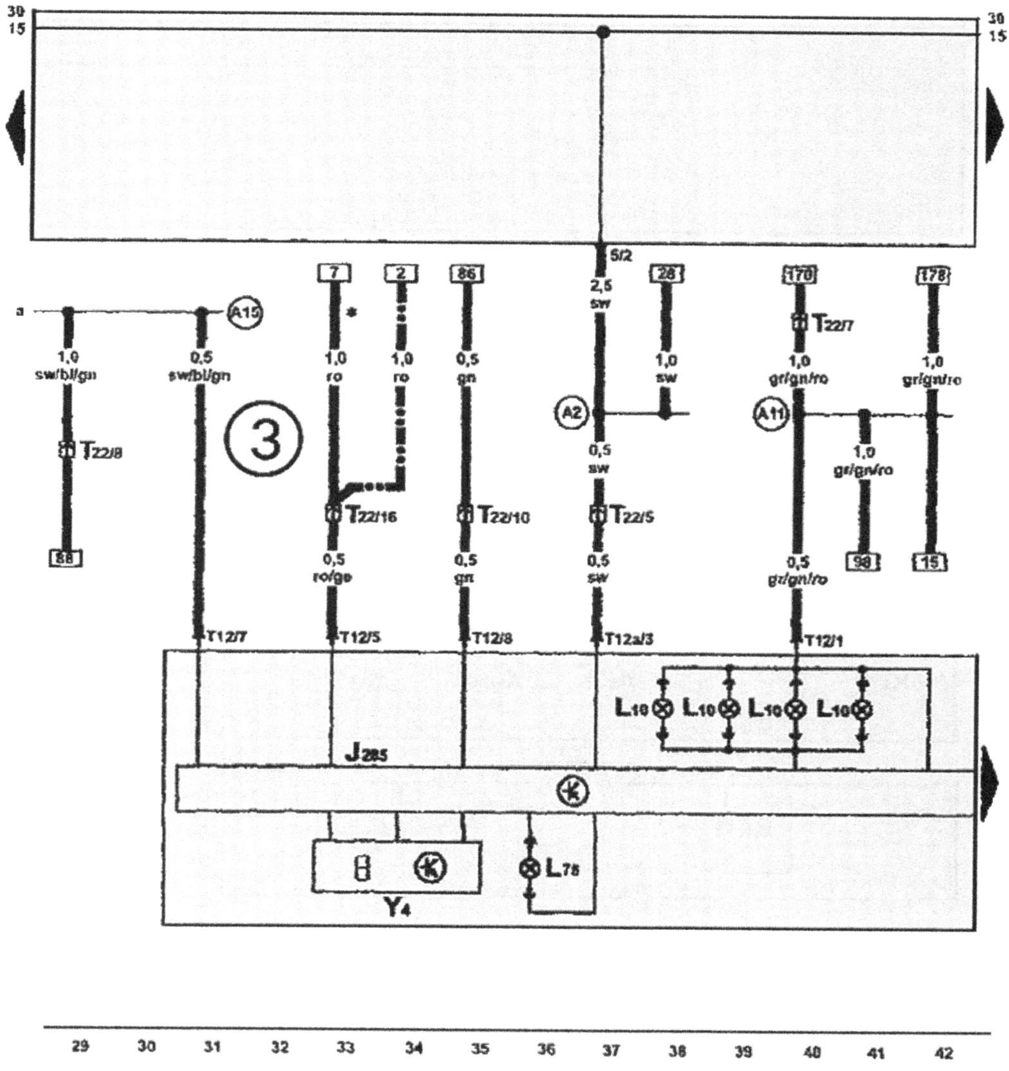

L285	Control unit with display in dash panel
L10	Bulb, dashboard lighting
L75	Bulb, digital display
T12	12-pin connector
T12a	12-pin connector
T22	22-pin connector, below dashboard, left
Y4	Distance recorder

In Circle:

A2	Positive connection (15), in dashboard cable harness
A11	Positive connection (58b), in dashboard cable harness
A15	Positive connection (15), in dashboard
*	Only from February 1999
**	Only to February 1999

Instrument Insert, Warning Lights for Flashers, Brake System, Handbrake

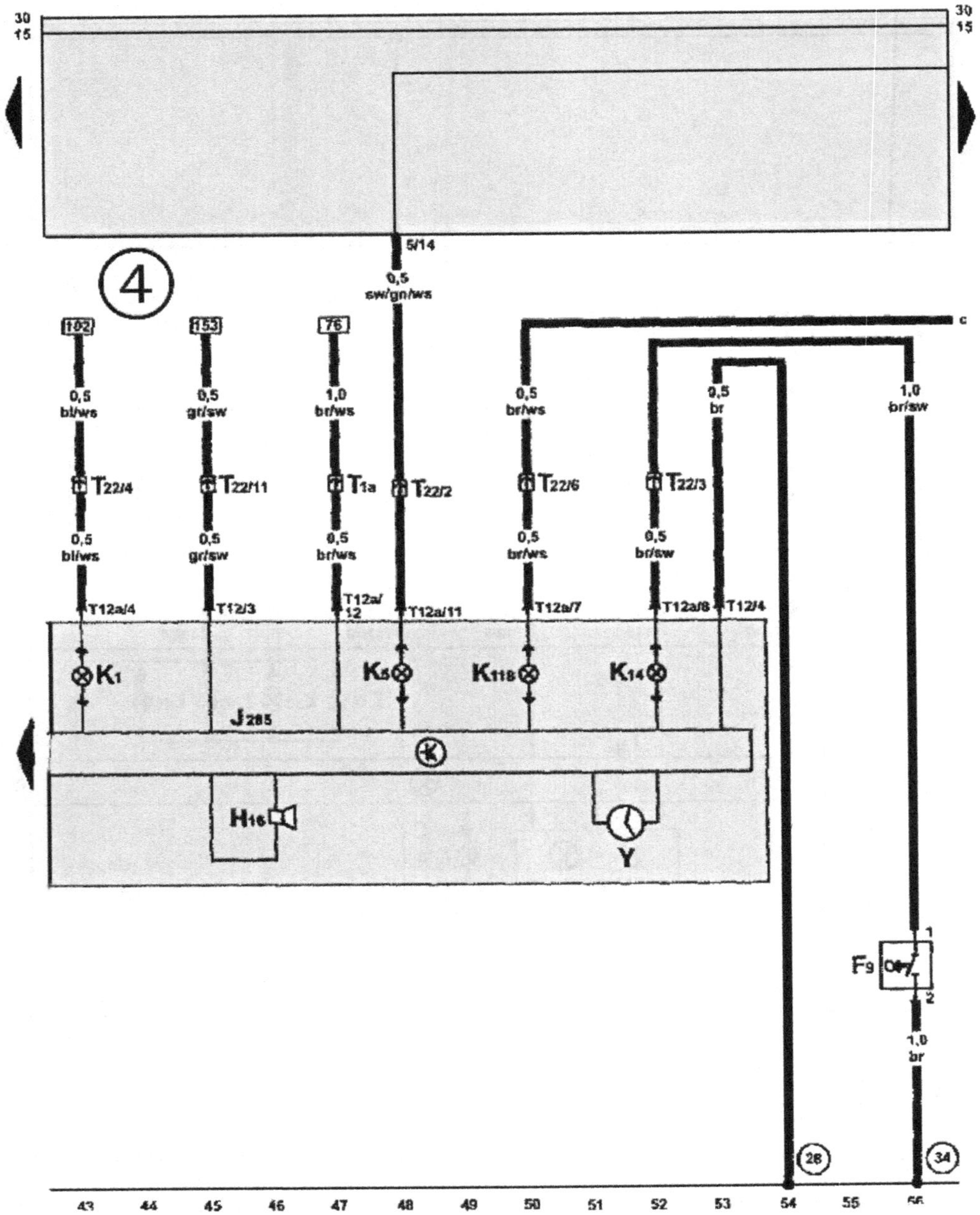

F9	Switch on handbrake lever	T12	12-pin connector
H16	Warning buzzer, "lights on"	T12a	12-pin connector
J285	Control unit with display in dashboard	T22	22-pin connector, below dashboard
K1	Main beam warning light	Y	Analog clock
K5	Flasher warning light	In Circle:	
K14	Handbrake warning light	28	Earth point, bulkhead, left
K118	Warning light, brake system	34	Earth point, below driver's seat
T1a	1-pin connector, below dashboard, left		

Brake Pad Wear Indicator, Brake Fluid level Warning

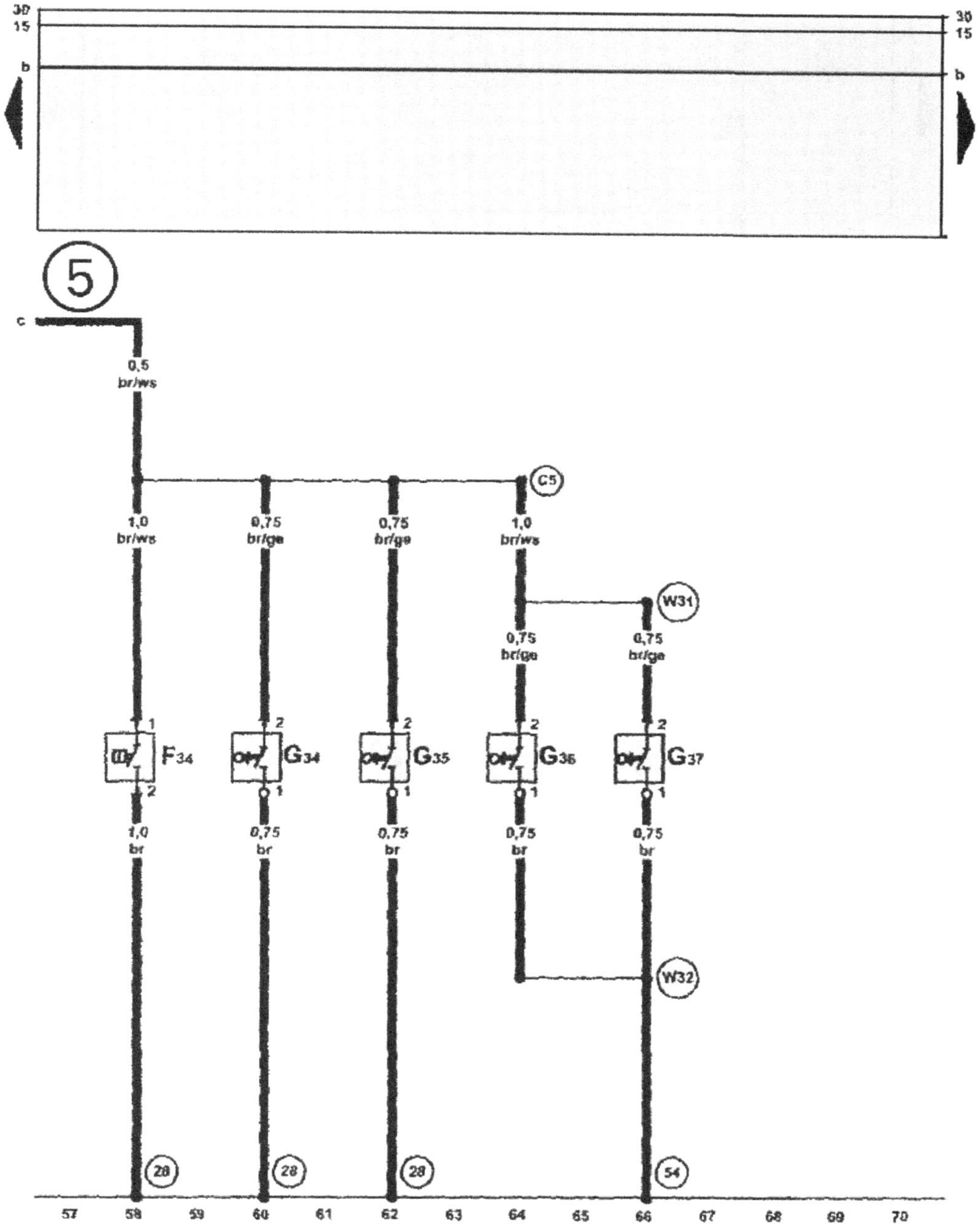

F34 Warning contact, brake fluid
G34 Sensor, pad wear indicator, front left
G35 Sensor, pad wear indicator, front right
G36 Sensor, pad wear indicator, rear left
G37 Sensor, pad wear indicator, rear right
In Circle:
28 Earth point, bulkhead, left
54 Earth point, cover panel, rear
C5 Connection 1, pad wear indicator, in cable harness, front
W31 Connection 1, pad wear indicator, in cable harness, rear
W32 Connection 2, pad wear indicator, in cable harness, rear

Door Contact Switches, Front and Rear Interior Lights

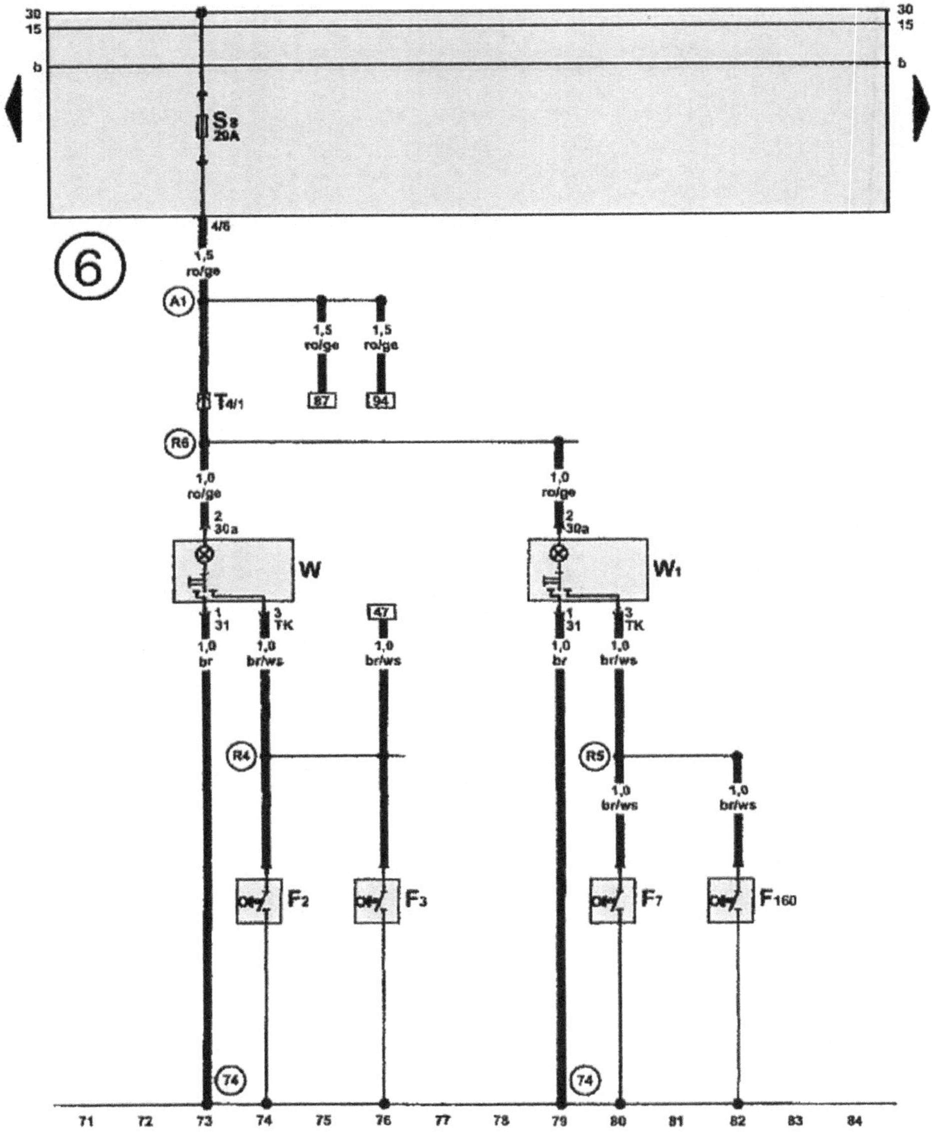

Door contact switch, driver's side
Door contact switch, passenger's side
Door contact switch, R.H. rear (side door)
Door contact switch, L.H. rear
4-pin connector, under dashboard, left
Interior light, front
Interior light, rear/cargo room

In Circle:

74	Earth point, under roof lining, front
A1	Positive connection (30a), in dashboard cable harness
R4	Connection 1, in interior light/door contact switch cable harness
R5	Connection 2, interior light harness
R6	Connection 1, interior light harness

Radio, Self-Diagnosis, Horn, Cigar Lighter

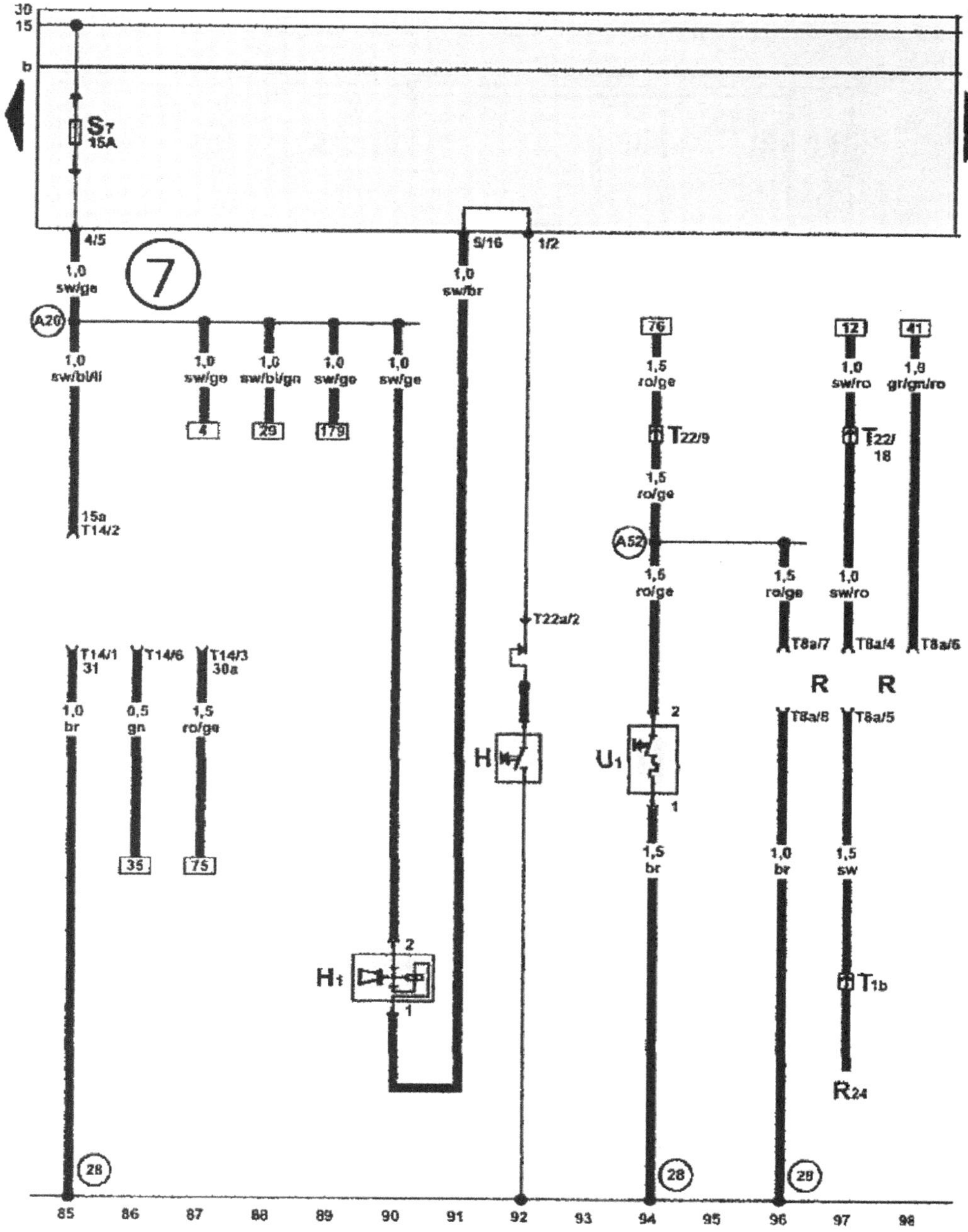

H	Horn button	T22	22-pin connector, under dashboard, left
H1	Horn	T22a	22-pin connector, on steering column switch
R	Radio	U1	Cigar lighter
R24	Arial amplifier	In Circle:	
T1b	1-pin connector, near interior light, front	28	Earth point, bulkhead, left
T14	14-pin connector, in foot well, front, For self-diagnosis control	A20	Connection, dashboard cable harness
		A52	Positive connection, dashboard harness

Headlamp Side Lights

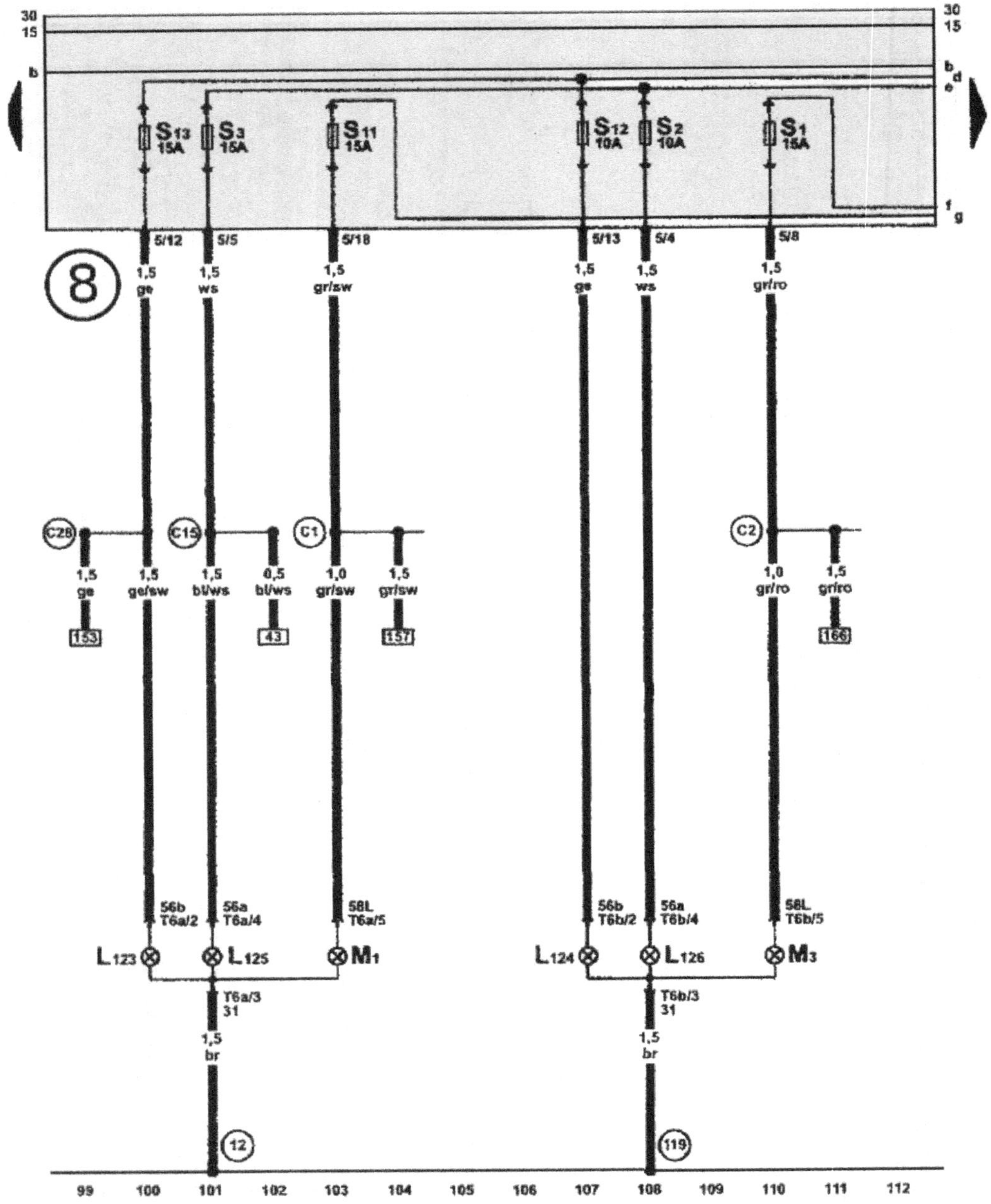

L123	Bulb, L.H. dipped beam
L124	Bulb, R.H. dipped beam
L125	Bulb, L.H. main beam
L126	Bulb, R.H. main beam
M1	Bulb, L.H. side light
M3	Bulb. R.H. side light
T6a	6-pin connector, L.H. headlamp
T6b	6-pin connector, R.H. headlamp

In Circle:

12	Earth point, engine comp., front left
119	Earth connection, headlamp cable harness
C1	Positive connection (58L), cable harness, front left
C2	Positive connection (58R), harness, front left
C15	Connection, in headlamp cable harness
C28	Connection (56b),in headlamp cable harness

Front Indicator Lights, Side Repeater Lights

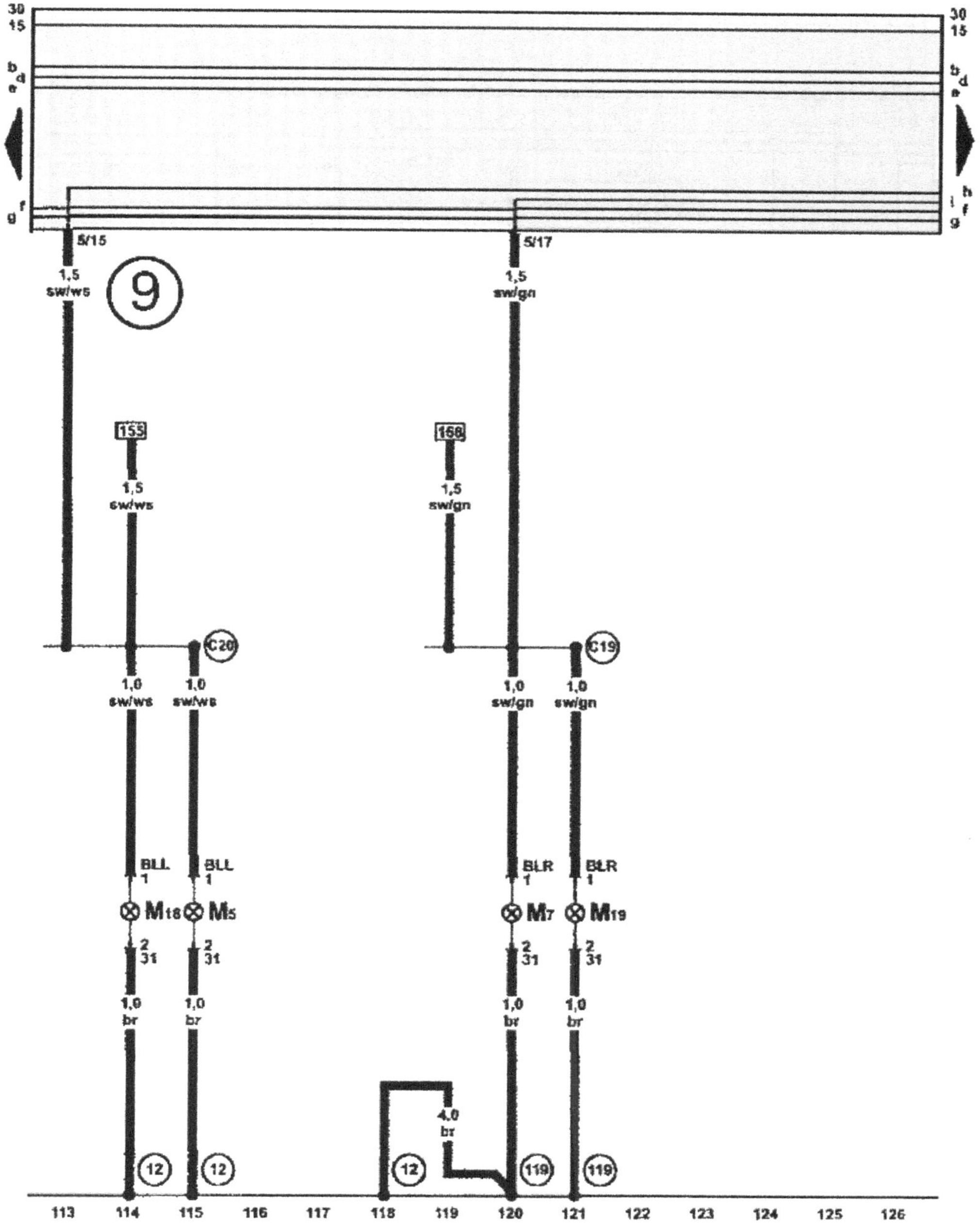

M5 Bulb, front, L.H. flasher lamp
M7 Bulb. Front R.H. flasher lamp
M18 Bulb, L.H. side repeater lamp
M19 Bulb, R.H. side repeater lamp
In Circle:
12 Earth point, engine compartment, L.H. front
119 Earth connection, headlamp cable harness
C19 Connection (L.H. flasher lamp), in headlamp cable harness
C20 Connection (R.H. flasher lamp), in headlamp cable harness

Dipped Beam, Headlamp Flasher, Flasher Switch, Light Switch, Sidelight Switch, Hazard Light Switch

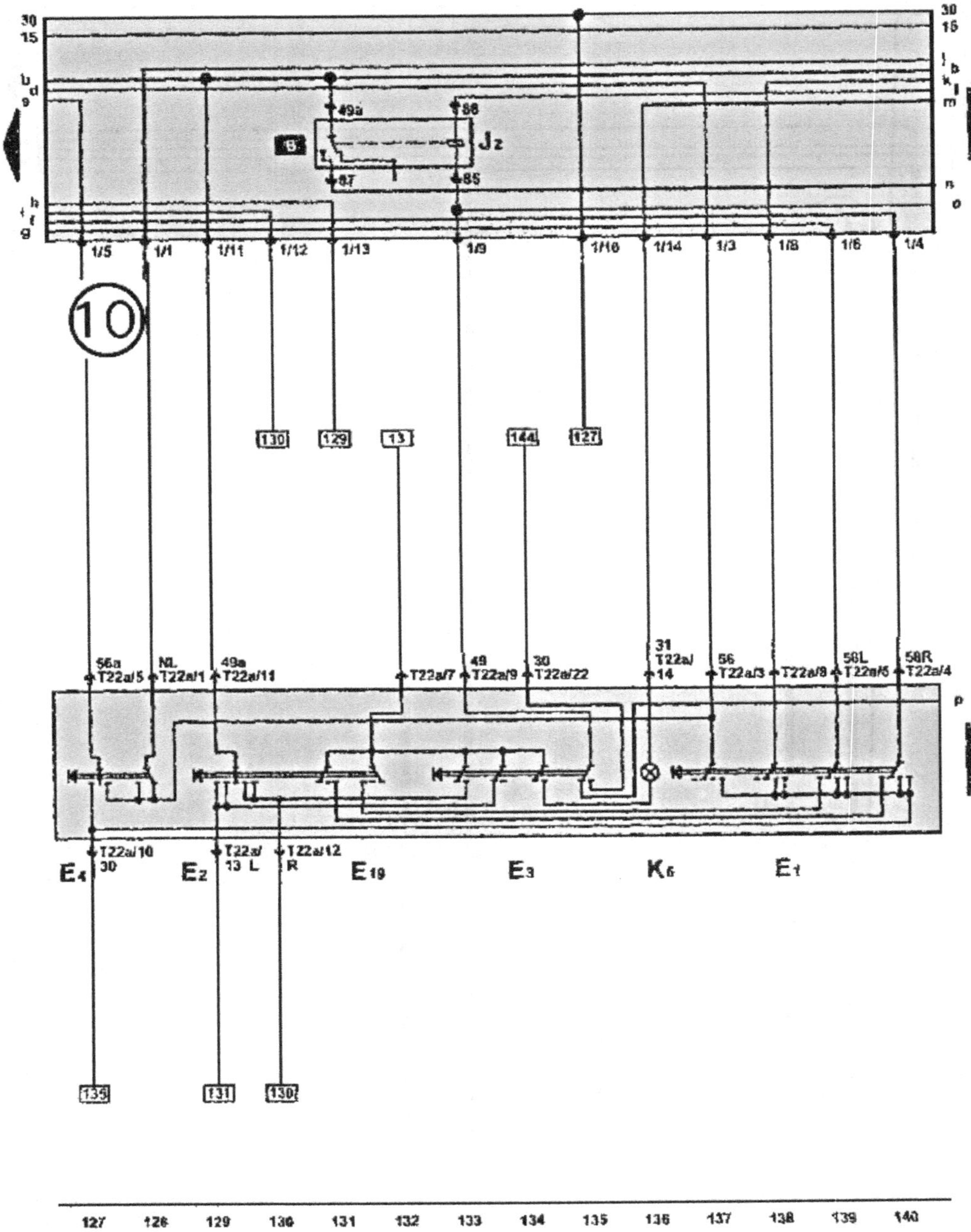

Lighting switch		E19	Side light switch
Flasher lamp switch		J2	Relays, hazard warning lights
Hazard warning light switch		K6	Warning light, hazard warning lights
Dipped beam and headlamp flasher switch		T22a	22-pin connector

Windscreen Wipers, Fog Lamp Switch

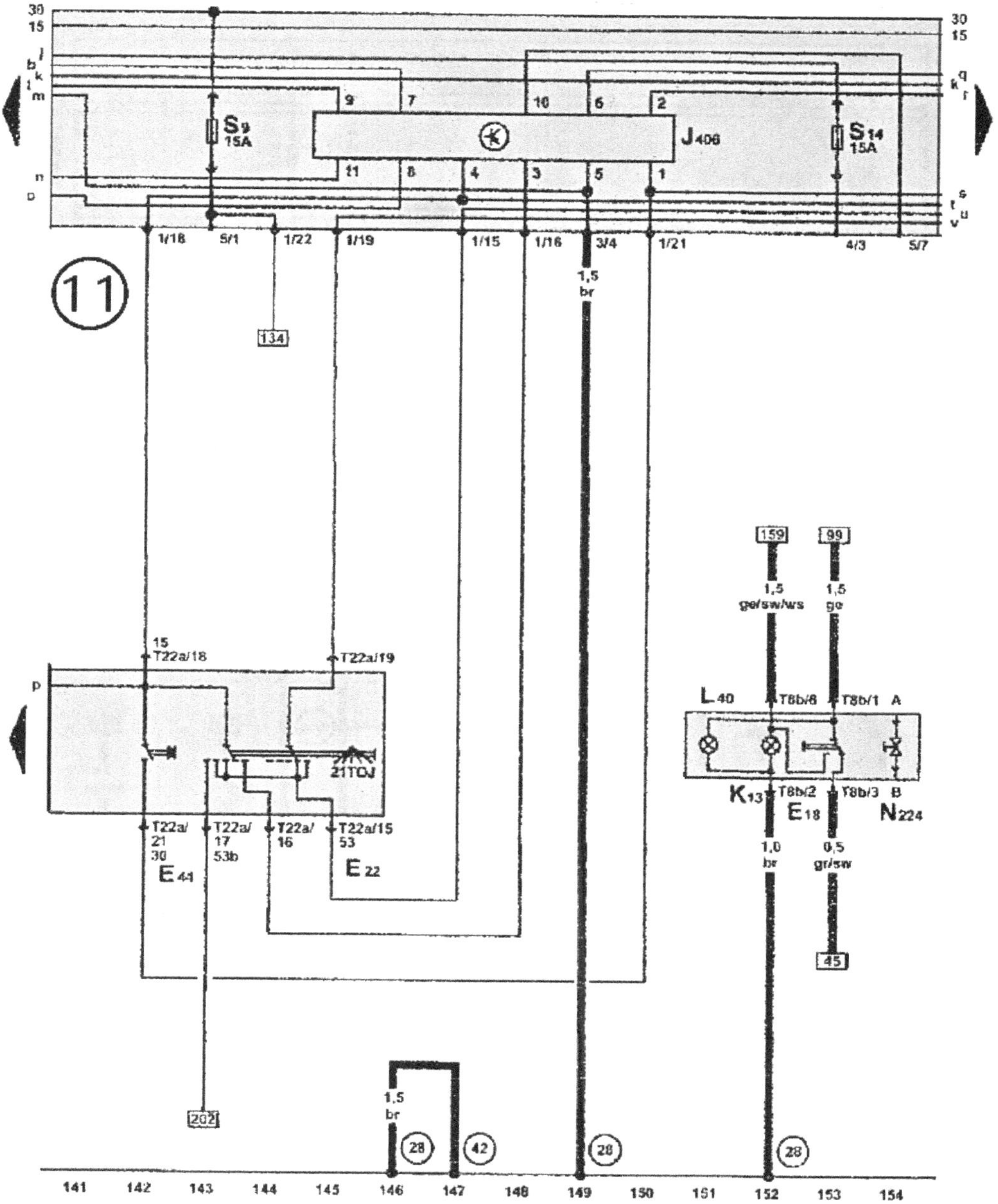

E18	Switch, rear fog lamp	K13	Warning light, rear fog lamp
E22	Windscreen wiper switch for	L40	Bulb, fog light switch, front and and rear fog lights
E44	Switch for washer pump (automatic Wash/wipe operation and headlamp Cleaning system)	N224	Valve for headlamp beam adjustment
		T8b	8-pin connector
J406	Control unit for wash/wipe automatic	T22a	22-pin connector

In Circle:

28	Earth point, bulkhead, left	42	Earth point on steering column

Rear Flasher Lamps,Brake Lights, Rear Lights, Reversing Brake Light Switch, Fog Lamps, High Brake Lights

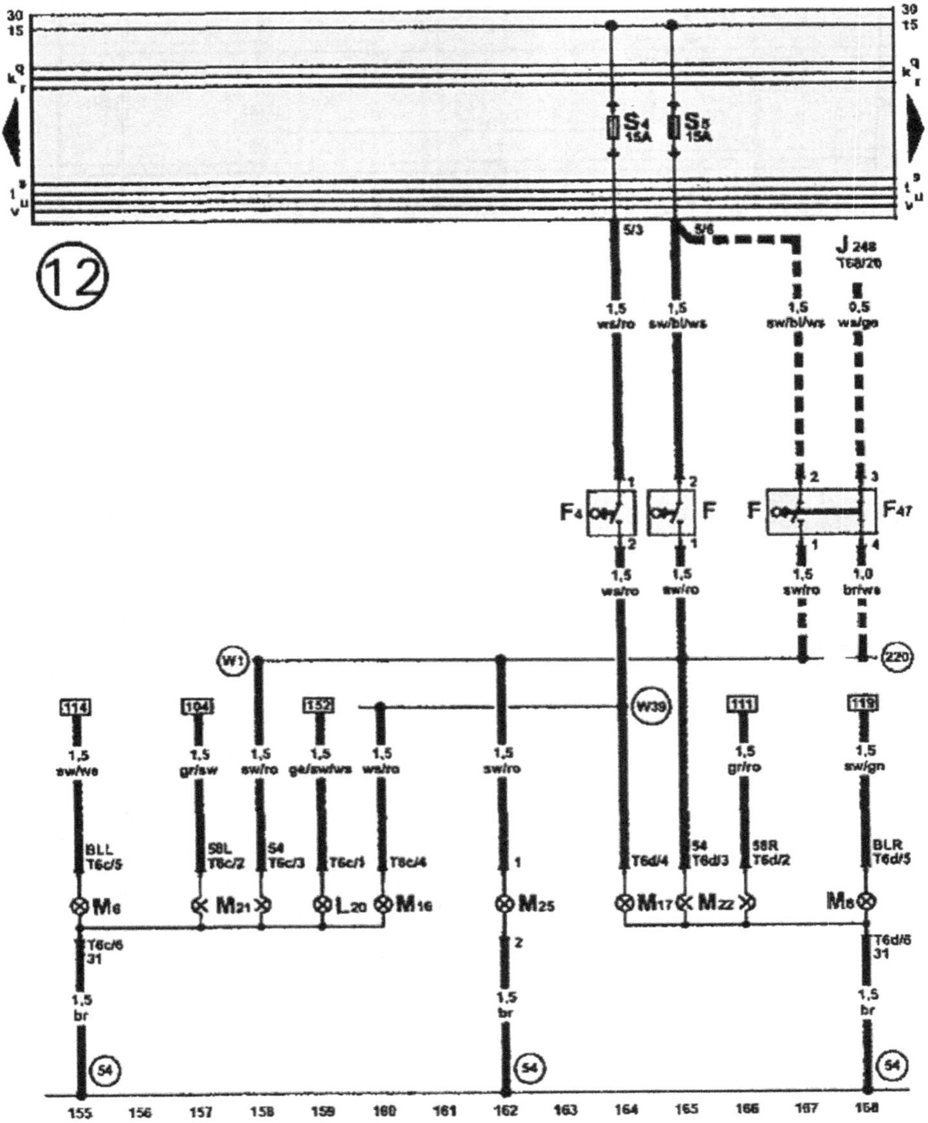

F	Brake light switch	M16	Bulb, L.H. reversing light
F4	Switch, reversing light	M17	Bulb, R.H. reversing light
F47	Brake pedal switch for diesel injection system, foot well	M21	Bulb, L.H. brake and rear light
J248	Control unit, diesel injection, near battery	M22	Bulb. R.H. brake and rear light
L20	Bulb, rear fog lamp	M25	Bulb, high brake light
M6	Bulb, L.H. rear flasher lamp	T6a	6-pin connector
M8	Bulb, R.H. rear flasher lamp	T6d	6-pin connector
In Circle:		T68	68-pin connector
54	Earth point, rear end panel	W1	Connection (54), rear cable harness
220	Earth connection, engine cable harness	W39	Connection (reversing light), in rear cable harness
			- - Only engine AHD and AGX

Number Plate Lights, Heated Rear Window, Left

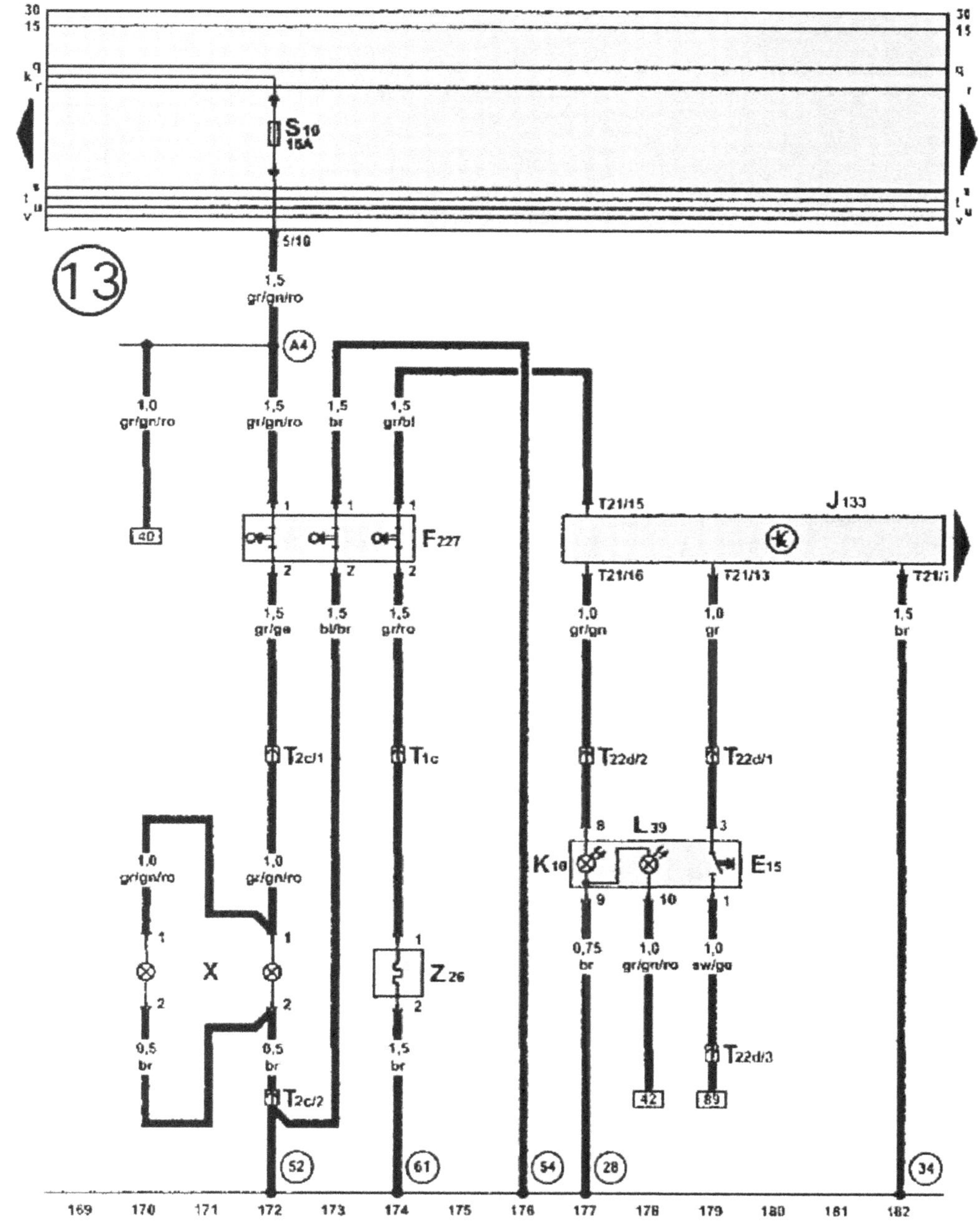

E15	Switch, heated rear screen	T22d	22-pin connector
F227	Door contact switch, left (L.H. rear door)	X	Number plate lighting
J133	Delay relay for heated rear screen	In Circle:	
K10	Bulb, warning light, heated rear screen	28	Earth point, bulkhead, left
L39	Bulb, heated rear screen switch	34	Earth point, under driver's seat
T1c	1-pin connector, in L.H. rear door	52	Earth point, L.H. rear swing door
T2c	2-pin connector, in L.H. rear door	54	Earth point, rear end panel
T21	21-pin connector	61	Earth point in L.H. rear swing door

Heated Rear Window, Right

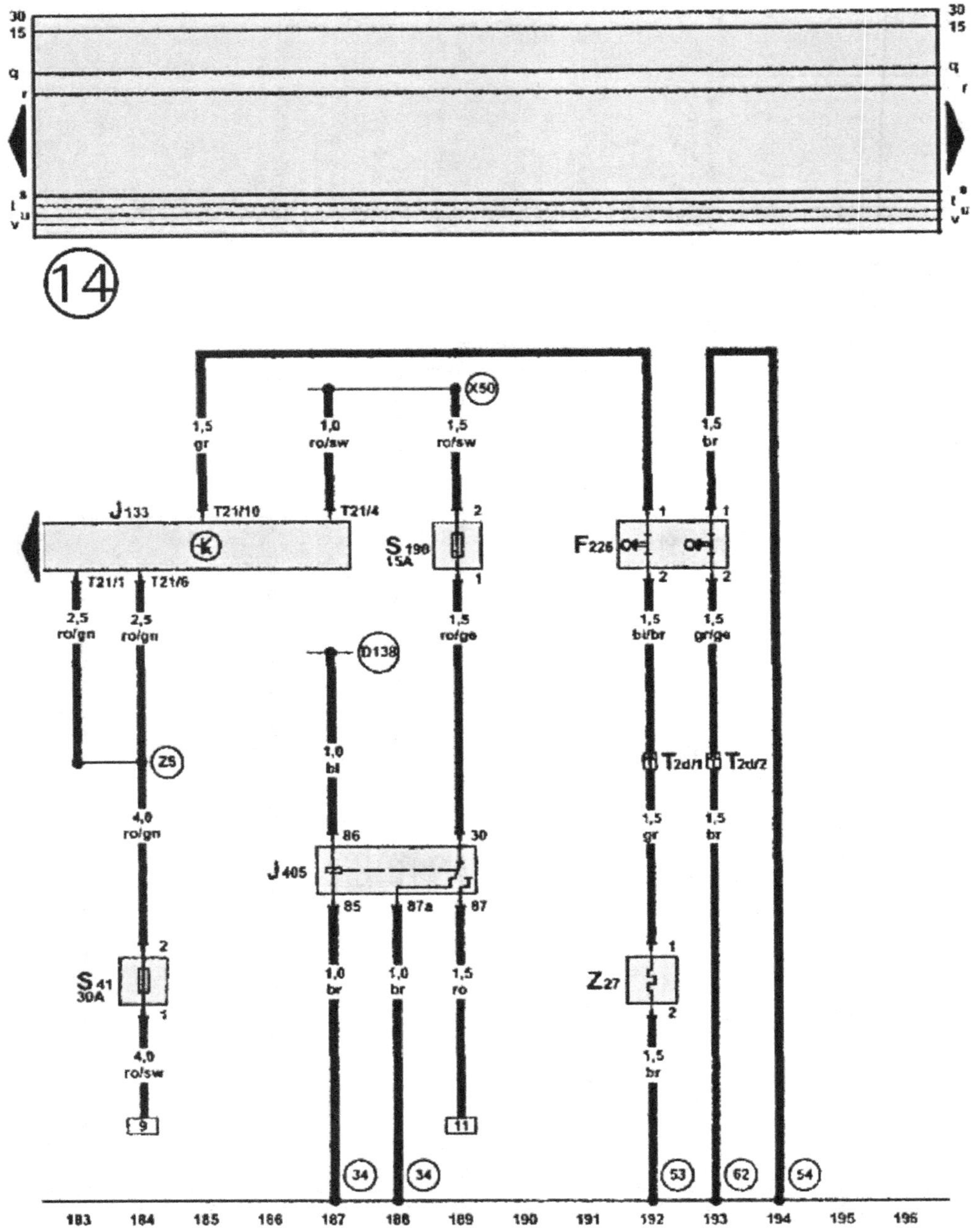

F226 Door contact switch, rear door
J133 Delay relay for heated rear screen
J405 Relief relay, under driver's seat
S41 Fuse, heated rear screen, below seat
S190 Fuse for voltage supply, below seat
T2d 2-pin connector, R.H. rear swing door
T21 21-pin connector
Z27 Heated rear screen, R.H. rear swing door

In Circle:
34 Earth point, below driver's seat
53 Earth point 1, R.H. rear swing door
54 Earth point, cover end panel, rear
62 Earth point 2, R.H. rear swing door
D138 Connection, engine comp. Cable harness
X50 Positive connection (30)
Z5 Connection, heated rear screen harness

Screen Wipers and Washers

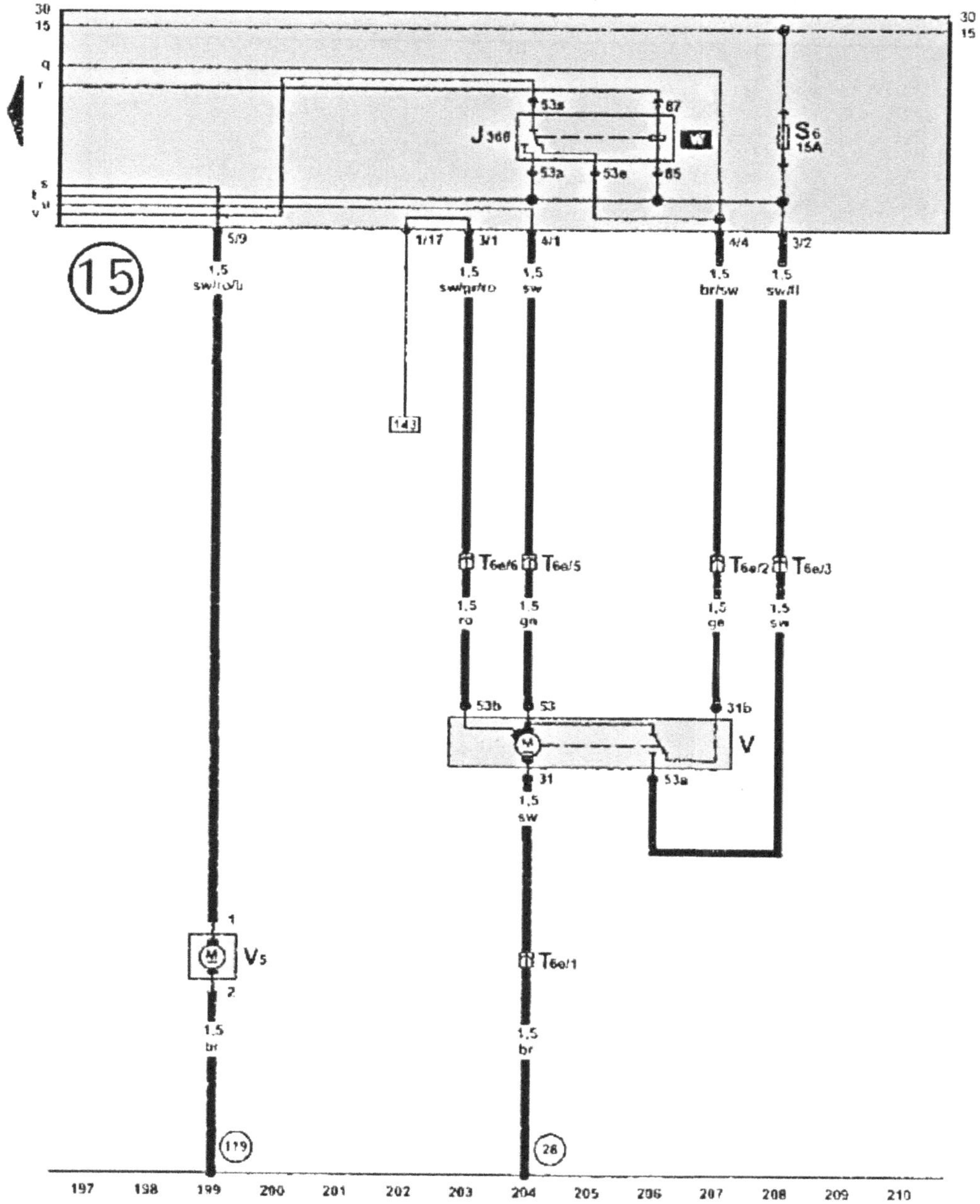

J368	Relay, windscreen wiper motor		In Circle:	
T6e	6-pin connector, near wiper motor		28	Earth point, bulkhead, left
V	Windscreen wiper motor		119	Earth connection, headlamp cable harness
V5	Pump, windscreen washer system			

Alternator, Starter Motor, 2.5 Litre Turbo Diesel from Sept. 2000

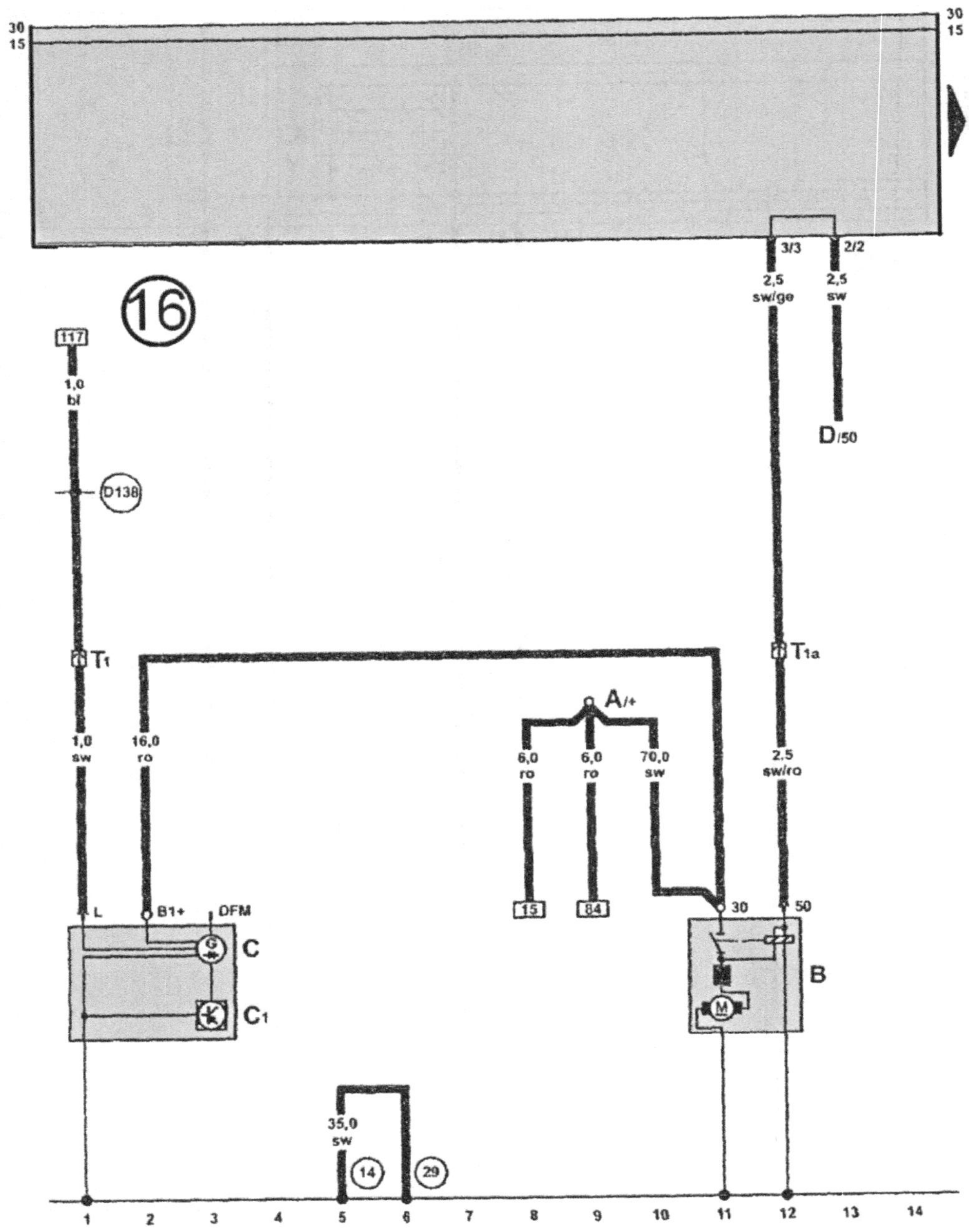

A	Battery	T1a	1-pin connector, near battery carrier
B	Starter motor	In Circle:	
C	Alternator	14	Earth point in gearbox
C1	Voltage regulator	29	Earth point, near gearbox tunnel
D	Ignition/starter switch	D138	Connection (D+), engine compartment cable harness
T1	1-pin connector, near battery carrier		

Alternator, Starter Motor, 2.5 Litre Turbo Diesel from Nov. 1999

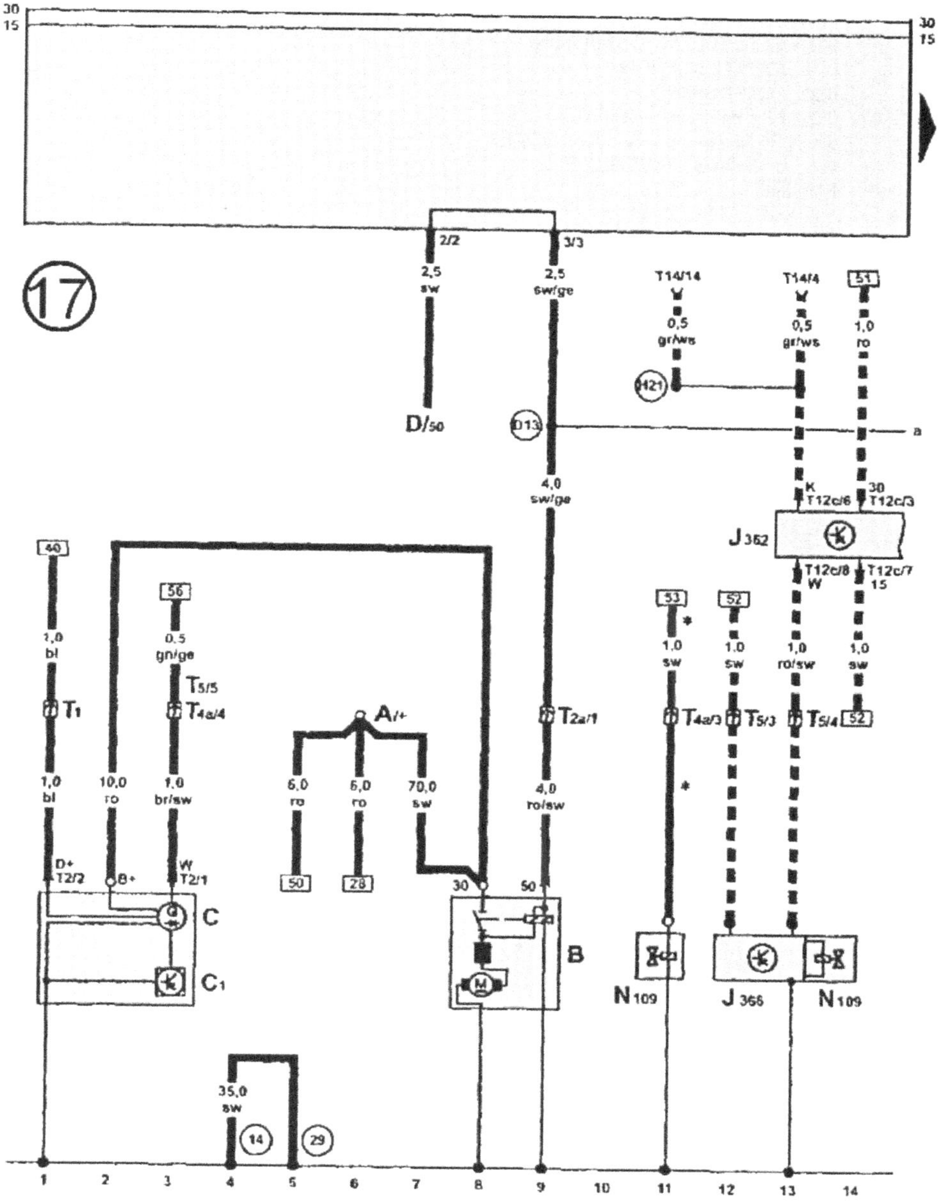

A Battery

B Starter motor

C Alternator

C1 Voltage regulator

D Ignition/starter switch

J362 Control unit, immobiliser

J366 Control unit for fuel shut-off valve

N109 Fuel shut-off valve

T1 1-pin connector, near battery carrier

T2/2a 2-pin connector, near battery carrier

T4a 4-pin connector, near battery

T5 5-pin connector, near battery

T12c 12-pin connector

T14 14-pin connector, foot well, L.H. front

In Circle:

14 Earth point, on gearbox

29 Earth point, near gearbox tunnel

D13 Connection (59), L.H. front cable harness

H21 Connection, diagnosis cable harness

* Only without immobiliser

--- Only with immobiliser

FULL CONTENTS

Made in United States
Orlando, FL
04 November 2026

90201281R00089